Communications and Computer Networks

Communications and Computer Networks

Edited by **Akira Hanako**

Clanrye International
New Jersey

Published by Clanrye International,
55 Van Reypen Street,
Jersey City, NJ 07306, USA
www.clanryeinternational.com

Communications and Computer Networks
Edited by Akira Hanako

International Standard Book Number: 978-1-63240-552-4 (Hardback)

Printed in China.

Contents

Preface

Every book is initially just a concept; it takes months of research and hard work to give it the final shape in which the readers receive it. In its early stages, this book also went through rigorous reviewing. The notable contributions made by experts from across the globe were first molded into patterned chapters and then arranged in a sensibly sequential manner to bring out the best results.

Communications and computer networks refer to a network of telecommunication which enables different computers and computerized peripherals to interconnect via a wire or wireless medium and communicate and share data with each other. There are different types of computer networks like personal area network, local area network, metropolitan area network, wide area network, internetwork, etc. This book will serve as resource guide for students and researchers. It will discuss in detail various techniques and advanced concepts related to computer networks and their use in communication. As this field is emerging at a rapid pace, this text will help readers to better understand the concepts of this area. The ever growing need of advanced technology is the reason that has fueled the research in the field of communications and computer networks. This book is appropriate for students seeking detailed information in this area as well as for experts.

It has been my immense pleasure to be a part of this project and to contribute my years of learning in such a meaningful form. I would like to take this opportunity to thank all the people who have been associated with the completion of this book at any step.

Editor

1

Job Scheduling Using Coupling in Grid

Avula Anitha

Department of Computer Science and Engineering, Keshav Memorial Institute of Technology, Hyderabad, India
Email: gamyav@yahoo.in

Abstract

The grid computing main concern is to use the resources efficiently. For achieving this many grid resource scheduling algorithms are used for the efficient use of unused resources, especially CPU. The scheduling algorithms assign single complete job to a single resource. Instead, if these algorithms consider the degree of dependency among the modules of a job, they can be allocated parallel to the different resources. This reduces the completion time of the job and the resources can be utilized to its maximum extent. Towards this, the job scheduling using coupling algorithm is proposed. This algorithm puts forward the idea of considering the coupling degree while allocating the modules of a job parallel to different resources. In this algorithm, resource selection is done by using both its functional and non-functional properties. The algorithm works in 3 phases. It groups the interdependent modules of a job into different sets using coupling in the first phase. It checks the non-functional property *i.e.* availability of a resource using echo procedure in the second phase and in the third phase, the sets created in first phase, are allocated parallel to different available and matching resources. From the simulation results it is observed that job scheduling using coupling algorithm gives better performance in terms of reduced turnaround time as compared to First Come First Served, Largest Task First and Minimum Execution Time scheduling algorithms.

Keywords

Grid Computing, Job Scheduling, Coupling, Functional and Non Functional Properties

1. Introduction

Grid computing organizes various geographically distributed resources into a virtual organization group for efficient utilization of idle resources [1] [2]. The grid is dynamic, in which resources are added or removed by the resource owners [2]. Hence, the grid resource discovery and scheduling become important. When the resource is requested, the scheduler should consider both functional properties such as computing capacity, operating system, and RAM size, and nonfunctional properties such as performance, availability and reliability. The allocation is done, if the requested resource properties and the advertised resource properties are matching to the expected degree. According to paper [3], there can be different types of matches such as exact match, plug-in match, subsume-match and no-match. Many existing resource scheduling algorithms such as Min-Min, Minimum Completion time algorithm (MCT), and $QPS_{Max\text{-}Min<>Min\text{-}Min}$ Scheduling, are considering only functional properties for resource matching [4]. Less importance is given to nonfunctional properties. The job cannot be executed on a resource if it is not working properly. The user has to wait for a long time, with a false impression that the job is being executed on some resource in the grid. This may have a greater impact on the trust and efficiency of the grid environment [5]. To avoid this, both functional and nonfunctional properties of a resource should be considered.

Generally, the resource scheduling algorithms select the best resource and assign a single complete job to it. If it is a computational grid, the maximum utilization of an idle CPU is given more importance. Rather than allocating a whole, completing job to a single resource, job's independent and dependent modules can be identified using coupling technique. Each set of dependent modules of a job is then allocated to different matching resources. Hence, various sets of dependent modules of a single job are executed parallel on different matching resources. This gives an improved impact on the job completion time and data flow between the resources can be reduced. The resource can be utilized to a maximum extent and if any further remaining availability of CPU power exists, it can be utilized for allocating next job's dependent modules.

Towards this approach, this paper proposes an idea by an algorithm named "job scheduling using coupling". In this, the degree of dependency among modules of a job using coupling, the functional and nonfunctional properties of resources are considered while allocating a resource. The job with a coupling degree of 1, indicates that only one set of dependent modules exists and the complete job should be allocated to a single resource. The job having low coupling degree less than 1, indicates that more than one set of dependent modules exists and each dependent set is allocated to a separate suitable resource. The number of dependent set's equals to the number of resources used for allocation. Further, in this paper, the related work is discussed in Section 2. The coupling concept, resource functional and non functional properties are discussed in Section 3. The pseudocode of the proposed algorithm, "job scheduling using coupling" is given in Section 4. Simulation design with an illustrative example and the performance comparison of the proposed algorithm with the existing algorithms are discussed in Section 5. Section 6 ends with conclusion and future work.

2. Related Work

Currently there are many grid resource scheduling algorithms which use different approaches while scheduling to reduce the make-span time and execution time of a job. Minimum execution time (MET) algorithm allocates the task with minimum execution time to the available resource without considering the current load of the resource [4]. The Minimum Completion Time [4] is similar to Minimum Execution Time where completion time for a task is calculated on all resources by adding the availability time of resources. LoBa-Min-Min [4] algorithm is based on load balance where Min-Min algorithm is used for preliminary scheduling and then, according to the standard of reducing make-span, the tasks on heavy loaded resources would be assigned to resources that need less time to load balance, raise resource utilization rate, etc. In First Come First Served (FCFS) scheduling algorithm, new set of jobs are added in the job scheduling queue. The resource is assigned based on queue order to maintain the fairness among grid users [6] [7]. Min-Min algorithm [8] belongs to batch mode scheduling in which a set of all unmapped tasks completion time is calculated on each resource by adding the expected execution time and the resource availability time. The task with the overall minimum completion time is selected and mapped to the resource which obtains the overall minimum completion time. Max-Min algorithm [9] schedules long tasks on resource with high computation power. Min-Mean algorithm [10] is an improved Min-Min scheduling algorithm based on average execution time of resource mean completion time. The Largest Task First (LTF) scheduler schedules the task from the task domain which has largest task size. Length comparator is used to identify the task size [11]. The New Threshold based job scheduling [12] attempts to minimize the cost of workflow execution time and provides service fairness. In this the jobs are scheduled in FCFS manner, if the user jobs execution time is less than the threshold. Scheduling algorithm implemented by [NGSched] provides distinct functionalities, such as dynamic workflow planning and just-in-time scheduling in a grid environment [13]. Serve On Time algorithm proposed in [14] is based on the idea of combining all the dynamic multi tasks so that all the tasks will obtain the rights to be served as soon as possible. Group based Job Scheduling algorithm [15] provides an optimize algorithm to queue of the scheduler using various scheduling methods like Shortest Job First, First Come First Serve, Round Robin algorithms for jobs. [16] discusses concepts of resource scheduling and application scheduling and presents a classification of scheduling algorithms. Opportunistic load balancing algorithm (OLB) is the job scheduling algorithm in which the next earliest available resource which is idle is selected randomly. The OLB algorithm will not take the execution time of task into consideration, hence resulting in poor throughput [17] [18]. The problem of allocating resources in Grid scheduling requires the definition of a model that allows local and external schedulers communicate in order to achieve an efficient management of the resources themselves. To this aim, tender/contract-net model for Grid resource allocation, showing the interactions among the involved actors is given in [19].

The above mentioned algorithms handles single job at a time and considers only functional properties. Reducing the make-span time and execution time by considering the dependent tasks and executing the dependent module sets of the job simultaneously and parallel across resources can be done using the proposed approach.

3. Coupling and Properties of a Resource

This section gives an overview of the concept on coupling which is used to find the interdependent modules and an overview functional and nonfunctional properties of a resource based on which the proposed algorithm is designed.

3.1. Coupling

Coupling [20] [21] is a measure of interconnection among modules in a software structure. Coupling depends on the interface complexity between modules, the point at which entry or reference is made to a module, and what data pass across the interface. In software design, we strive for lowest possible coupling. Simple connectivity among modules results in software that is easier to understand and less prone to a "ripple effect", caused when errors occur at one location and propagates through the system. There are different types of coupling. In Content Coupling one module refers to the content of another module. Common Coupling exists when two or more modules will have access to same global data. Two modules are controlled coupled when one module controls can directly affect the execution of another module using control parameter. Data coupling exhibits the properties that all parameters to a module are either simple data types, or in the case of a record being passed as a parameter, all data members of that record are used/required by the module. That is, no extra information is passed to a module at any time. A variation of data coupling, called *stamp coupling*, is found when a portion of a data structure (rather than simple arguments) is passed via a module interface.

The following is the equation used for measuring the coupling degree of a module.

Let

di represents the number of input data parameters;

ci represents the number of control parameters;

do represents the number of output data parameters;

co represents the number of output control parameters;

gd represents the number of global variables used as data;

gc represents the number of global variables used as control;

w represents the number of modules called from the current module *i.e.* fan-out;

r represents the number of modules calling the current module *i.e.* fan-in.

$$C = 1 - 1/\left(di + 2*ci + do + 2*co + gd + 2*gc + w + r\right) \quad (1)$$

Coupling degree "*C*" makes the value larger the more coupled the module is. This number ranges from approximately 0.67 (low) to 1.0 (high).

3.2. Functional and Nonfunctional Properties of a Resource

Each resource will have its own functional and nonfunctional properties. The scheduling is done using these properties. The resources which can be shared in the Grid environment are registered and their property information is maintained by Grid Information Server (GIS) in the Grid System. The grid user submits the job request to the scheduler. Then the scheduler schedules the job based on scheduling algorithms and assign resources to the grid job with the help of GIS. After the job is completed, the available resource is updated with GIS and the job result is sent to the grid user [11]. The resources differ from each other in functional and non functional properties.

The functional properties of a resource are system processing speed, processing element ids, architecture, software platform and other associated factors, such as memory, storage and connectivity [22]. They also can be of different types like supercomputers, workstations, databases, storages, networks, software, and special instruments, advanced display devices, cache, CPU cycle and so on. Among these, resource as CPU cycles is considered in the proposed "Job Scheduling Using Coupling" Algorithm.

The nonfunctional properties of a resource are its availability, reliability and performance. If the resource allocated to a job is not reliable or it is of low performance or if it is not available, *i.e.* it is "down" at the moment, then the make-span, execution time and the turn around time of a job will vary to a considerable amount. Among the nonfunctional properties mentioned, the availability of a resource is considered by our proposed algorithm. The Job Scheduling Using Coupling algorithm will check whether the resource is available or not by using the "echo" procedure mentioned in the next section. If it is available and working properly, then the resource is considered while scheduling the job on it.

4. Job Scheduling Using Coupling (JSC) Algorithm

In Job Scheduling Using Coupling algorithm, when the user submits the job with desired resource parameters, it will consider the resources whose advertised parameters are matching with the requested parameters by the user. This algorithm works in three phases. 1) Coupling Phase; 2) Check Availability Phase; and 3) Job Scheduling Phase.

1) Coupling Phase: In this phase, it is assumes that the user while requesting for a resource, will submit not only the desired resource parameters, but also the information about the interdependent modules of a job. In this algorithm the modules are referred to as tasks of a job. Using the given information, this phase will create the discrete dependency sets. Each set will hold interdependent modules.

2) Check Availability Phase: Before scheduling is done, this phase checks whether the resource is available and working properly or not by sending the "echo message". If the resource is available, then the acknowledgment is received from the resource, *i.e.* echo message is returned. The time taken for receiving the echo message back is calculated. If the time taken is less than the predefined threshold time value, then the resource is considered for allocation.

3) Job Scheduling Phase: From the first phase, the dependency sets are receive and the scheduling is done on the sets. Each dependency set is allotted with a separate available resource. Different sets of a job are executed in parallel on different resources. The remaining CPU power of the resource is utilized for allotting the next jobs' dependency set. This phase sees that the job is executed parallel on different resources by decreasing the make-span, execution time and thereby, turn around time of the job. Here, the sequential execution of a job is done within the set, but parallel execution is done between the sets.

Following is the Job Scheduling Using Coupling algorithm.

Let Job's be J1, J2, ···, Jn.

Let Task's in each job Ji be T1, T2, ···, Tk.

1. Find the dependent modules or tasks of Job Ji using coupling, *i.e.* by considering user given information about modules and their interdependency.

2. Let m sets are obtained, where each set is having interdependent modules/tasks

$$\text{S1} = \{\text{T1}, \text{T2}, \cdots, \text{Tx1}\},\ \text{S2} = \{\text{Tx1}+1, \text{Tx1}+2, \cdots, \text{Tx2}\}, \cdots, \text{Sm} = \{\text{Tx2}+1, \text{Tx2}+2, \cdots, \text{Tk}\}$$

The sets S1, S2, ···, Sm are discrete.

3. Check for the availability of resources, *i.e.* checks whether requested resource properties and advertised resource properties are matching or not and whether the resources are working properly or not.

4. If they are matching and available, Let R1, R2, ···, Rnum are the resources which can be utilized for assigning the job.

5. If the number of sets are equal to or less than the number of available resources, then GOTO step 6 otherwise GOTO step 7.

6. Assign each set to each resource. GOTO step 11.

7. If the number of sets are greater than the number of resources, then assign the sets from S1 to Snum to resources R1 to Rnum.

8. Assign the remaining sets to the existing resources in round robin fashion, *i.e.* assigns remaining sets Snum + 1 to Sy to resources R1 to Rnum in rotation.

9. Repeat step 8 until all the sets Sy + 1 to Sm are assigned to existing resources. Using the steps from 7 to 9, each resource will be allocated with more than one dependency set of job Ji.

10. Repeat Step 1 to 9, till all n jobs are allocated to the existing available resources.

11. Exit.

The following **Figure 1**, shows the flow chart of Job Scheduling Using Coupling Algorithm.

The Pseudocode of Job Scheduling Using Coupling is given in Appendix A.

5. Simulation Design

Grid Simulator (Gridsim) is used for simulation. It is a tool which is used to simulate the Grid Computing Environment. This simulation contains entities for users, resources, information service, etc. Jobs are described with information like job's execution time, necessary machine architectures. In this simulation, the dataset for job's configuration is created randomly. Six jobs are initially used for checking the performance of the proposed algorithm. Then, worked out with 25 jobs and slowly increased and checked with 175 jobs. The simulation experiments are done 8 times and in each experiment the scheduling algorithms FCFS, MET, LTF and proposed JSC algorithm is worked out on the specific number of jobs by using 4 to 8 matching resources. The Turn Around time is calculated in every simulation experiment for different scheduling algorithms. For implementing and evaluating proposed algorithm, random number of modules are generated for each job and they are grouped in the form of sets. Each set holds the interdependent modules. The results and performance is evaluated using the performance metrics mentioned in Section 5.1.

The Simulation is performed using the simulation parameters mentioned in **Table 1**.

5.1. Performance Metrics

Turn Around Time: Let the total number of jobs be N and the completion time of job ji is ci and the job arrival time is ai. The Total Turn Around Time is defined by the following equation

$$\text{TAT} = \frac{1}{N}\sum_{i=1}^{N}(ci - ai) \tag{2}$$

Makespan: It is defined as the time spent from the beginning of the first task to the end of the last task in the schedule. The assumptions made are resources are available continuously and jobs are ready to be scheduled. The Makespan is given as

$$\text{M} = \text{Max}\{\text{C1}, \text{C2}, \text{C3}, \cdots, \text{Cn}\} \tag{3a}$$

where Ci is the completion time of task i. Lesser the makespan means more efficient is the algorithm, *i.e.* less time is taken to execute the algorithm.

For the proposed algorithm the Makespan of Job ji is

Table 1. Simulation parameters used.

Parameter	Value
No. of jobs	6 - 175
No. of resources	4 - 8
Job workload	50 - 500 (billion instructions)
Node processing speed	100 MIPS
Network bandwidth	2 - 6 Mbps

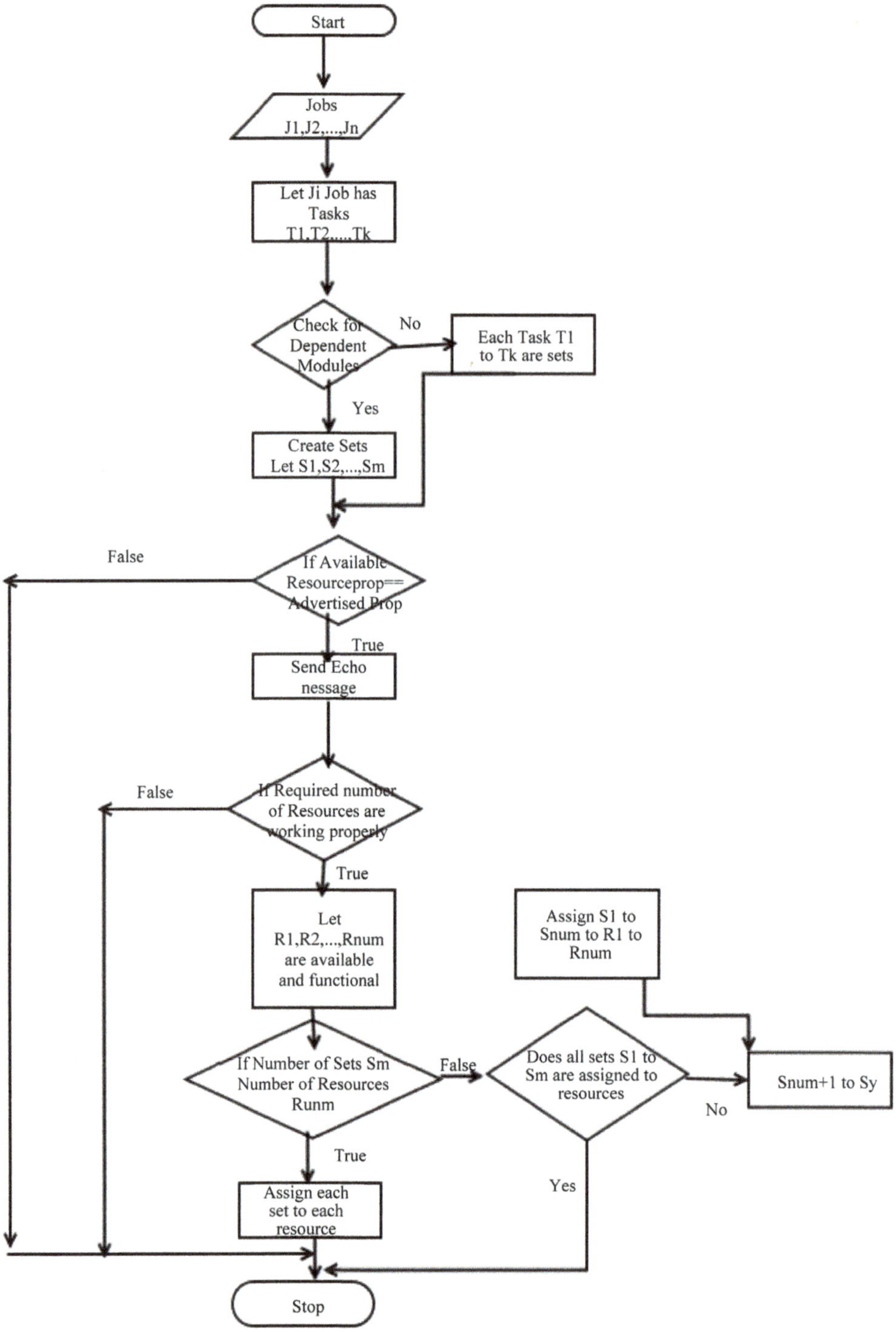

Figure 1. Flow chart of JSC algorithm.

$$\begin{aligned} M &= \text{Max}\{t1, t2, \cdots, tk\} = \text{Max}\{\text{dset1}, \text{dset2}, \cdots, \text{dsetp}\} \\ &= \text{Max}\left\{\{C11, C12, \cdots, C1x\}_1, \{C21, C22, \cdots, C2y\}_2, \cdots, \{Ck1, Ck2, \cdots, Ckz\}_p\right\} \end{aligned} \tag{3b}$$

where ti, t2, ···, tk are the modules or tasks of the job ji. The tasks are coupled using the Coupling phase of the proposed algorithms and dependent sets are created, dset1, dset2, ···, dsetp are the sets. Each set is a group of interdependent modules or tasks and C11, C12, ···, C1x are the completion times of tasks in dset1, C21, C22, ···,

C2y are the completion times of tasks in dset2 and Ck1, Ck2, ···, Ckz are the completion times of tasks in dsetp.

5.2. An Illustrative Example

This section describes the working procedure of the proposed algorithm using 6 jobs, each with a different number of interdependent modules. Four resources are considered for illustrating this example. The following are the assumptions made while illustrating the example.

1) It is assumed that all 4 resources are working properly and available.
2) The resources advertised properties are matching with the user requested properties.
3) The interdependency among the modules for each job is given by the user.
4) The arrival time of jobs is 0 *i.e.* all jobs arrived at the same time.

Let 6 jobs are submitted by 6 users are {J1, J2, J3, J4, J5, J6} and the data used for each job is as follows.

Let Job J1 has 3 modules/tasks, *i.e.*, t1, t2, t3 each with execution time 2, 4, 4 secs. Tasks t1, t2 is interdependent and discrete from t3. Therefore, 2 dependent sets are created using the coupling phase of JSC algorithm and let, they are referred as J1S1 = {t1, t2} and J1S2 = {t3}. The total execution time of J1S1 is 6 secs and J1S2 is 4 secs. The total execution time of the job J1 is 10 secs.

Let Job J2 has 1 module/task, *i.e.*, t1 with execution time 5 secs. Only one dependent set is created with only one task in it and let, it is referred as J2S1 = {t1} with the execution time of 5 secs.

Let Job J3 have 5 modules/tasks, *i.e.*, t1, t2, t3, t4, t5 each with execution time 5, 2, 4, 4, 5 secs. Tasks t1, t2, t4 are interdependent and t3, t5 are interdependent. Therefore, 2 dependent sets are created using the coupling phase of JSC algorithm and let, they are referred as J3S1 = {t1, t2, t4} and J3S2 = {t3, t5}. The total execution time of each dependent set is 11 and 9 secs respectively. The total execution time of job J3 is 20 secs.

Let Job J4 have 5 modules/tasks, *i.e.*, t1, t2, t3, t4, t5 each with execution time 4, 3, 2, 2, 3, 5 secs. Tasks t1, t4 is interdependent and t2, t3, t5 are interdependent. Therefore, 2 dependent sets are created using the coupling phase of JSC algorithm and let, they are referred as J4S1 = {t1, t4} and J4S2 = {t2, t3, t5}. The total execution time of each dependent set is 7 and 10 secs respectively. The total execution time of job J4 is 17 secs.

Let Job J5 have 3 modules/tasks, *i.e.*, t1, t2, t3 each with execution time 3, 2, 3 secs. Tasks t1 is discrete from other two interdependent tasks t2, t3. Therefore, 2 dependent sets are created using the coupling phase of JSC algorithm and let, they are referred as J5S1 = {t1} and J5S2 = {t2, t3}. The total execution time of each dependent set is 3 and 5 secs respectively. The total execution time of job J5 is 8 secs.

Let Job J6 have 6 modules/tasks, *i.e.*, t1, t2, t3, t4, t5, t6 each with execution time 4, 3, 2, 3, 1, 2 secs. Tasks t1, t4 is interdependent, t2, t6 are interdependent and t3, t5 are interdependent. Therefore, 3 dependent sets are created using the coupling phase of JSC algorithm and let, they are referred as J6S1 = {t1, t4}, J6S2 = {t2, t6} and J6S3 = {t3, t5}. The total execution time of each dependent set is 7, 5 and 3 secs respectively. The total execution time of job J6 is 15 secs.

Using the above details the Job Scheduling algorithm third phase will allocate the resources to each dependent set of each job in parallel.

Figure 2 shows the order of allocation of each dependent set of each job on 4 resources simultaneously along with the completion time of each job.

It can be observed from **Figure 2** that the completion times of Jobs {J1, J2, J3, J4, J5, J6} are {6, 5, 13, 16, 17, 20} respectively. The turn around is calculated using Equation (2) as

$$\text{TAT} = \frac{(6+5+13+16+17+20)}{6} = 12 \text{ secs.}$$

5.3. Performance

The simulation results are shown in **Figure 3**. It gives the simulation results of Turn Around Time and performance of the FCFS, MET, LTF and JSC. This is simulated under random job failure condition. It is observed from the results that Turn Around Time of the LTF and FCFS are high. As they either assign largest task first, making shortest execution time jobs waiting or assigns First Come First Serve basis without considering resource capability of executing the assigned job in time. The MET performs better than LTF and FCFS as the resources assigned first with minimum execution time. Large jobs need not starve for a long time as resources are released within a short time as soon as the job completes its task and can be available for assigning to next largest job in

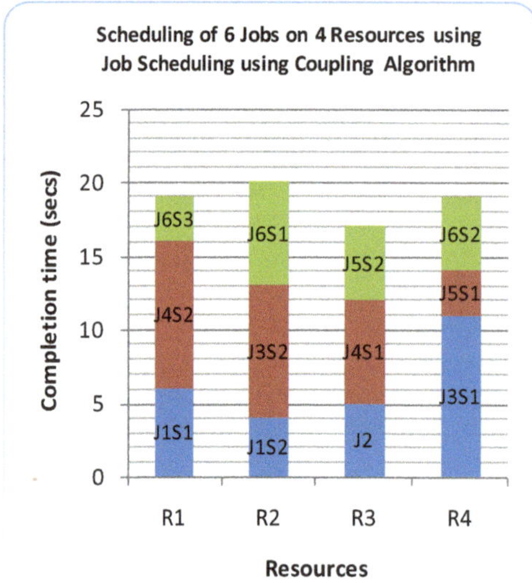

Figure 2. JSC algorithm for 6 jobs 4 resources.

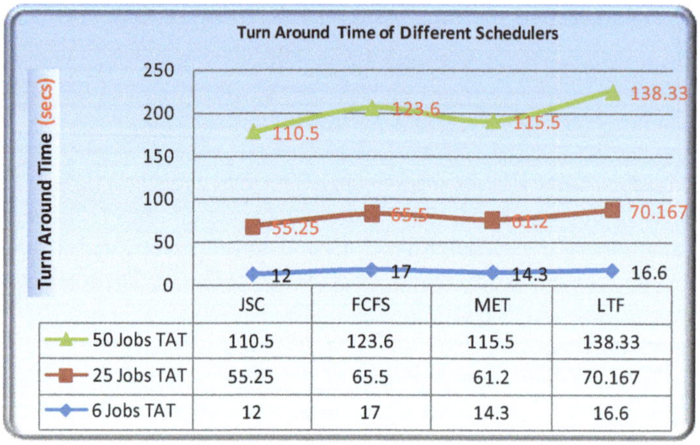

Figure 3. Turn around time of different schedulers.

the queue. These algorithms executes the job sequentially and the complete job is allocated to a single resource. In these algorithms the availability of the resource and its capability of executing is not considered. The proposed algorithm considers nonfunctional properties such as availability and resource capability using the "echo" procedure described in Section 4. It considers only available and high performance resources and assigns the sets to the resources and sees that the sets of a job are executed simultaneously or in a parallel manner. With this approach, each job's makespan is reduced, thereby reducing the execution time and decreasing the total turnaround time. The performance of the proposed algorithm is observed better than the existing algorithm such as FCS, MET and LTF scheduling algorithms. The total Turn Around Time of proposed algorithm is observed lesser than other algorithms considered for comparison.

In **Figure 3**, the three curves are the results of performing the simulation using 6, 25 and 50 jobs. Each curve shows the Turn Around Times of JSC, FCFS, MET and LTF.

It is observed in **Figure 3** that the Turn Around Time of LTF and FCFS increases with an increase in the number of jobs. When MET and JSC are compared, the Turn Around time decreases with the increase in the number of jobs, but JSC's Turn Around Time is considerably lesser than MET Turn Around Time.

6. Conclusion

In this paper, job scheduling using coupling algorithm is proposed in which the job is scheduled parallel on the resources by considering the groups of dependent tasks. Each dependent set will have set of dependent modules/tasks. Each dependent set of a job is scheduled on different resources rather than on a single resource to avoid sequential execution. The simulation is done on GridSim, and the make-span time and total turn around time of the proposed algorithm (JSC) are compared with FCFS, MET and LTF algorithms. The JSC algorithm will select only the resources which are available and performing well, *i.e.* the non functional properties of a resource are considered in selecting the resources from the matching set before proceeding with the actual allocation of a dependent set of a job on a resource. As JSC algorithm is executing the job's tasks simultaneously in a parallel manner on different matching resources, the turn around time and make-span time are observed to be reduced considerably when compared to the FCFS, MET, and LTF. The coupling concept mentioned in Section 3 measures whether a module is dependent on another module or not. But, it will not give the information about the module details on which the current module is dependent. Therefore, in the proposed algorithm the dependency sets are created using the user given details about the modules or tasks of a job, *i.e.*, the JSC algorithm depends on the user given information about the details of the interdependent modules or tasks. If this dependency check and identification of dependent tasks are done by the Grid Resource Manager rather than depending on the user's input, the user's burden will be reduced. Therefore, identification of the interdependent tasks after the job is submitted to grid remains the future work of this proposed work.

References

[1] Kesselman, C., Foster, I. and Steven, T. (2001) The Anatomy of the Grid-Enabling Scalable Virtual Organizations. *The International Journal of Super Computer Application*, **15**, 200-222. http://dx.doi.org/10.1177/109434200101500302

[2] Andrew, R. and Krzysztof, P. (2004) Grid Computing: The Current State and Future Trends (in General). University of Canterbury, Christchurch, TR-CoSc 01/04.

[3] Paolucii, M., Kawamura, T., Payne, T. and Sycara, K. (2002) Semantic Matching of Web Service Capabilities. *Proceedings of* 1*st International Semantic Web Conference* (*ISWC*2002), Berlin.

[4] Zhang, Z.P., Wen, L.J. and Wang, Z.P. (2013) LoBa-Min-Min-SPA: Grid Resources Scheduling Algorithm Based on Load Balance Using SPA. *The Open Automation and Control Systems Journal*, **5**, 87-95. http://dx.doi.org/10.2174/1874444301305010087

[5] Anitha, A. (2015) Trust Management in Grid-Trust Assessment and Trust Degree Calculation of a Resource—A Novel Approach. *Journal of Computers and Communications*, *Scientific Research Publications*, **3**, 34-41. http://dx.doi.org/10.4236/jcc.2015.36005

[6] Ahmed, I., Kwok, Y.K., Wu, M.Y. and Li, K. (2004) Experimental Performance Evaluation of Job Scheduling and Processor Allocation Algorithms for Grid Computing on Metacomputers. *Proceedings of the* 18*th International Symposium Parallel Distributed Processing*, IEEE Xplore Press, 170-177.

[7] Daulamis, N.D., Daulamis, A.D., Varvarigos, E.A. and Varvarigou, T.A. (2007) Fair Scheduling Algorithms in Grid. *IEEE Transactions on Parallel Distributed System*, **18**, 1630-1648. http://dx.doi.org/10.1109/TPDS.2007.1053

[8] Yu, X.P. and Yu, X.G. (2009) A New Grid Computation-Based Min-Min Algorithm. 6*th International Conference on Fuzzy Systems and Knowledge Discovery*, Tianjin, 14-16 August 2009, 43-45. http://dx.doi.org/10.1109/fskd.2009.81

[9] Ni, L.N., Zhang, J.Q., Yan, C.G. and Jiang, C.J. (2005) A Heuristic Algorithm for Task Scheduling Based on Mean Load. 1*st International Conference on Semantics*, *Knowledge and Grid* (*SKG*'05), Beijing, 27-29 November 2005. http://dx.doi.org/10.1109/skg.2005.13

[10] Kamalam, G.K. and Muralibhaskaran, V. (2010) A New Heuristic Approach: Min-Mean Algorithm for Scheduling Metatasks on Heterogenous Computing Systems. *International Journal of Computer Science and Network Security*, **10**.

[11] Menasce, D., Saha, D., Porto, S.C.D., Almeida, V.A.F. and Tripathi, S.K. (1995) Static and Dynamic Processor Scheduling Disciplines in Heterogenous Parallel Architectures. *Journal of Parallel and Distributed Computing*, **28**, 1-18. http://dx.doi.org/10.1006/jpdc.1995.1085

[12] Ramaparvathy, L. (2014) A New-Threshold Based Job Scheduling for Grid System. *Journal of Computer Science*, *Science Publications*, **10**, 1069-1076. http://dx.doi.org/10.3844/jcssp.2014.1069.1076

[13] Madhuri, B. and Pradhan, S.N. (2011) NGSched—An Efficient Scheduling Algorithm Handling Interactive Jobs in Grid Environment. *International Journal of Grid and Distributed Computing*, **4**, 1.

[14] Ajith, A., Rajkumar, B. and Baikunth, N. (2000) Nature's Heuristic for Scheduling Jobs on Computational Grids. *Proceedings of* 8*th IEEE International Conference on Advanced Computing and Communications* (*ADCOM*2000).

[15] Pinky, R., Ravinder, S., Payal, S. and Dilip, S. (2012) Group Based Job Scheduling Algorithm Using Priority Queue and Hybrid Algorithm in Grid Computing. *International Journal of Grid Computing and Applications*, **3**, 55-64.

http://dx.doi.org/10.5121/ijgca.2012.3405

[16] Prajapathi, H.B. and Vipul, A.S. (2014) Scheduling in Grid Computing Environment. *4th International Conference on Advanced Computing & Communication Technologies* (*ACCT*), Rohtak, 8-9 February 2014, 315-324. http://dx.doi.org/10.1109/ACCT.2014.32

[17] Armstrong, R., Hensegen, D. and Kidd, T. (1998) The Relative Performance of Various Mapping Algorithms Is Independent of Sizable Variances in Run-Time Productions. *7th IEEE Heterogenous Computing Workshop* (*HCW*'98), Orlando, 30 March 1998, 79-87. http://dx.doi.org/10.1109/HCW.1998.666547

[18] Freund, R.F. and Siegel, H.J. (1993) Heterogenous Processing. *IEEE Computer*, **26**, 13-17.

[19] Massimiliano, C. and Stefano, G. (2008) Resource Allocation in Grid Computing. *WSEAS Transactions on Computer Research*, **3**.

[20] Jefferson, O.A., Mary, J.H. and Priyadarshan, K. (1996) A Software Metric System for Module Coupling. *Journal of Systems & Software*, **20**, 295-308.

[21] Allen, E.B. and Khoshgoftaar, T.M. (2001) Measuring Coupling and Cohesion of Software Modules: An Information Theory Approach. *Proceedings of 7th International Conference on Software Metric Symposium*, London, 4-6 April 2001, 124-134. http://dx.doi.org/10.1109/METRIC.2001.915521

[22] Ferreira, L., Berstis, V., Armstrong J., Kendzierski, M., Neukoetter, A., Takagi, M., Bing-Wo, R., Amir, A., Murakawa, R., Hernandez, O., Magowan, J. and Bieberstein, N. (2003) Introduction to Grid with Globus. 2nd Edition, IBM Redbook.

Appendix A

Pseudocode

```
For i = 1 to n
Begin
S[]=CoupledSets(Ji)
R[]=CheckAvailable_Res(parameters);
Rnum=R.length;
If S.length <= Rnum
      For j = 1 to m
      Begin
          Assign(S[j],Rj)
      End For
Else
      For j = 1 to m
      Begin
   Assign(S[j],Rj)
   If j==Rnum
   Rnum=1
      End For
End If
End For
```

Pseudocode for CheckAvailable_Res procedure

```
CheckAvailable_Res(parameters user_requested_parameters[])
Begin
For each registered resource Ri
Begin
If match(user_requested_parameters,resource_advertised_parameters of Ri)==1)
Begin
ack=send echo "hello" to resource Ri
If(ack>=1and ack.time<=threshold_set_time)
Begin
//Select resource and consider it as available and working properly. Add to the selected resources
Add(R[ ], Ri)
End if
End if
End for
Return R
End CheckAvailable_Res
```

The CheckAvailable_Res procedure sends an echo message "hello" to all registered resources. If the resource is working properly and it is available, it will send the acknowledgement (ack). If acknowledgement is received within the threshold time, which is already set, then the resource is added to the set of available resources array R [].

Pseudocode for CoupledSets(j)

```
CoupledSets(J)
Begin
For all tasks of Job J
Begin
```

//Identify the tasks which are interdependent using user given information about the interdependency among modules
//Group them into a set, referred here as "dependent set" and add it to S [].
End for
Return S //*i.e.* Dependent sets
End CoupledSets

2

Usability Analysis of Touch Screen for Ground Operators

Jungwan Hong[1], Seungju Baek[1], Heesoo Jung[1], Suhwan Kim[2], Yongjin (James) Kwon[1*]

[1]Department of Industrial Engineering, Ajou University, Suwon, South Korea
[2]1st Division, 3rd Department, Agency for Defense Development, Daejeon, South Korea
Email: *yk73@ajou.ac.kr

Abstract

The importance of the aircraft is increasing gradually; among them it may be called Ground Control Station (GCS) receiving the highest expectations. In aspect of operation, operator's workload and working speed are one of the most important factors. Thus, we need to compare Touch mode and Non Touch mode in order to improve workload and working speed. In this paper, we analyzed the differences of Touch mode and Non Touch mode about workload, working speed in operation of aircraft. Through NASA-TLX and Measured working time, quantified data were collected and an analysis of variance (ANOVA) was performed. According to experimental result, working speed and workload according to input mode showed significant difference. In ground operation, Touch mode is better than Non Touch mode about working speed and workload. In the Touch mode, the optimal value of lower display angle was selected as 60° angle.

Keywords

Touch Screen, Work Load, Ground Operator, Working Time, NASA-TLX, Ground Control Station

1. Introduction

The importance of the aircraft is increasing gradually, among them it may be called anti-air ground control station (GCS) are receiving the highest expectations. This can be regarded as worldwide trend. A GCS is defined as a "powered, aerial vehicle that does not carry a human operator, uses aerodynamic forces to provide vehicle lift, can fly autonomously or be piloted remotely, can be expendable or recoverable, and can carry a lethal or nonlethal payload". The biggest advantage of the GCS is people do not ride with no loss of life can do long-term mission. Also, ground control center has undergone a process of change and evolve over time. As time goes on, in order to maximize the performance of the battlefield situation, a study was conducted for each of the elements such as chairs, consoles, pedals. So ground control center has been evolving in the direction of improvement for the convenience and usability.

However, research on the touch screen usability is likely to not actively make. In the future, the touch screen is considered important to maximize the ability to operation of GCS. In operation as a GCS, it is necessary to

*Corresponding author.

compare the performance of mouse and touch screen. Therefore, in this study, we measured the response time, workload in operating GCS simulator. And we analyzed the results shown by measurements.

2. Related Study

2.1. NASA-TLX Analysis Method

The NASA Task Load Index (NASA-TLX) is a widely-used, subjective, multidimensional assessment tool that rates perceived workload in order to assess a task (see **Table 1**) [1]. Through various industrial accidents and disasters, Workload has been found as one of the factors that have the greatest impact on the safety of the system. So in order to prevent the workload reduction, a number of studies were carried out. As a result, it was found that the main cause of Workload reduction is the complexity of the task. While operators are performing complex task, operators will experience a significant amount of the cognitive load [2]. Due to the complex task, if required cognitive load exceed operations ability, rapid task ability will occur [3]. This fact is supporting that assessing and managing task difficulty will prevent accident caused by task load decrease. The most effective method of assessing task ability is giving direct question to the operator. Therefore, various question investigations to assessing task ability were developed, such as NASA-TLX (task load index), Cooper-Harper scale, SWAT (subjective workload assessment technique) are known [4]. Of these, the most stable subjective task difficulty assessing method up to now is NASA-TLX Method [5].

NASA-TLX is configured to be able to quantify the degree of difficulty through the entire. There is a description for each of these subscales that the subject should read before rating. They are rated for each task within a 100-points range with 5-point steps. These ratings are then combined to the task load index. Providing descriptions for each measurement can be found to help participants answer accurately. If you perform a task difficulty assessing through the NASA-TLX, you can know reason about what is being affected. Through the effective management of the task difficulty level, in order to inhibit workload, it evaluates the task difficulty. Identifying efficient factor is very important, so more efficient management of the task difficulty level is possible.

2.2. Previous Studies

Many studies have compared touchscreens to other selection devices for various tasks Hall, Cunningham, Roache, and Cox (1988) investigated the effects of various factors on touchscreen performance. They reported that accuracy varied from 66.7% for targets 10 mm per side, to 99.2% for targets 26 mm per side, and that accuracy was maximized once targets were approximately 26 mm per side [6]. Ostroff and Shneiderman (1988) compared a touchscreen, mouse, number keys, and arrow keys. The results were similar to those of most other studies comparing the touchscreen and the mouse, indicating that the touchscreen was faster. They found no significant difference between error rates for the mouse and the touchscreen [7]. Ahlstrom and Lenman (1987) compared a conductive touchscreen and mouse for the selection. This study indicated that the touchscreen was faster, but resulted in a much higher error rate [8]. Karat, McDonald, and Anderson (1986) compared a touchscreen, mouse, and keyboard for selection tasks. The results indicated that the touchscreen was the preferred device for the task without the typing subtask, while the keyboard was preferred when the subtask was included. The touchscreen was the fastest for both tasks [9].

Table 1. Evaluation factors and description of NASA-TLX.

Evaluation Factors	Description
Mental Demand	How mentally demanding was the task?
Physical Demand	How physically demanding was the task?
Temporal Demand	How hurried or rushed was the pace of the task?
Performance	How successful were you in accomplishing what you were asked to do?
Effort	How hard did you have to work to accomplish your level or performance?
Frustration	How insecure, discouraged, irritated, stressed, and annoyed were you?

3. Experiment Design

3.1. Experiment 1

In this study, we have conducted experiments to maximize the operational capability as the operation in the GCS. Factors were set to lower display angle (45˚, 60˚, 80˚) and input mode (Touch, Non Touch). In this regard, we measured the working time and the workload according to the factor by using the Flight Simulator X and to measure the workload, it was carried out NASA-TLX. (The baseline of NASA-TLX is 45˚/Non touch.) In Experiment 1, 2, both of the samples were determined 10.

In this experiment, including the Main Panel, Trim panel, Overhead Panel, Throttle Panel were used (see **Figures 1** and **2**).

Scenario is as follows.

1. Sequentially perform the setting for take-off on the GCS (As a given number). (Change the settings of 5 kinds at Main Panel, 2 kinds at Trim panel, 2 kinds at Overhead Panel, 4 kinds of Throttle Panel.)
2. Perform the take-off.
3. When climbed to a certain safe trajectory, perform the setting for normal flight (As a given number).

(Change the settings of 5 kinds at Main Panel, 2 kinds at Trim panel, 2 kinds at Overhead Panel, 4 kinds of Throttle Panel).

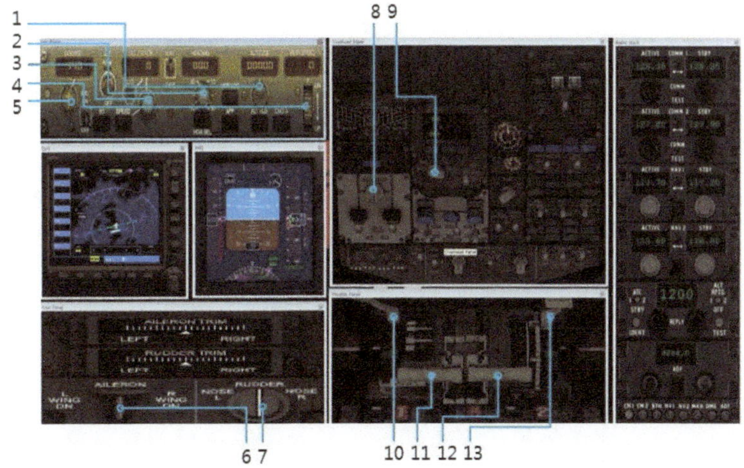

Figure 1. Interface of Experiment 1.

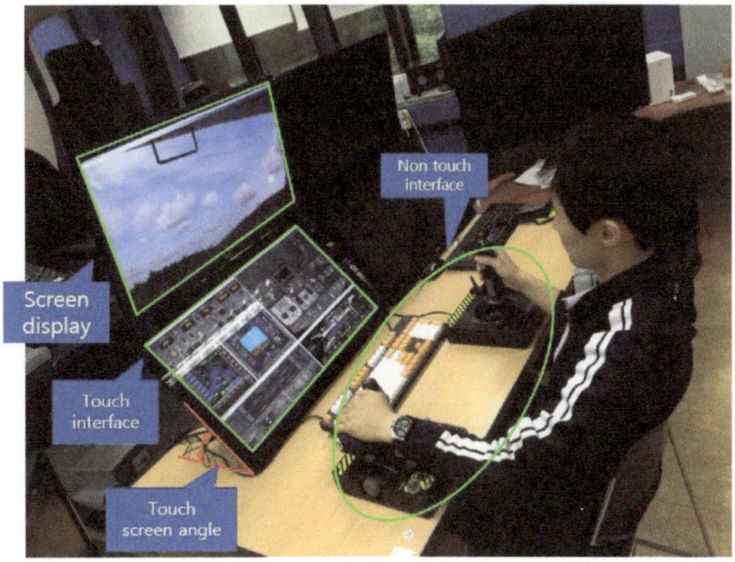

Figure 2. The example of Experiment 1.

3.2. Experiment 2

Using the quantitative results obtained by experiments, it was performed ANOVA. The optimum lower display angle and input method according to the ANOVA were adopted (see **Table 2**). This experiment 2 was performed on the reaction rate and the workload measurement based. During GCS operating, we measured the reaction rate to a target is displayed at randomly (see **Figure 3**). We measured the workload by the NASA-TLX. Accordingly, we adopted optimum lower display angle and input mode in operating GCS. In Experiment 1, 2 both of the samples was determined 10.

4. Result Analysis

4.1. Analysis of the NASA-TLX

In Experiment 1, NASA-TLX showed that workload was optimal at lower display angle 60°, touch mode. Also, in significance level 0.05, N = 10. All of result about NASA-TLX was significant. (Mental Demand ($p = 0.005$), Physical Demand ($p = 0.000$), Temporal Demand ($p = 0.014$), Performance ($p = 0.000$), Effort ($p = 0.000$), Frustration ($p = 0.006$).) In Experiment 1, it was determined to be the most significant in 60° angle (see **Figure 4**). And Experiment 2, the workload was measured according to the input mode in the 60° angle (see **Figure 5**). Also, in significance level 0.05, N=10. As a result, all of results are optimal at touch mode. Also like the previous Experiment 1, in significance level 0.05, all of result about NASA-TLX was significant (Mental Demand ($p = 0.000$), Physical Demand ($p = 0.010$), Temporal Demand ($p = 0.002$), Performance ($p = 0.000$), Effort ($p = 0.000$), Frustration ($p = 0.004$)).

4.2. Analysis of the Working Time

In Experiment 1, Also, in significance level 0.05, N = 10. Working time was optimal at 60° angle and Touch mode. But, significant value is 0.928. So, It showed that the results are not significant (see **Figure 6**). Because, Flight Simulator X is regarded as not being optimized for the touch screen, In Experiment 2, Also, in significance level 0.05, N = 10. the experiment was conducted in a different environment. As a result, Touch mode is faster than Non Touch mode. Also, significant value is 0.034. So, the results were significant (see **Figure 7**).

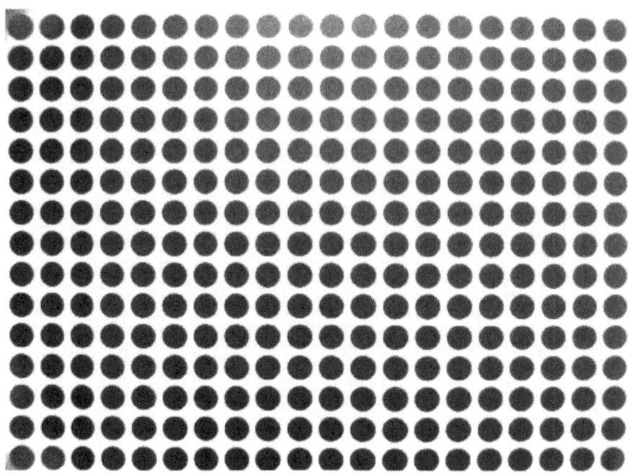

Figure 3. Interface of Experiment 2.

Table 2. Experiment design.

Scale	Input Mode	
	Non Touch	Touch
Lower Display angle	45°, 60°, 80°	45°, 60°, 80°
Time	Measured in Seconds	Measured in Seconds
NASA-TLX	Rating from given 6 questions (100 points range with 5-point steps)	Rating from given 6 questions (100 points range with 5-point steps)

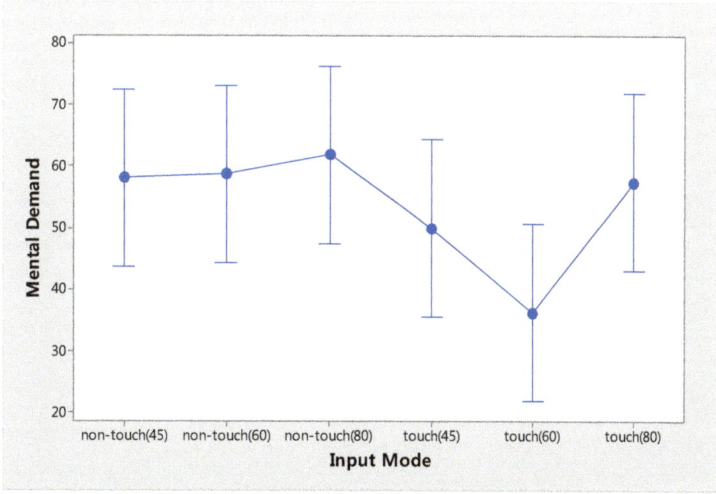

Figure 4. NASA-TLX according to input mode (in Experiment 1).

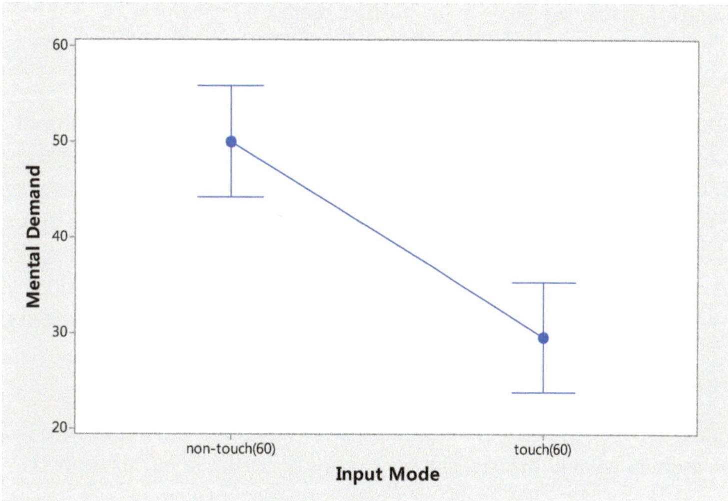

Figure 5. NASA-TLX according to input mode (in Experiment 2).

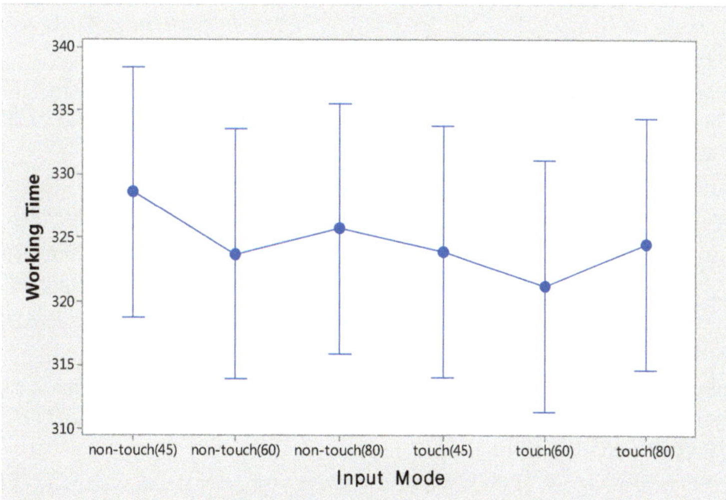

Figure 6. Working time according to input mode (in Experiment 1).

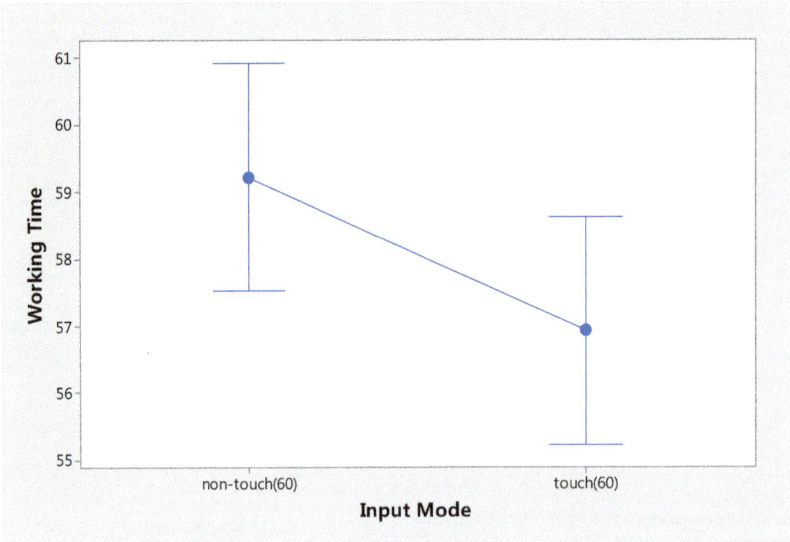

Figure 7. Working time according to input mode (in Experiment 2).

5. Conclusion

As a result, in Experiment 1, it was found that Touch mode was the best results of 60° angle. Also, in Non Touch mode, 80° angle is optimal but, in NASA-TLX, Touch mode is far more efficient than Non Touch mode. Therefore, it was found that Touch mode is optimal and in Touch mode, 60° angle is optimal. But, Flight Simulator X using in Experiment 1 is regarded as not being optimized for the touch screen, the significance level showed that the working speed is not significant. So, secondarily we performed different experiment about working speed and workload. In different experiment, it was found that touch mode is faster than Non Touch mode. As a result, we conclude that Touch mode is more optimal about working speed and workload.

Acknowledgements

This work was supported by the Agency for Defense Development (ADD) under the Contract No. UD140066CD. The authors wish to express sincere gratitude for the financial support.

References

[1] Hart, S.G. and Staveland, L.E. (1988) Development of NASA-TLX (Task Load Index): Results of Empirical and Theoretical Research. In: Hancock, P.A. and Meshkati, N., Eds., *Human Mental Workload*, Elsevier Science Publisher B. V., North-Holland, 139-183. http://dx.doi.org/10.1016/s0166-4115(08)62386-9

[2] Campbell, D.J. and Gingrich, K.F. (1986) The Interactive Effects of Task Complexity and Participation on Task Performance: A Field Experiment. *Organizational Behavior and Human Decision Processes*, **38**, 162-180. http://dx.doi.org/10.1016/0749-5978(86)90014-2

[3] Woods, D.D. (1990) On Taking Human Performance Seriously in Risk Analysis: Comments on Dougherty. *Reliability Engineering and System Safety*, **29**, 375-381. http://dx.doi.org/10.1016/0951-8320(90)90022-F

[4] Miyake, S. (2001) Multivariate Workload Evaluation Combining Physiological and Subjective Measures. *International Journal of Psychophysiology*, **40**, 233-238. http://dx.doi.org/10.1016/S0167-8760(00)00191-4

[5] Hill, S.G., Iavecchia, H.P., Byers, J.C., Bittner, Jr., A.C., Zaklad, A.L. and Christ, R.E. (1992) Comparison of Four Subjective Workload Rating Scale. *Human Factors*, **24**, 429-439.

[6] Hall, A., Cunningham, J., Roache, R. and Cox, J. (1988) Factors Affecting Performance Using Touchentry Systems: Tactal Recognition Fields and System Accuracy. *Journal of Applied Psychology*, **73**, 711-720. http://dx.doi.org/10.1037/0021-9010.73.4.711

[7] Ostroff, D. and Shneiderman, B. (1988) Selection Devices for User of an Electronic Encyclopedia: An Empirical Comparison of Four Possibilities. *Information Processing & Management*, **24**, 665-680. http://dx.doi.org/10.1016/0306-4573(88)90004-0

[8] Ahlstrom, B. and Lenman, S. (1987) Touch Screen, Cursor Keys, and Mouse Interaction. Work with Display Units, Elsevier, North-Holland, 831-837.

[9] Karat, J., McDonald, J. and Anderson, M. (1986) A Comparison of Selection Techniques: Touch Panel, Mouse, Keyboard. *International Journal of Man-Machine Studies*, **1**, 73-92. http://dx.doi.org/10.1016/S0020-7373(86)80034-7

3

Foreign Fiber Image Segmentation Based on Maximum Entropy and Genetic Algorithm

Liping Chen, Xiangyang Chen, Sile Wang, Wenzhu Yang, Sukui Lu

School of Computer Science and Technology, Hebei University, Baoding, China
Email: mc_chenlp@sina.com

Abstract

In machine-vision-based systems for detecting foreign fibers, due to the background of the cotton layer has the absolute advantage in the whole image, while the foreign fiber only account for a very small part, and what's more, the brightness and contrast of the image are all poor. Using the traditional image segmentation method, the segmentation results are very poor. By adopting the maximum entropy and genetic algorithm, the maximum entropy function was used as the fitness function of genetic algorithm. Through continuous optimization, the optimal segmentation threshold is determined. Experimental results prove that the image segmentation of this paper not only fast and accurate, but also has strong adaptability.

Keywords

Foreign Fibers, Image Segmentation, Maximum Entropy, Genetic Algorithm

1. Introduction

The foreign fibers in cotton refer to those non-cotton fibers and dyed fibers, such as polypropylene fiber silk, hemp, feathers, colored line, colored cloth, hairs and so on. Though very low content of foreign fibers in cotton, the presence of foreign fibers will seriously affect the quality of the final cotton textile products, as they may debase the strength of the yarn, they are not easy to be dyed, and this will lead to great economic loss for the cotton textile enterprises [1]. Therefore, fast and accurate measurement of foreign fibers in lint cotton enterprises is an urgent problem. Using machine vision technology to identify and measure the foreign fiber is an effective and feasible solution. The basis and key of the measurement of the foreign fiber content is the foreign fiber's classification, and the image segmentation is the premise and guarantee of the classification of the foreign fiber.

Image segmentation is one of the primary stages in image processing and machine vision system, and it is also the precondition of image analysis. The objective of image segmentation is to partition the image into meaningful connected-components to extract the features of objects [2]. According to the means of image segmentation, the image segmentation method can be divided into threshold method, boundary detection method, area method and so on. Among them, the threshold method is the earliest studied and used. At present, due to the characteristics of clear physical meaning, obviously effect, easily implementing and good real-time, the threshold method became one of the most commonly used image segmentation method in image analysis, image recognition and machine vision systems.

The current threshold segmentation method, according to the threshold selection criterion function types, can be divided into maximum entropy method [3], maximum inter-class variance (Otsu) method, cross entropy method [4], minimum error method, fuzzy entropy method and so on [5]. How to determine the optimal threshold is the key of the threshold segmentation method. Since the maximum entropy method does not require a priori knowledge, it is widely used in image segmentation. But the maximum entropy method is through exhaustive search of global optimal solution, it is very complex and slow. In recent years, due to the fast random search capability, genetic algorithm (GA) has been successfully applied in many fields, such as image processing, machine learning, parameter optimization and so on [6] [7]. In this paper, through combine GA with maximum entropy method, we realized the fast segmentation of the foreign fibers.

The remainder of this paper is organized as follows. Section 2 presents the maximum entropy method and the GA briefly. Section 3 provides the proposed algorithm in detail. Results and discussions are followed in Section 4. Section 5 declares the conclusion.

2. Methodology

2.1. Maximum Entropy

Maximum entropy algorithm is a segmentation method which based on the histogram of the image [8]. Let x be the gray level of the image (from 0 to L), $p(x)$ is the probability of the pixel which gray level is x in the image, then the entropy of the image is represented as:

$$H = -\sum_{x=1}^{L} p(x) \lg p(x) \tag{1}$$

Let the segmentation threshold is t, which will divide the image into two parts: target A and background B. Assuming that the gray values which less than the threshold of the pixels belong to the target A, the gray values which greater than or equal to the threshold of the pixels belong to the background B, then the probability of pixels in the image are related to target A and background B respectively is:

Target A:

$$p_a = \sum_{i=0}^{t-1} p_i \tag{2}$$

Background B:

$$p_b = \sum_{i=t}^{255} p_i \tag{3}$$

The entropy of the target region and the background region are respectively defined as:

$$H_a(t) = -\sum_{i=0}^{t-1} (p_i/p_a) \lg (p_i/p_a) \tag{4}$$

$$H_b(t) = -\sum_{i=t}^{255} (p_i/p_b) \lg (p_i/p_b) \tag{5}$$

The entropy of the whole image is defined as:

$$H(t) = H_a(t) + H_b(t) \tag{6}$$

The greater the value of function $H(t)$, the more information of the target and background represent, the more accurate the target and the background area. The global optimal threshold (let it be t^*) is the value which make $H(t)$ take the maximum value, that is, t^* can be expressed as:

$$t^* = \arg\max H(t) \tag{7}$$

2.2. Genetic Algorithm

Genetic algorithm (GA) is a parallel random search optimization method which simulates natural genetic mechanism and biological evolution theory. It is stochastic rather than exhaustive. It can obtain the global optimal solution with large probability. In addition, GA does not require the evaluation function to be monotone, so GA is very suitable for real-time processing, which is not only easy to implement, but also has strong robustness. The basic idea of the algorithm is first encoding the problem to be solved into a string (named chromosome). Each chromosome is a solution to the being solved problem. Next, GA calculates the fitness value of each individual in a group of chromosomes, then chromosomes with higher fitness are selected for crossover and mutation operation and then produce a new generation with higher fitness. At last, the relative optimal solution is obtained by continuous iteration.

3. Foreign Fiber Image Segmentation Based on Maximum Entropy and Genetic Algorithm

Through analysis the cotton foreign fiber images, it is found that the background of the cotton layer has the absolute advantage in the whole image, while the target is very small (**Figures 2(a)-5(a)**). Furthermore, the brightness and contrast of the image are all poor. Using the traditional image segmentation method (e.g. Otsu), the segmentation results are very poor. Therefore, we adopt the maximum entropy threshold method in this paper. However, the essence of the maximum entropy method is to obtain the maximum value of the objective function (Equation (6)) in the gray space, the computation of the algorithm is very large, and it takes much more time. In order to reduce the complexity of the algorithm, we combine GA with the maximum entropy in the foreign fiber image segmentation. The maximum entropy function was used as the fitness function of GA. By continuous optimization, the optimal segmentation threshold is determined. The flow chart of the algorithm was shown as **Figure 1**.

3.1. Encoding and Design Fitness Function

Encoding a solution of a problem into a serial of genes, called chromosome, is very important when using GA. Various encoding methods have been proposed for particular problems to provide effective implementation of GA, such as binary encoding, real number encoding, integer or literal permutation encoding, or general data structure encoding [9]. In our research, we chose the binary encoding method. Since the gray level of the foreign fiber image has 256 gray level, so encoding the segmentation threshold with 8 binary string, and take the entropy function of Equation (6) as the fitness function.

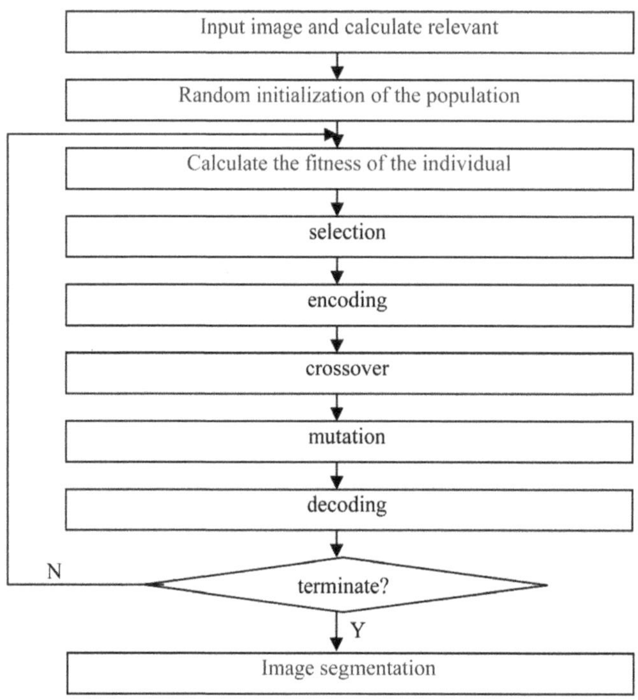

Figure 1. Flowchart of the proposed method.

3.2. Population Initialization

The initial population for evolution is created randomly. Individual in the initial population is equivalent to the candidate solution in the solution space. If the population size is too large, the computational complexity is high; if the scale is too small, the search space is limited, the search may stop at the immature stage. Therefore, the population number is set to 20, the maximum number of iterations is set to 100.

3.3. Selection Operation

The selection operation, also known as the copy operation, is to determine which individual is genetic and which individual is eliminated. The common methods include roulette wheel method, the elite preservation strategy and the order selection method. As the crossover and mutation operations are performed, the optimal solution can be easily lost in an intermediate step. So we adopted the method of 10% elite strategy and 90% roulette wheel. Individuals with the largest fitness directly into the next generation, which can guarantee that the algorithm converges to the global optimal solution; the remaining 90% of the individuals are selected according to the fitness proportionate. The larger the fitness of the individual, the greater the probability of being selected. The probability of the individual i is selected to the next generation group is:

$$P_i = F_i \Big/ \sum_{i=1}^{n} F_i \tag{8}$$

where n is the population size, F_i is the fitness of the individual i.

3.4. Crossover Operation

Commonly used crossover operations include single-point crossover, double-point crossover, multi-point crossover, etc. In our research, we chose the single-point crossover operator, the crossover probability is 0.8 and 0.6 respectively.

3.5. Mutation Operation

According to the mutation probability, the individual's value is replaced by some other gene values, which can increase the diversity of the population, and can also improve the local search ability of GA. Various mutation operations have been introduced by researchers, such as inversion mutation, neighbor exchange mutation, etc. We chose inversion mutation in our research.

3.6. Probability of Crossover and Mutation

The probabilities of crossover and mutation, denoted by P_c and P_m, affect seriously the search ability and convergence speed of GA [10]. For the standard GA, P_c and P_m are all fixed. But in the early stage of evolution, the individual difference is bigger, so we should increase P_c and reduce P_m in order to accelerate the evolution process and at the same time reducing the amount of calculation. In the later stage of evolution, the individual difference is small, so P_m should be reduced and P_c should be increased to avoid local optimal solution.

For the proposed algorithm, P_c is 0.8 during the first 50 generations of the evolution, so as to search the optimal solution as soon as possible. In the second part of the evolution, P_c is 0.6. The mutation probability P_m is calculated by:

$$P_m = \begin{cases} P_{m0} / \log_2 (g+1) & f \geq \bar{f} \\ P_{m0} & f \leq \bar{f} \end{cases} \tag{9}$$

where P_{m0} is the initial mutation probability; f is the fitness of the parent chromosome; $\bar{f}$ is the average fitness of the chromosomes in the current population [11].

3.7. Terminate Criterion

We set two terminate criterions in our GA: 1) maximum generation number. When the maximum number of iterations is reached, the algorithm terminates; 2) set the minimum number of iterations is 10, when the iteration number is larger than the value, check the difference between the average fitness of the current population and the average fitness of the last generation whether less than a minimum value ε (here $\varepsilon = 0.0001$), once the condition is satisfied, the iteration terminates.

4. Results and Discussions

In order to verify the effectiveness of the paper's algorithm, we collected 60 foreign fiber images which include 5 foreign fibers such as hair, feather, hemp rope, polypropylene, cloth and so on. The image is 128×4096 size. The experiments are performed on over Matlab7.1 platform. We compared the results of the paper's method with the results of the Otsu's method. **Figures 2-5** show a part of the experimental results.

It can be seen from **Figures 2(a)-5(a)**, for the original image, the cotton layer background occupies the absolute advantage in the whole image, while the foreign fiber only accounts for a very small part, and the contrast of the image is very low. Therefore, in many cases, the target cannot be separated from the background by using the Otsu method (e.g. **Figures 2-4**). For **Figure 2(c)** and **Figure 3(c)**, although the target was segmented, but part of the background was divided into the target too. The foreign fiber of **Figure 4** was not extracted at all. So the reliability of the Otsu is very poor. In contrast, this paper's method works well in most cases. The extraction of the target is accurate and complete. **Table 1** lists the statistical segmentation results of 60 foreign fiber images with two different methods.

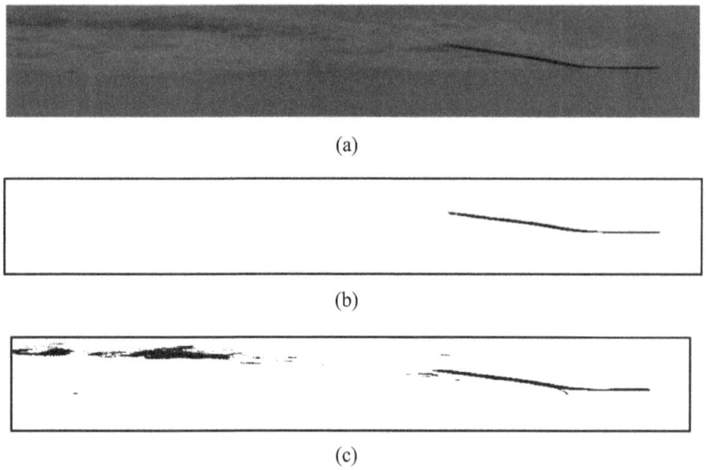

Figure 2. (a) Color image of polypropylene fiber; (b) The result of the paper's method; (c) The result of Otsu.

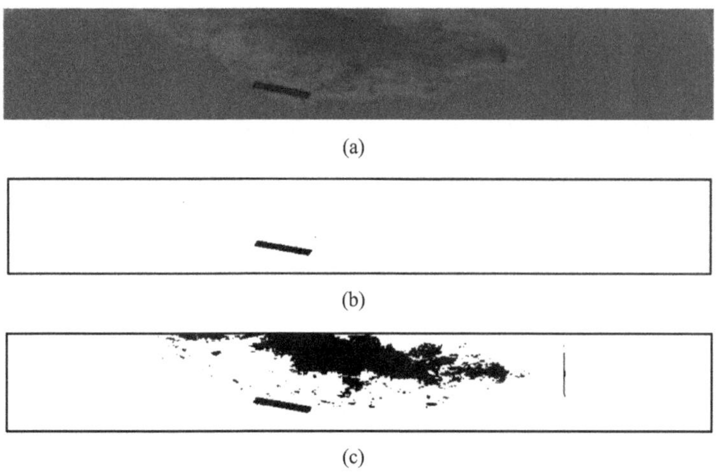

Figure 3. (a) Color image of cloth piece; (b) The result of the paper's method; (c) The result of Otsu.

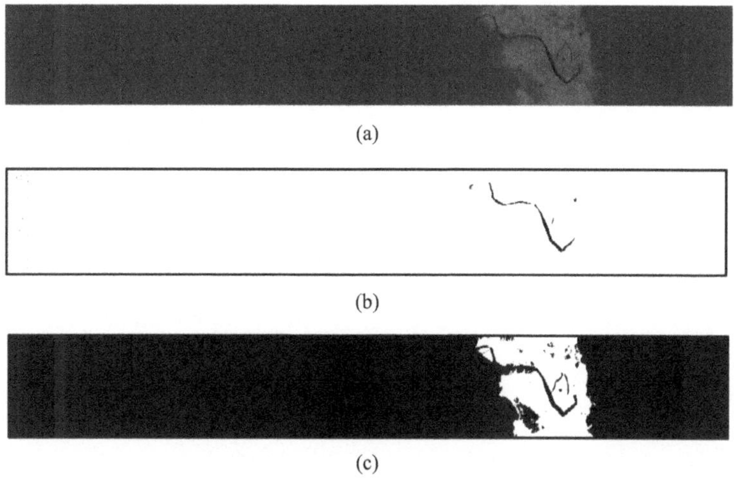

Figure 4. (a) Color image of hemp rope; (b) The result of the paper's method; (c) The result of Otsu.

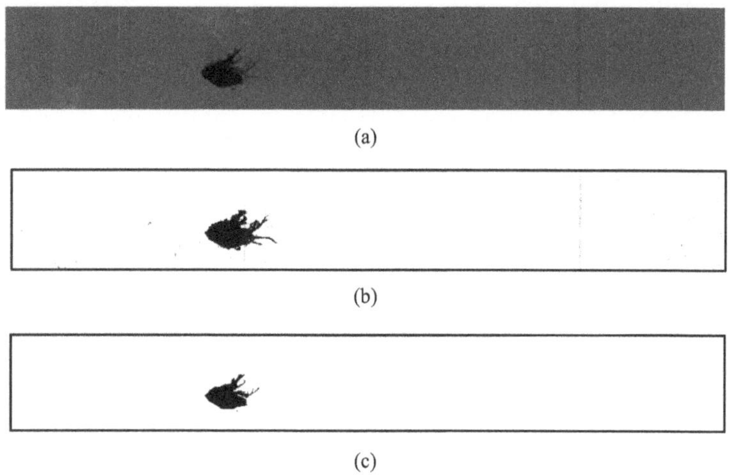

Figure 5. (a) Color image of feather; (b) The result of the paper's method; (c) The result of Otsu.

Table 1. Caparison of the segmentation results by two different methods.

Image name	Image number	Segmentation accuracy	
		The proposed method	Otsu
Hair	10	90%	20%
Feather	15	100%	100%
Polypropylene	10	100%	50%
Cloth	15	100%	60%
Hemp rope	10	100%	20%

The results shown in **Table 1** indicated that for different foreign fiber images, the segmentation accuracy of the proposed method is very high, it has strong adaptability. While for the Otsu method, the segmentation accuracy is higher when the area of the target is larger (such as the feather images, the segmentation accuracy is 100%). When the foreign fiber's area is small, the segmentation accuracy is very low (such as hair and twine images, the accuracy rate is only 20%). So, for low contrast images, the method of combining the maximum entropy and genetic algorithm can effectively improve the image segmentation accuracy.

5. Conclusions

Due to the low brightness, low contrast and small proportion of the target of the foreign fiber images, we adopted the image segmentation based on maximum entropy and genetic algorithm. Experimental results show that the adopted algorithm is more accurate than the traditional Otsu method, and the speed is fast and adaptable, which meets the real-time requirements of the foreign fiber detection system.

Next work: for the detection of white foreign fibers (such as plastic sheeting, plastic film, etc.), the accuracy of the segmentation algorithm is poor, there is still to be further improved.

Acknowledgements

The authors thank Hebei Natural Science Foundation (F2015201033), the Ministry of Science and Technology of the People's Republic of China (2013DFA11320), for their financial support.

References

[1] Yang, M.H. (2014) Hazards and Prevention of Foreign Fiber in Cotton. *Jiangsu Textile*, **4**, 44-46. (In Chinese with English Abstract)

[2] Zhang, Y.J. (2013) Image Engineering (II) Image Analysis. Tsinghua University Press, Beijing.

[3] He, Z.J. and Wang, H.F. (2010) Chinese Word Sense Disambiguation Based on Maximum Entropy Model with Feature Selection. *Journal of Software*, **21**, 1287-1295. http://dx.doi.org/10.3724/SP.J.1001.2010.03591

[4] Wu, Y.Q. and Zhang, X.J. (2011) Two-Dimensional Symmetric Cross-Entropy Image Thresholding. *Journal of Image and Graphics*, **16**, 1393-1401.

[5] Wu, Y.Q., Meng, T.L. and Wu, S.H. (2015) Research Progress of Image Threshold Methods in Recent 20 Years (1994-2014). *Journal of Data Acquisition and Processing*, **30**, 1-23.

[6] Ou, P. and He, D. (2011) 2-D Maximum Entropy Method of Image Segmentation Based on Genetic Algorithm. *Journal of Computer Simulation*, **1**, 294-297.

[7] Guo, M.S. and Liu, B.H. (2008) 2-D Maximum Entropy Method in Image Segmentation Based on Chaos Genetic Algorithm. *Computer Technology and Development*, **18**, 101-104.

[8] Cao, J.N. (2012) Review on Image Segmentation Based on Entropy. *Pattern Recognition and Artificial Intelligence*, **25**, 958-959.

[9] Zhang, C.Q., Zheng, J.G. and Qian, J. (2011) Comparison of Coding Schemes for Genetic Algorithms. *Application Research of Computers*, **28**, 819-822.

[10] Cao, D.Y. and Cheng, J.X. (2010) A Genetic Algorithm Based on Modified Selection Operator and Crossover Operator. *Computer Technology and Development*, **20**, 44-51.

[11] Yang, W.Z., Li, D.L. and Zhu, L. (2010) An Improved Genetic Algorithm for Optimal Feature Subset Selection from Multi-Character Feature Set. *Journal of Agricultural Machinery*, **38**, 2733-2740.

4

An Application of Canny Edge Detection Algorithm to Rail Thermal Image Fault Detection

Libo Cai[1], Yu Ma[1], Tangming Yuan[1*], Haifeng Wang[2*], Tianhua Xu[3]

[1]Department of Computer Science, University of York, Heslington, York, UK
[2]National Engineering Research Center of Rail Transportation Operation and Control Systems, Beijing Jiaotong University, Beijing, China
[3]Laboratory of Rail Traffic Control and Safety, Beijing Jiaotong University, Beijing, China
Email: *tommy.yuan@york.ac.uk, *hfwang@bjtu.edu.cn

Abstract

The paper discusses an application for rail track thermal image fault detection. In order to get better results from the Canny edge detection algorithm, the image needs to be processed in advance. The histogram equalization method is proposed to enhance the contrast of the image. Since a thermal image contains multiple parallel rail tracks, an algorithm has been developed to locate and separate the tracks that we are interested in. This is accomplished by applying the least squares linear fitting technique to represent the surface of a track. The performance of the application is evaluated by using a number of images provided by a specialised company and the results are essentially favourable.

Keywords

Fault Detection, Rail Thermal Image, Canny Edge Detection, Linear Least Squares

1. Introduction

Rail track failures may directly lead to train derailment which may cause destruction and threaten people's lives. Detecting rail track defects in advance can improve the reliability of the railway transport systems. In addition, the low efficiency and high error rate of manual detection is no longer appropriate for the fast moving railway industry, *i.e.* high speed rail. An application for automatic and effective fault detection is needed and the aim of this paper is to contribute to addressing this.

Currently, the most popular non-destructive method to detect rail defects is ultrasonic techniques [1]. Ultrasonic testing needs the direct contact between the probe and the railhead in order to detect the rail squat, but the false and missed detections are unavoidable. Infrared thermal imaging is a relatively new but widely used non-destructive inspection technology, and it can be used to detect the fault of the track surface rather than rail squats

*Corresponding authors.

[2].

In order to detect the faults of the track surface, a vision based inspection technique [3] is adopted in this paper. Since an edge consists of a great deal of information for an image, edge detection is treated as the key process for image processing and analysis. The information provides the foundation for image segmentation, character determination, image categorization and pattern recognition. Canny edge detection algorithm [4] is one of the most classical methods for edge detection and it specifies three principles, namely, low error rate, good localization and low spurious to a single edge. Although Canny edge detection is a good tool for edge detection, it may not be directly applicable to every situation. This paper represents an improved version of the Canny algorithm to the processing of rail track thermal image.

The rest of the paper is organised as follows. Section 2 discusses the method of incorporating histograms with Canny edge detection algorithm in order to process rail thermal image. Section 3 provides the method of applying the improved Canny algorithm to the rail track fault detection. Section 4 discusses the software application that has been specially developed for this purpose. An evaluation of the algorithm and the software tool has been conducted and documented in Section 5.

2. Incorporate Histograms with Canny Edge Detection Algorithm

Canny algorithm utilizes a mask generated by Gaussian function as Equation (1) to reduce the noise of the original image and the first derivative Gaussian is shown in Equation (2) which is used for computing gradient. According to the features of Gaussian function and the consideration of the computation cost, one-dimensional instead of two-dimensional mask is adapted to the solution as the convolution computation [5].

$$G(x) = e^{-\frac{x^2}{2\sigma^2}} \quad (1)$$

$$G'(x) = \left(-\frac{x}{\sigma^2}\right) e^{-\frac{x^2}{2\sigma^2}} \quad (2)$$

In some original thermal images, the rail tracks are difficult to be distinguished from the background due to the low contrast. Therefore, before reducing the noise by using Gaussian kernel, it is necessary to enhance the contrast of the raw thermal image, and the reason for this is that the value of gradient will be enhanced when calculating the potential edge by using the algorithm. As a result, the edge will be recognized easier. As an efficient approach to the contrast enhancement, histogram equalization (HE) [6] is incorporated to the Canny algorithm for image pre-processing as discussed below.

For an image with width N and length M, the total number of pixels is $M * N$, and let n_i be the occurrence times in the total pixels with grey level x_i, then p_i, the probability of occurrence times of grey level x_i, can be defined as Equation (3). L is the number of grey levels and in the current application its value is always 256 and this means the range of grey level is from 0 to 255.

$$p(x_i) = \frac{n_i}{MN}, 0 \le i < L \quad (3)$$

If let $s(x_i)$ is a transformation of $p(x_i)$ that is defined as Equation (4) below, the histogram generated by $s(x_i)$ is the cumulative histogram. For example, there is a histogram of one original thermal image demonstrated as **Figure 1**. The percentage of normal histogram is displayed with a bold line, while the cumulative histogram is drawn thinner than normal histogram line and presented as a monotonically increasing line.

$$s(x_i) = T(x_i) = \sum_{j=0}^{i} p(x_j) \quad (4)$$

Then, $x_{\min}$ can be defined as the minimum level of grey when the value of transformation is first time greater than 0, $x_{\max}$ can be defined as the maximum level of grey when the value of transformation is first time reach up to $M * N$ respectively. Then let the equalized level of grey be r_i, the definition of r_i is shown in Equation (5).

$$r_i = equalize(x_i) = Round\left[\frac{x_i - x_{\min}}{x_{\max} - x_{\min}}(L-1)\right] \quad (5)$$

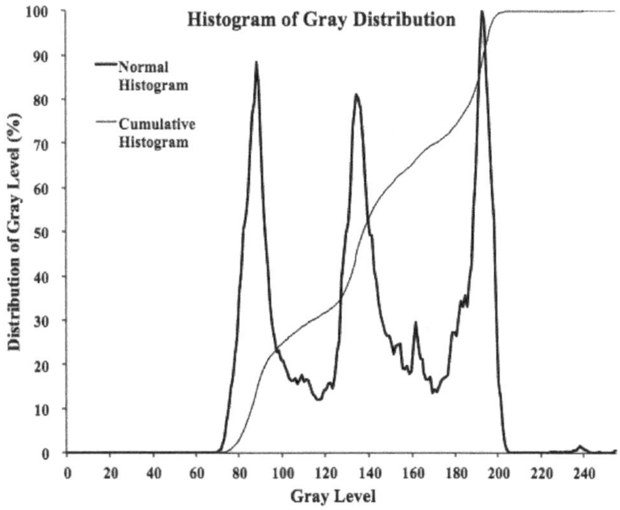

Figure 1. Histogram of gray distribution.

After histogram equalization, the contrast of original image is strengthened and the new histogram should be distributed from 0 to 255 (the max level of grey). However, sometimes if the distribution of pixels which belong to noise or other objects is far from that of the rail, the effect of HE will be greatly reduced, or even completely disappeared. Therefore, in this case, utilizing HE to enhance the contrast with starting and ending percentage instead of adopting $x_{\min}$ and $x_{\max}$ is a possible means to suppress the influence.

3. Rail Track Detection

The most significant step of rail detection is the determination of edges of rail from all edges detected by the Canny algorithm discussed in Section 2 above. This is one of the key requirements for rail thermal image detection. Before the recognition of the edges, the set of points for each edge can be collected by an iterative algorithm from eight directions for each edge point. The algorithm is described below:

Define *an integer array with the size of the total number of pixels*

Initialize *the current edge index as* 0

Scan *from point* P_0 (0, 0) (*top left of the image*) ***until*** P_1 (*width*-1, *height*-1) (*bottom right of the image*):

for *each point, check whether it is an edge point or not,*

if *it is not* ***then*** *check the next point.*

if *the point is an edge point* ***and*** *the point is not marked with any index of edge,*

then *set current edge index as the value of the element that can represent the point from integer array defined earlier,* ***and*** *check all eight directions around the point (northwest, north, northeast, west, east, southwest and southeast) respectively. This checking should be iterative until meet the ending condition and should be marked with the same edge index. The current index number should be transferred as parameter. The current edge index increases by* 1 *after each iteration.*

The output from the algorithm above represents the set of edge points. By analysing a large number of detected edge point results, it can be found that rail edges have three features in rail thermal images. Firstly, since the thermal cameras are positioned at the bottom of a train, the current rail track will always be in the lower half of the image. Secondly, in most of images with crossing over junction, the edges of rails are always displayed as a group of single lines, which are similar to horizontal lines but with a little lean. Thirdly, the edges of rail should be longer than other edges, or it should be combined with some edges which have the similar angle of tilt. Combining the above three features, the linear function with slope and intercept is selected as the characterization function.

Since the collection of points for each detected edge is not difficult to be obtained, the horizontal and vertical coordinates for each point that belongs to one edge can be used for calculating the fitted line by Linear Least Squares Fitting Algorithm [7] and all the edges can be expressed by a linear function with slope and intercept as shown in Equation (6),

$$y = f(x) = kx + b \tag{6}$$

where k is slope and b is intercept.

4. Railway Track Thermal Analyzer

By applying the algorithms and methods discussed in Section 2 & 3 above, a railway track thermal analyzer System has been implemented by a team of MSc Software Engineering students at the University of York [8]. The tool can be used to analyze the proprietary LSJPI/LSJPG video files to automatically identify thermal hotspots in the rail captured by the camera. The system is divided into eight modules, as shown in **Figure 2**, which also details the flow of data between the modules.

Thermal Analyzer is the main driver class for the application. It sets up the logging object, then obtains an instance of Controller and passes it the path supplied to the program by the user. The *Controller module* has the job of coordinating all the other main functional modules. If requested to run a single video file, the Controller obtains an instance of the FileReader module linked to that video file and requests a sample of frames from the video. It also obtains an instance of the EdgeDetector module and, for each sample frame, the Controller requests edge detection results from the EdgeDetector. Once it has all the edge detection results, the Controller requests the RegionCalculator module to analyze them and calculate the region of the video frame images that should be analyzed for hotspots. The aim is to restrict the analysis to the rail of interest to reduce processing overhead and to ignore hotspots in a second railway track often visible in the upper portion of the video image.

The *FileReader* module is mainly responsible for reading and retrieving information from the video files. The video files are a proprietary format, comprising metadata and a series of grey scale, thermal JPG images. The *FileReader* reads and retrieves metadata from the LSJPI file and uses it to access the image data in the related LSJPG file. The *EdgeDetector* module is responsible for analyzing a video frame image and identifying the position of the edges that bound the rail of interest for analysis. It uses the Canny Edge Detection algorithm discussed in Section 2 above. The *Region Calculator* takes the results of the *EdgeDetector* module and determines the area of the video image that should be analysed for hotspots, *i.e.* where in the image the rail of interest appears throughout the video. The Region Calculator implements the algorithm described in Section 3.

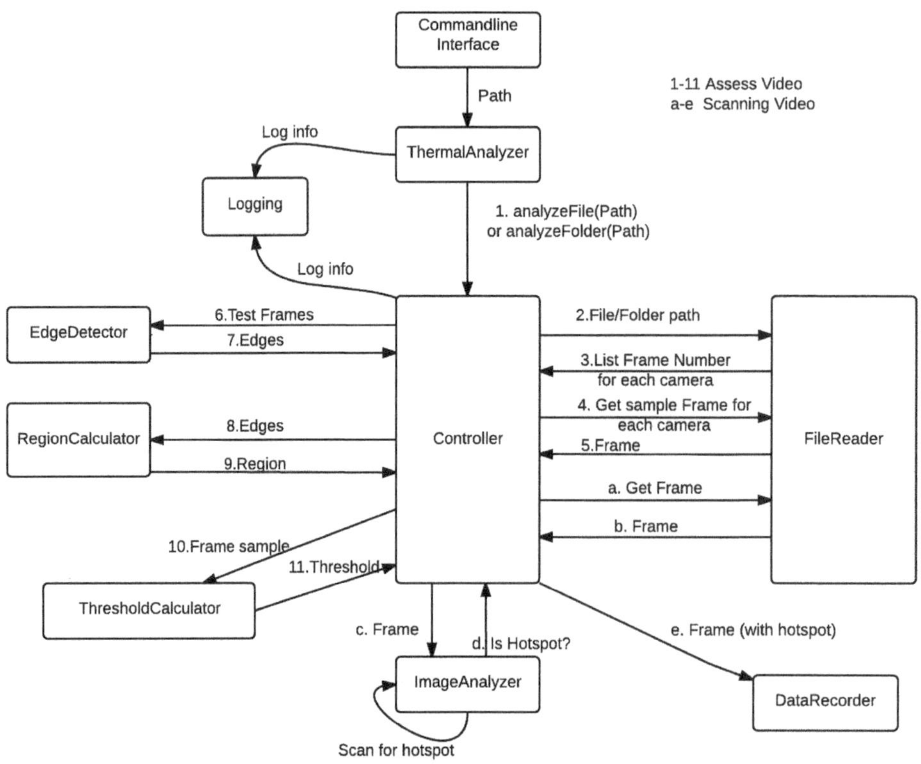

Figure 2. Architecture of railway track thermal analyzer.

The *Threshold Calculator* module is a simple implementation that returns a fixed value, either as specified in the configuration file or the default value of 255. This value is used by the *ImageAnalyzer* to determine whether a frame image pixel is hot and thus potentially part of a hotspot. A desirable approach might be to calculate the appropriate threshold based on the qualities of each video. However, there was not time to investigate such a method and using a fixed value of the maximum possible brightness was suitable for the data observed so far. This functionality is included in a module (rather than a simple hard coded value) to enable easy modification should there be further work around threshold calculation.

The *Image Analyzer* module scans the specified region of a frame's image to find hotspots. Within the set region, it compares each pixel's brightness to the set threshold to determine whether it is bright enough to be considered hot. It keeps count of the hot pixels found and if there are more than a specific value (this value can be set in the configuration values; the default is 20), then the frame is deemed to have a hotspot. A more complex algorithm that checks locality of the hot pixels found (to prevent a hotspot being declared when the hot pixels are spread throughout the regions, for example) were desired, but this feature is not implemented in the current version for the system. We are catering this in the future. The *Data Recorder* module records and retrieves data relating to video frames to and from a database. The *Logging* module provides logging facilities, recording messages passed with a timestamp in text files.

5. Evaluation

The Algorithms and method discussed above have been evaluated by using the tool developed in Section 4 above. The data files for evaluation are provided by a specialised railway software company at York. **Figure 3** illustrates the results for the processing with HE. The original image is shown in **Figure 3(a)** with its histogram in **Figure 3(b)**. In **Figure 3(c)**, the original image equalized with 5% as starting and 95% as ending point, which greatly enhance the contrast of the image. In addition, the histogram of the equalized image in **Figure 3(d)** can be clearly seen that the grey level of the image has been equalized. On the other hand, it should be noticed that although the weak edges can be obtained more effectively, some part of data are also lost which probably leads to image distortion.

Figure 4 shows the result for railway track detection. The original image is shown in **Figure 4(a)**, the edges with red lines that are generated by Canny edge detection are displayed based on original image, which is shown

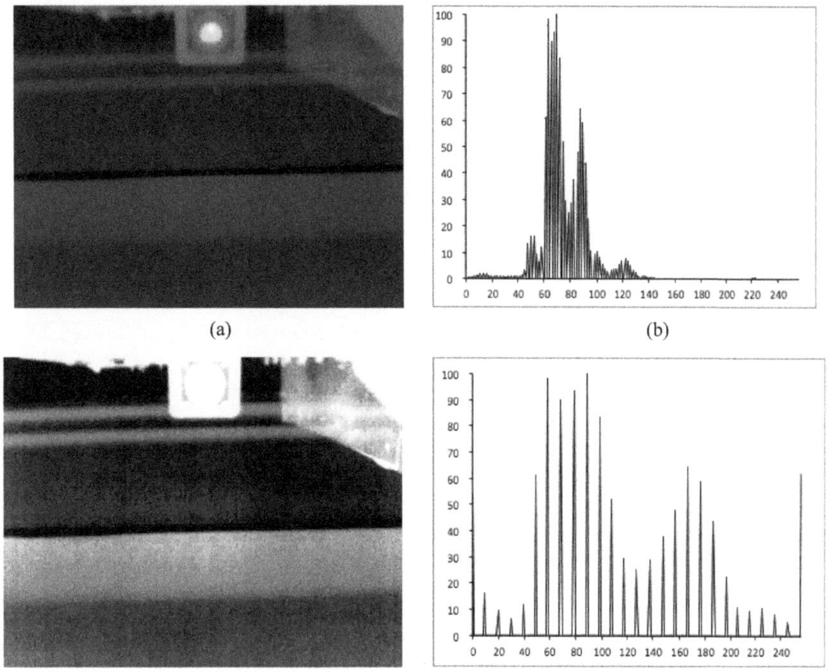

Figure 3. Histogram equalization processing.

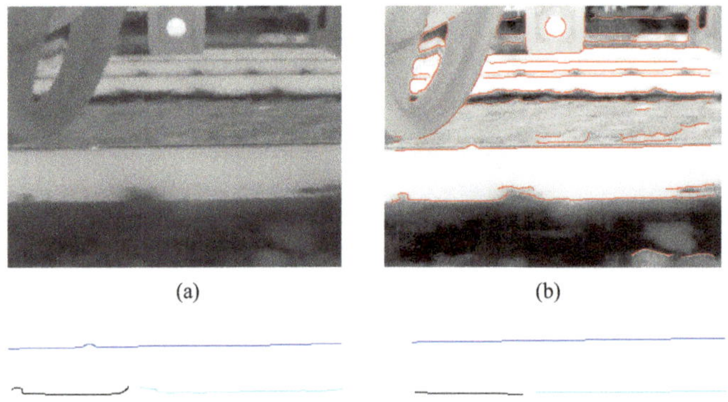

(a) (b)

(c) (d)

Figure 4. Rail track detection.

in **Figure 4(b)**. Note that the histogram equalization processing is performed before Canny edge detection, and the low threshold and high threshold are 2.5 and 7.5 respectively. The edges need to be filtered by a series of rules to get the region that the rail track belongs to. Using the algorithm described in Section 3 to collect all the points of edges. The test image with detected edges is shown in **Figure 4(c)**, each of colours represents an edge. The points in an edge are used to fit a straight line, and the result of Linear Least Squares Fitting Algorithm is shown in **Figure 4(d)**.

6. Conclusion

A railway track thermal analyzer has been presented. The original image is pre-processed by using histogram equalization prior to Canny edge detection. An iterative algorithm is used to obtain appropriate edges of rail tracks. These edges are then filtered by some conditions to get the appropriate region of a track. The pixels in the rail track region are scanned to judge whether hotpots are present. The system is able to generate an output description file with hotspot information. In the future, more complicated situations other than hotspots need to be considered, for example, the recognition of the region of a rail joint. The threshold values in Canny edge detection are currently fixed and they can only be adjusted manually. It would be necessary to tune these thresholds for Canny edge detection and the adaptive thresholds would be useful to get a better result of edges.

Acknowledgements

This work was supported by the National Natural Science Foundation of China (Grant No. 61473029), and the Science and Technology R&D program of China Railway Corporation (Grant No. 2014X008-A).

References

[1] Clark, R. (2004) Rail Flaw Detection: Overview and Needs for Future Developments. *NDTE International Independent Nondestructive Testing and Evaluation*, **37**, 111-118. http://dx.doi.org/10.1016/j.ndteint.2003.06.002

[2] Peng, D. and Jones, R. (2013) NDI of Rail Squats and Estimating Defect Size and Location Using Lock-In Thermography. *Engineering*, **5**, 29-38. http://dx.doi.org/10.4236/eng.2013.51005

[3] Deutschl, E., *et al.* (2004) Defect Detection on Rail Surfaces by a Vision Based System. *IEEE Intelligent Vehicles Symposium*, 507-511. http://dx.doi.org/10.1109/ivs.2004.1336435

[4] Canny, J. (1986) A Computational Approach to Edge Detection. *IEEE Transactions on Pattern Analysis and Machine Intelligence*, **8**, 679-698. http://dx.doi.org/10.1109/TPAMI.1986.4767851

[5] Parker, J.R. (2010) Algorithms for Image Processing and Computer Vision. 2nd Edition, John Wiley & Sons.

[6] Gonzalez, R. and Woods, R. (2008) Digital Image Processing. 3rd Edition, Prentice Hall.

[7] Freedman, D. (2009) Statistical Models: Theory and Practice. Revised Edition, Cambridge University Press. http://dx.doi.org/10.1017/CBO9780511815867

[8] Cai, L., *et al.* (2005) Railway Track Thermal Analyzer System Maintenance Manual. MSc Software Engineering Team Project Group Report, University of York.

5

Robot Control System Based on Electrooculography and Electromyogram

Minoru Sasaki[1], Muhammad Syaiful Amri Bin Suhaimi[1], Kojiro Matsushita[1], Satoshi Ito[1], Muhammad Ilhamdi Rusydi[2]

[1]Department of Mechanical Engineering, Gifu University, Gifu, Japan
[2]Department of Electrical Engineering, Andalas University, Padang, Indonesia
Email: sasaki@gifu-u.ac.jp, u3128026@edu.gifu-u.ac.jp, kojirom@gifu-u.ac.jp, satoshi@gifu-u.ac.jp, rilhamdi@yahoo.com

Abstract

This paper describes a man-machine interface system using EOG and EMG. A manipulator control system using EOG and EMG is developed according to EOG and EMG. With the eye movement, the system enabled us to control a manipulator. EOG is using for moving the robot joint angles and EMG is using for object grasping. The EOG and EMG discrimination method is used to control the robot. The robot arm joint movements are determined by the EOG discrimination method where the polarity of eye gaze motion signals in each Ch1 and Ch2. The EMG discrimination method is used to control arm gripper to grasp and release the target object. In the robot control experiment, we are successfully control the uArm™ robot by using both EOG and EMG discrimination method as the control input. This control system brings the feasibility of man-machine interface for elderly person and handicapped person.

Keywords

EOG, EMG, Man-Machine Interface, Robot Control

1. Introduction

Recently, machine and robot technology has progress rapidly, improving human live and making human work become easier. Unfortunately most of the conventional machine and robot are usually controlled by hand manipulation and have complicated control system. By the year 2014, the world's population has reach above 7 billion human but in that 7 billion human there are quite a percentage of people with disabilities and aged people. For example, in Japan alone, based on year 2009 report, the people who are with disabilities and need supports are about 4,690,000. For those aged people and people with disabilities, the ever growing machine and robot technology seem unfeasible to make use of it. The people with disabilities especially those who have physical disabilities, the hand manipulation based control system and the complicated control system are unusable for them. As for aged people, the human aging is closely related to the decrease of the mass and the voluntary contractile of muscle and this condition often causes a disability condition among older people. To overcome this problem with the machine and robot control system, human machine interface technology has being research progressively to solve the problem. In the human machine interface there many ways of control system exists and one of the prominent ways is by using the bio-signal.

The bio-signals are a type of signal that could be obtained by any living being. In the bio-signal, the bio-signal that are usually used as medium of communication for human and machine is electrooculography (EOG) and Electromyography (EMG). Electrooculography (EOG) is a method for measuring the cornea-retinal electrical potential that exists between the front and the back of the eye. Electromyography (EMG) in other hand is a method for obtaining the electrical potential produced by human muscles.

Performing some special tasks using EOG and EMG in daily activities is being developed in various areas. These inventions improve the life quality of people in human machine interface. EOG is recorded by attaching electrodes to the skin around the subject's eyes. As for EMG, the signal is recorded by attaching electrodes on effected muscle area that have relationship with human mastication or biting motion. Both of the EOG and EMG research used NF Instrument as the signal recording device. This device has an amplifier (EEG head box), a processors box, and four electrodes. The electrodes consist of one ground electrodes, one reference electrodes, channel 1 (Ch1) and channel 2 (Ch2) for recording both EOG and EMG signal simultaneously. Three digital filters were given by the processor box, which are 1.6 Hz High Pass Filter, 30 Hz Low-Pass Filter and 60 Hz Ham Filter. The 30 Hz Low-Pass Filter was added to reduce the electricity equipment noise beside effectively record the EMG signal produced from biting motion. The EMG signal theoretically has higher amplitude than EOG signal, the NF Instrument EEG head box used for the research could not record the EMG signal effectively unless the low-pass filter is set to lower value. There are three basic steps in human machine interface using both bio-signals, amplifying and acquisition of signal from human, pattern recognition of the signals and action to control a machine using the signal. In previous studies [1]-[11], we have worked on electrooculography (EOG) and applied EOG-based-control for a robot arm. However, EOG is suitable for directional control, but unsuitable for grasping objects. Therefore, we propose robot-arm control based on both EOG and EMG signals: EOG is for moving the robot joint angle and EMG is for object grasping.

2. Type of Bio-Signal

2.1. Electrooculography (EOG)

The electrooculography (EOG) is a type of bio-signal that have relationship with eye movement. Electrooculography (EOG) is a method for measuring the cornea-retinal electrical potential that exists between the front and the back of the eye. The frontal part of the eye cornea produced positive electrical potential and the back part of the eye retina produced the negative electrical potential. One of the merit of the signal is the signal is easily to be recorded if the electrode is positioned around one's eye. The demerit of the signal produced a lot of noises, the signal is need to be filtered to get good signal for the EOG. Besides that, if a subject doing EOG for a certain long period, the signal produced will also have many noises.

2.2. Electromyogram (EMG)

The electromyogram (EMG) is a bio-signal that associated with the electrical potential from human muscle. There are many ways to get the EMG, from moving one's hand to moving one's head, depending on a motion that associated with the use of one's muscle. In my research, I have used mastication or biting motion as a way to produce EMG signal. To determine the EMG for the study, we need to understand which muscle used for biting motion. The muscle used primarily is around one's mouth. Usually, the muscle used Masseter, Buccinator, Zygomaticus minor/major and depressor anguli oris as shown in **Figure 1**. Human mastication or biting motion also depend on one age, race, area of reside and eating habit.

3. Research Equipment

3.1. Bio-Signals Instrument

NF Instrument is an equipment to detect and record signals. It can be used to get both EOG and EMG signal. NF instrument as shown in **Figure 2** have 2 boxes, one is processor box and other is EEG head box. Both of the boxes have different function. Head box is used to amplify signal to send the data from operators to processor box. Processor box is used to provide some specification of filters that could be used for the research.

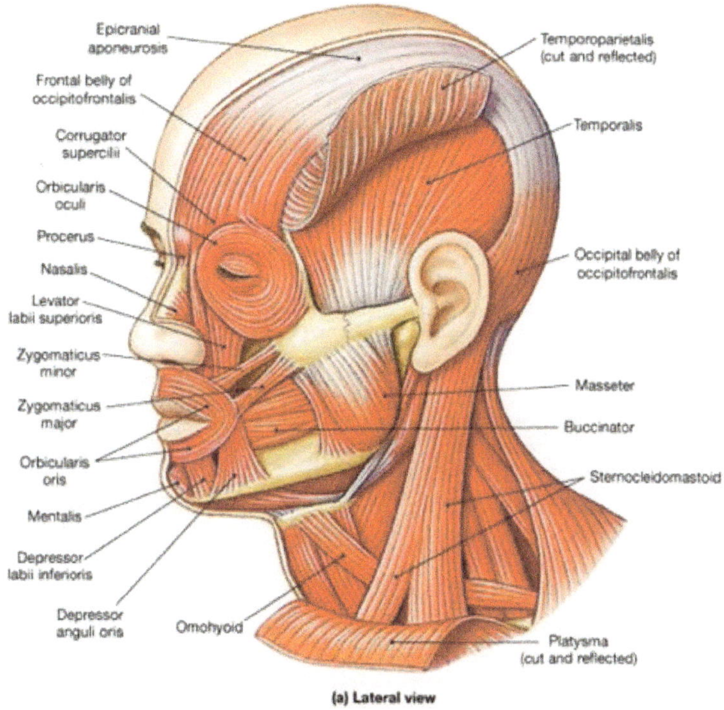

Figure 1. Human face muscles (Ref.: http://www.advocurenf2.org/livingwithnf2_ailments+care_face.php).

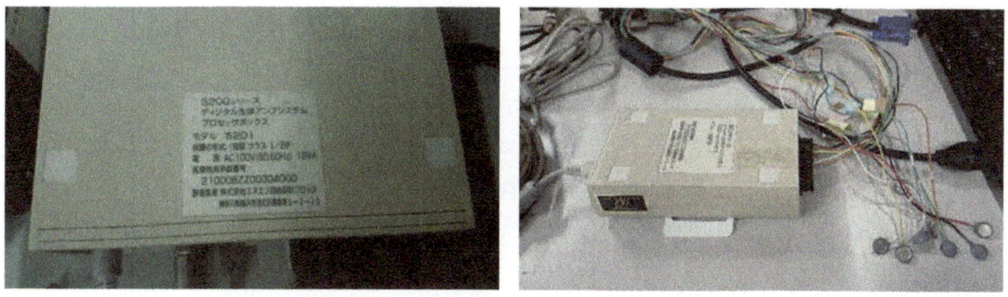

Figure 2. NF instrument processor box (Left picture) and EEG head box with electrodes (Right picture).

3.2. uArm™

The robot used for the research is uArm™ robot as shown in **Figure 3**. The robot is made from u Factory company from China. The robot used 5 servo motor each motor use to control the motion function of the robot. The motion function consists of moving up and down the robot arm joint, stretch forward and backward robot arm joint, robot base rotation, robot gripper part rotation and gripping and releasing the robot gripper.

3.3. Visual Studio 2008 and Bio-Signals

Visual studio 2008 and 2010 express is a software that be used to program a 2010 Express instrument and device based on C/C++ programming language. In this study, the Visual Studio is used as a medium for programming the NF instrument for EOG and signal detection and recording. In addition the visual studio is used to program the uArm™ for EOG and EMG robot control system in C/C++ programming via serial communication.

4. Research Method

4.1. EOG Electrode Position

In previous study, we have determined the position for detecting and recording the EOG signal. As the EOG has relationship with eye activities, the electrode is positioned around the as shown in **Figure 4**. In this study, we are searching for the EOG signal characteristic and patterns when doing up, down, left and right gaze motion.

Figure 3. uArm^{TM} robot.

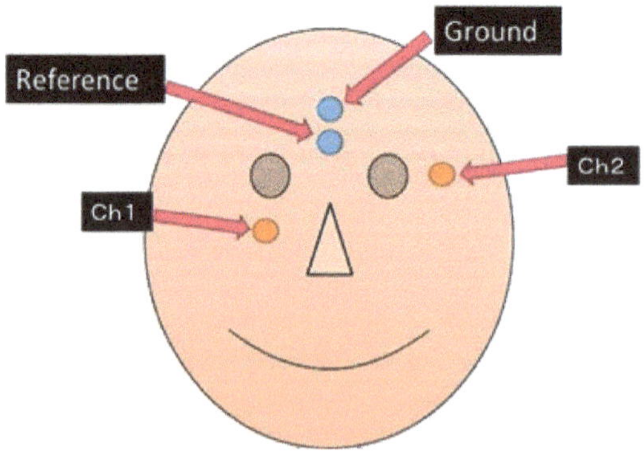

Figure 4. EOG electrodes position.

4.2. Experimental EMG Electrode Position

EMG signal could be derived by many way regarding human activities that using any muscle in one's body. In this study we are using mastication or bite motion as a method to obtain the EMG signals as shown in **Figures 5-8**. In order to obtain those signal, the electrodes are positioned consequently with the face muscle that be used during the biting motion. The target face muscle for this study is Masseter, Buccinator, Depressor anguli oris, and Zygomaticus major/minor. The electrode will position on face skin proportional to the muscle mentioned. The electrode positions are changed for each section and the number of biting motion for each section is three times. The respective result of each muscle electrode position will be derived to find signal characteristic and pattern.

4.3. EOG and EMG Electrode Position

As we have determined the location of best location on EOG and EMG respectively, the next step of our study is to determine whether the both EOG and EMG electrode positions could be combined together and get the same signals result as done independently. Basically, 4 electrodes are used in the experiment, the electrode of Ch1 and Ch2 are specialize to detect and record EOG and electrode Ch3 and Ch4 are specialize for EMG. Each electrode position is given a code name perpendicular to their number of electrodes used to discriminate them. The method of obtaining the EOG and EMG signals are the same, eye gazes motion and biting motion. Besides, the sampling rate is 100 Hz and the low-pass filter used for the experiment is 60 Hz. Each results are used to determine the signals characteristic and pattern.

4.4. Robot Control System

The control system for the uArm robot arm is shown in **Figure 9**. The NF Instrument of processor box and EEG head box are used in detecting and recording the EOG and EMG signal from eye gaze motion and biting motion of a subject. The computer is used to discriminate the EOG and EMG signals respectively and use it as a control input for the robot arm. In addition, the computer also is used for programming the robot arm to do motion functions such as joint arm movement and gripping for each EOG and EMG signals control input.

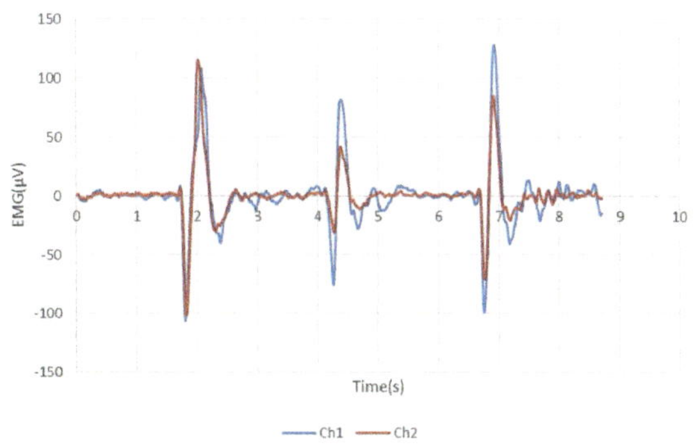

Figure 5. Masseter EMG signal.

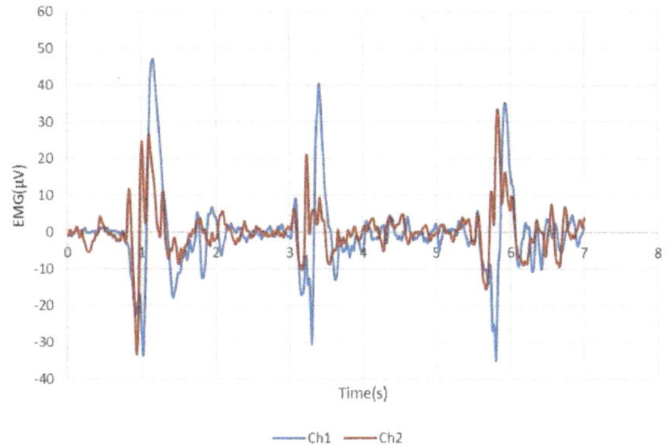

Figure 6. Buccinator EMG signal.

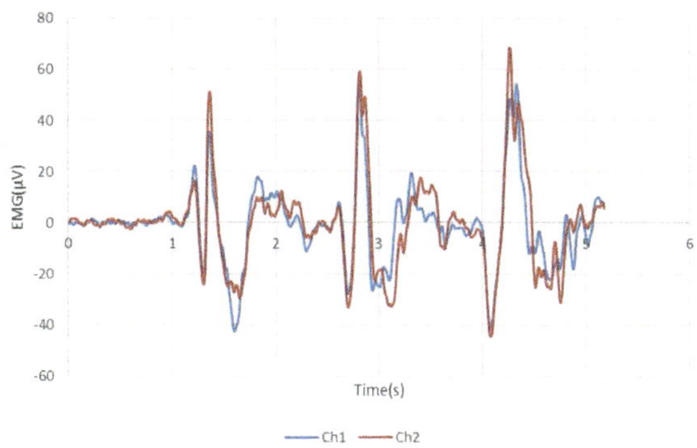

Figure 7. Depressor anguli oris EMG signal.

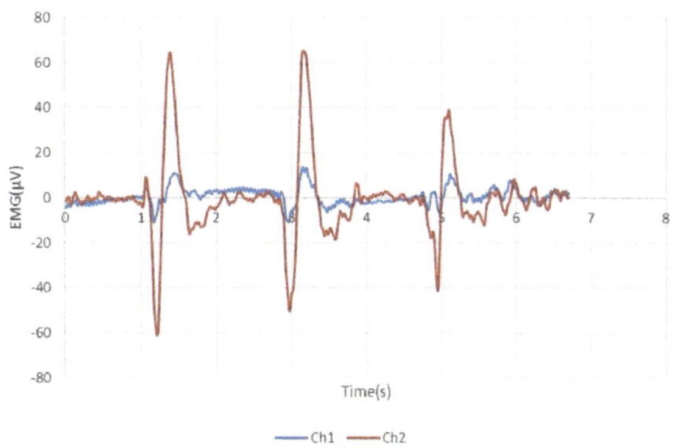

Figure 8. Zygomaticus major/minor EMG signal.

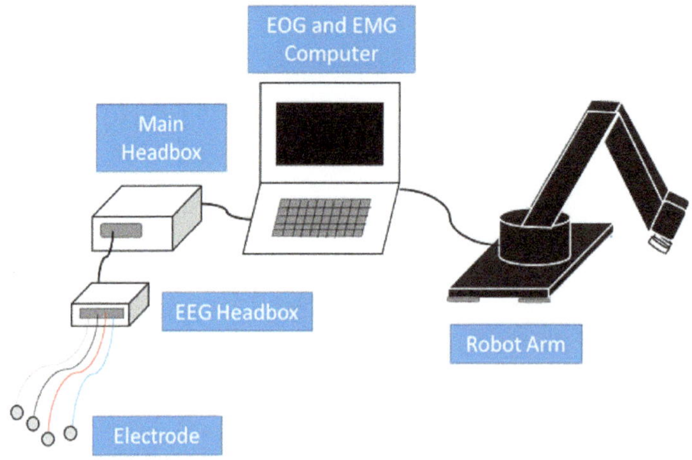

Figure 9. Robot control system.

5. Result and Discussion

5.1. EOG and EMG Electrode Position

In the combination of EOG and EMG electrode position, the signal recorded especially in Ch3 and Ch4 is unstable and hard to be analyzed. In all experiment, the biting motion signal obtained similar to noise signal. In addition, during doing eye gaze motion, all specialized electrode which used for detecting and recording EOG and EMG reacted to the motion. It is the as for biting motion, all specialized electrode reacted to the motion. Thus, the EOG and EMG signals are hard to discriminate for the robot control system.

From signal result obtained for all the electrode position, we realized that we do not many electrode to detect and record EOG and EMG signal. It is possible if only using EOG electrode position to detect and record both EOG and EMG signal as shown in **Figures 10-12**. The cut-off frequency for the low-pass filter should be lowered to 30 Hz in order to record EMG signal effectively.

5.2. EOG Discrimination Method

Based on the signals obtained, we could discriminate each gaze motions. The discrimination method for each gaze motions could be determined by the polarity of the first peak amplitude in Ch1 and Ch2 of the electrodes as shown in **Tables 1** and **2**.

5.3. EMG Discrimination Method

As for EMG signal we decided to use strong biting motion as medium to get EMG signal. Strong biting motion has higher magnitude of amplitude than other and has consistent signal result than other. The higher magnitude of amplitude and consistency of EMG signal are essential to discriminate the EMG than EOG signal.

The discrimination method for EMG signal, we done several strong biting motion experiment to investigate the signal characteristic and pattern. By the experiment done, the EMG signal can be discriminate using the Ch2 EMG signal as the Ch2 gave stable and consistent signal.

To discriminate the EMG signal as control input for robot control, we found that the sum and ratio of the signal can be used for this. The sum and ratio of Ch2 EMG signal is much higher in compare with the Ch2 EOG. **Table 3** shows the polar peak amplitude of Ch2 EMG signal relationship with sum and ratio of the signal obtained. The equation for the sum and ratio of signal is derived as below.

$$\text{Sum} = | 〖Ch2〗\_max| + | 〖Ch2〗\_min| \quad (1)$$

$$\text{ratio} = | 〖Ch2〗\_max/ 〖Ch2〗\_min| \quad (2)$$

where

〖Ch2〗 _max = Positive Peak Amplitude

〖Ch2〗 _min = Negative Peak Amplitude

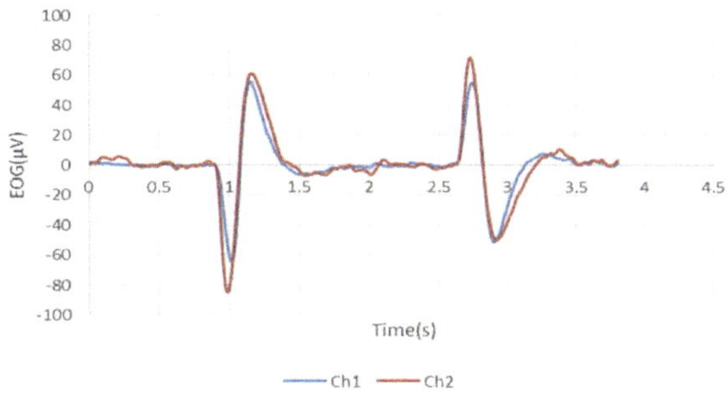

Figure 10. Up and down gaze motion signal.

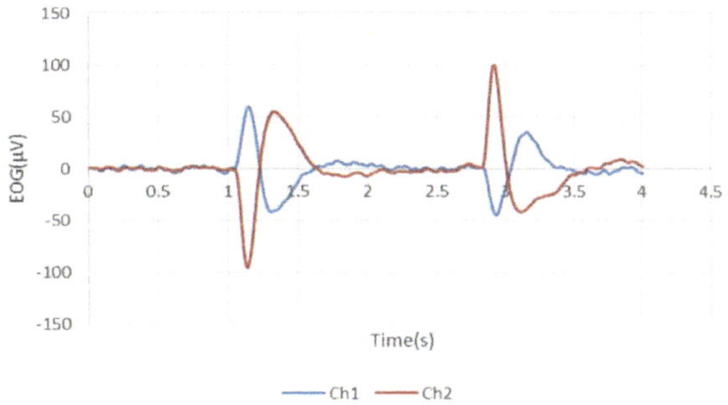

Figure 11. Right and left gaze motion signal.

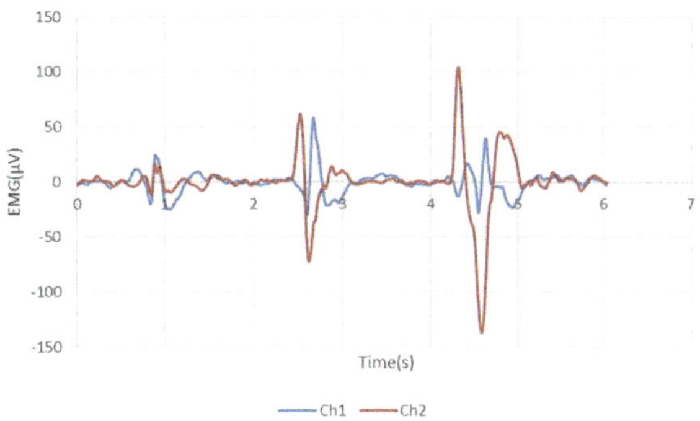

Figure 12. Biting motion strength signal. Left to right; small, normal and strong biting motion signal.

5.4. Robot Control

In this part, the EOG and EMG discrimination method is used to control the robot. The robot arm joint movements are determined by the EOG discrimination method where the polarity of eye gaze motion signals in each Ch1 and Ch2. The EMG discrimination method is used to control arm gripper to grasp and release the target object. In the robot control experiment, we are successfully control the uArm robot by using both EOG and EMG discrimination method as the control input.

6. Conclusions

In this study we conclude that EOG and EMG signal can be detected and recorded using the EOG electrode position as shown in **Figure 4**. We successfully archived to measure EMG signals with EOG electrodes by setting the low-pass filter (cut off freq.: 30[Hz]). Besides we were able to classify the EOG and EMG into 5 signal patterns as shown in **Figure 13**, and use those for robot control input. We also found that the Ch2 sum for EMG signal from the strong biting motion is between 220 - 180 (µV) and with a signal ratio of between 0.8 and 1.1. The patterns are as shown in **Figure 13**.

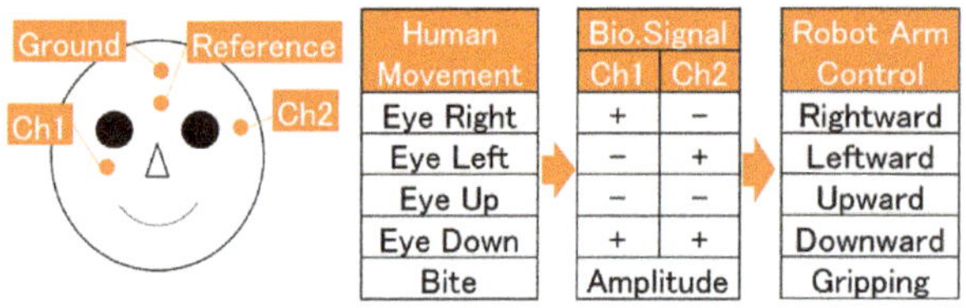

Figure 13. EOG and EMG 5 signal patterns discrimination.

Table 1. EOG gaze motions discrimination result.

	Up	Down	Right	Left
Ch1	+	−	−	+
Ch2	+	−	+	−

Table 2. EOG peak amplitudes from 4 EOG gaze motions.

Gaze motion	Up			Down			Left			Right		
|Ch2_max|	55.28	60.60	63.90	71.97	76.69	66.17	99.67	98.92	103.27	54.70	41.91	41.97
|Ch2_min|	82.66	88.17	99.49	53.67	51.56	46.16	41.61	40.60	41.15	95.53	105.68	108.15
Sum	137.94	148.77	163.39	125.64	128.25	112.33	141.28	139.52	144.42	150.23	147.59	150.12
Ratio	0.67	0.69	0.64	1.30	1.49	1.43	2.39	2.44	2.51	0.57	0.40	0.39

Table 3. EMG peak amplitudes from strong bite EMG.

No. of Experiment	1st	2nd	3rd	4rd	5th	6th
|Ch2_Max|	94.84	101.13	102.02	76.71	97.76	94.22
|Ch2_Min|	99.76	109.87	117.20	115.19	112.37	102.59
Sum	194.6	211.00	219.22	191.90	210.13	196.81
Ratio	0.95	0.92	0.87	0.67	0.87	0.92

Together with the signal patterns, we now can create a control system using EOG and EMG signal respectively. The EOG signals are used as robot joint angle movement control and EMG is used for controlling the degree of gripping.

References

[1] Cristian-Cezar, P., Florin, G. and Doru, T. (2012) EOG-Based Visual Navigation Interface Development. *Expert Systems with Application*, **9**, 10857-10866.

[2] Deng, Y.L., Hsu, C.L., Lin, T.C., Tuan, J.S. and Chang, S.H. (2010) EOG-Based Human-Computer Interface System Development. *Expert Systems with Application*, **37**, 3337-3343. http://dx.doi.org/10.1016/j.eswa.2009.10.017

[3] Lledo, L.D., Ubeda, A., Ianez, E. and Azorin, J.M. (2013) Internet Browsing Application Based on Electrooculography for Disabled People. *Expert Systems with Applications*, **40**, 2640-2648. http://dx.doi.org/10.1016/j.eswa.2012.11.012

[4] Bera, R., Boquete, L., Mazo, M. and Lopez, E. (2002) Wheelchair Guidance Strategies Using EOG. *Journal of Intelligent and Robotic Systems*, **34**, 279-299. http://dx.doi.org/10.1023/A:1016359503796

[5] Eduardo, I., Jose, M.A., Eduardo, F. and Andre, U. (2010) Interface Based on Electrooculography for Velocity Control of a Robot Arm. *Applied Bionics and Biomechanics*, **7**, 199-207. http://dx.doi.org/10.1155/2010/813184

[6] Sasaki, M., Ito, S., Takeda, K., Okamoto, T. and Rusydi, M.I. (2013) Developing a Two-Link Robot Arm Controller Using Voluntary Blink. *Proceedings of the 22nd MAGDA Conference*, Miyazaki, December 2013.

[7] Rusydi, M.I., Okamoto, T., Ito, S. and Sasaki, M. (2014) Rotation Matrix to Operate a Robot Manipulator for 2D Analog Tracking Object Using Electrooculography. *Robotics*, **3**, 289-309. http://dx.doi.org/10.3390/robotics3030289

[8] Rusydi, M.I., Mori, Y., Okamoto, T., Sasaki, M. and Ito, S. (2012) Using EOG Signal to Control Manipulator. *Proceedings of the 7th Asia Pacific Symposium on Applied Electromagnetics and Mechanis*, Ho Chi Minh City, July 2012.

[9] Rusydi, M.I., Sasaki, M. and Ito, S. (2014) Affine Transform to Reform Pixel Coordinate of EOG Signals for Controlling Robot Manipulators Using Gaze Motions. *Sensor*, **14**, 10107-10123. http://dx.doi.org/10.3390/s140610107

[10] Okamaoto, T., Sasaki, M., Ito, S., Takeda, K. and Rusydi, M.I. (2013) Using Gaze Point and Blink Detection to Control Robot Arm by Use of the EOG Signal. *Proceedings of the 22nd MAGDA Conference*, Miyazaki, December 2013.

[11] Rusydi, M.I., Sasaki, M. and Ito, S. (2014) Calculate Target Position of Object in 3-Dimensional Area Based on the Perceived Location Using EOG Signals. *Journal of Computer and Communications*, **2**.

6

Research on Application of Mobile Agent Based on Immune Algorithms in Ad Hoc Network

Qinggong Ma, Bo Yang

Changzhou University Huaide College, Taizhou, China
Email: hd.xxzx@cczu.edu.cn

Abstract

Ad Hoc network is a kind of multi hop, self-organizing wireless network without center. Each node in network can be used as host as well as router and it can form any network topology through wireless connections. Because of characteristics of itself, many new service items and application fields appear; meanwhile, it is also faced with many new security threats. Using immune agent can perceive the change of network node and make corresponding decision behavior, find the misbehavior nodes in the network as soon as possible and reduce the network attack and improve the immune competence of network.

Keywords

Immune Network, Mobile Agent, Ad Hoc Network Security

1. Introduction

Mobile Ad Hoc network [1] is a collection of multiple mobile nodes that can communicate with each other without fixed infrastructure and network topology can arbitrarily change. As a kind of special wireless network, nodes in network generally consist of mobile terminal. Because Ad Hoc network has characteristics of open media, dynamic topology, lack of center authentication mechanism, and limited node ability, compared with traditional network, it has more network security problems. Security problems have become choke point to hinder the development of Ad Hoc network. Attacks in Ad Hoc network can be divided into “active attack” and “passive attack”. Active attack can be detected easily but passive attack has high concealment and it is not easy to be found [2]. While mobile agent based on immune system can well monitor each node in the network, timely find its misconduct and make use of the immune ability of the network to restore the health of network nodes or eliminate it out of the entire network. This article will focus on how to apply immune algorithms and mobile agent to improve network security performance in Ad Hoc network.

2. Immunology Principle

2.1. Immune System

Immune system is a highly collateral distributed, adaptive information processing learning system [3]. The system can recognize self and non-self materials, eliminate and defend foreign invasive viral substances or molecules. In order to recognize self and non-self materials, in biological systems, it usually has T cells that specially detect antigen and B cells that produce antibody. There are two types of T cells. One kind is the helper T cell Th that is in the majority to be responsible for promoting B cells to secrete antibodies and strengthen the immune function of T cells and macrophage; another kind is suppressor T cell Ts, which activates under the action of cytokines to form effector cell and eliminate bad cells.

The immune response can be divided into three phages: immune recognition phase, antibody production stage, antigen elimination stage.

1) Immune recognition phase: when external antigen enters in the system, at first, the T cell does immunological recognition. If it recognized similar antigen and retain the information before, the memory part of T cell will combine with the antigen to realize immunological recognition;

2) Antibody production stage: if T cell has recognized the antigen, T cell will be activated and stimulate B cell to secrete large amounts of antibodies;

3) Antigen elimination stage: antibodies produced by B cell combine with antigen, which destroy the activity of antigen and it excretes the antigen out of the body. The antigen disappears, and the immune system returns to normal.

2.2. Immune Algorithms

Immune algorithm is the learning algorithm based on the immune system [4], and there are four basic elements, antigen recognition, antibody formation, immune selection and immune memory respectively. In the algorithm, regard the objective function of the optimal solution as the antigen, including problems to be solved and various constraint conditions. Solve the fitness function $f(x_i)$ of x_i and regard it as antigen, search the direction and the distribution of solution group according to concentration adaptive modulation of antibody. The representation of immune selection is $\dfrac{\sum_{j=1}^{N}\left|f(x_i)-f(x_j)\right|}{\sum_{i=1}^{N}\sum_{j=1}^{N}\left|f(x_i)-f(x_j)\right|}$. Immune memory refers to reserving the solving result of a specific problem and characteristics and parameters of the problem as initial solutions to solve the similar problems next time so as to accelerate the solving speed and effect.

The process to realize immune algorithms is as follows:

1) Initialization, randomly generate initial B cell mass P_0.

2) Calculate the fitness of each B cell to generate antibodies $f(x_k)$.

3) Save B cell with the optimal antibody as immune memory cell $x_k^{\max}$.

4) Calculate the concentration of antibody and generate B cell mass P_k according to the probability of immune selection.

5) P_k forms the next generation P_{k+1} of B cell through clone and differentiation.

6) If immune memory cell hasn't completely recognized antigen, return to step 2).

3. The Working Mechanism of Mobile Agent

Agent technology derives from artificial intelligence domain [5], can simulate human behavior and relationships and it is a procedure that can run independently and provide specific service. It can sense the environment change, make judgment and reasoning for changes and form the decision to control corresponding behavior and complete a task. Compared with traditional distributed computing model, mobile agent technology has many advantages, such as save network bandwidth, support off-line calculating, unrelated to platform and balance computation load, etc. In the network environment, under the self-control, the mobile agent can move from one computer to another computer, hang up at any point, and it continues execution from the hanging point when moving to a new machine. Under normal circumstances, mobile agent system (MAS) includes mobile agent MA and mobile agent platform MAE. Mobile agent platform is responsible for establishing safe and correct operation environment for mobile agent, providing basic services and forming restriction mechanism, fault-tolerance strategy and security control for MA. The mobility of mobile agent and the solving ability of problems depend largely on the services provided by mobile agent platform.

4. The Application of Immune Agents in Ad Hoc Network

Because the Ad Hoc network topology is changeful, the establishment of a connection mode completely depends on the location of each node and coverage area of their transceiver. Each node in the network is equivalent to a router, realizing the retransmission of information packet and participating in the discovery and maintenance of route. Ad Hoc network brings convenience to people and is faced with serious security threat. Attacks from network can be divided into passive attacks and active attacks. Under passive attack mode, attackers don't interfere with the communication of network nodes and their purpose is to obtain sensitive information in the network; active attack refers to that the attackers interfere with or destroy the communication of network nodes and cause that the network nodes cannot work normally. Flooding DDOS attack is a typical attack type. In order to early find misbehavior nodes in network and maintain normal communication of the system, this article combines immune algorithm with mobile agent and applies them to Ad Hoc network, makes network node can dynamically monitor change situation of other nodes in network, timely remove or restrict the behavior of misbehavior nodes and realize autonomous management of network nodes under the circumstances of no one involved.

4.1. Working Model of Immune Agent

It is easy to find the corresponding relation between immune system and Ad Hoc network through analyzing the division of labor of lymphocyte, T cell, B cell and the working mechanisms of mobile agent in immune system. The security architecture of Ad Hoc network is equivalent to the entire immune system; misbehavior nodes in Ad Hoc network are equivalent to the antigens in immune system; normal nodes in Ad Hoc network are equivalent to healthy cells in immune system; mobile agent in Ad Hoc network is equivalent to lymphocyte in immune system and responsible for transporting other agents with specific function. In view of the function of T cell and B cell, three types of agents have been set in the system, and they are monitoring agent, coordinating agent and blocking agent respectively. Monitoring agent is equivalent to T cell in immune system, distributes in all the nodes of the whole system, and it is responsible for collecting behavior information of neighbor nodes and regularly reports it to coordinating agent; coordinating agent distributes in different areas of the network and it is equivalent to B cell in immune system and responsible for arranging the information of nodes in the charge of it in the network and storing it into information memory bank. When coordinating agent receives a new report, it will compare the report with the information in memory bank and strategy library. If it is normal information, the agent will ignore it; if it isn't normal behavior information (if a node receives routing information and doesn't transpond), judge the time difference of bad report at the same node in last time. If the interval is nearer, then accumulate. Otherwise, remove the bad report number of this node. If it exceeds the threshold value, activate the blocking agent. Blocking agent is equivalent to antibody in immune system and can clone itself. Its task is to eliminate the misbehavior nodes [6]. Coordinating agent needs certain tolerance and can tolerate a small amount of bad behavior of individual nodes (some bad behaviors may be caused by network interference).

The safety verification of antibody moving from point h_0 to h_{i+1} is described as follows:

$$O_i = E_0\left(H_i, S_i\left(o_i\right), Uid\right), h_i, 0 \le i \le n$$

$$h_0 = h\left(r_0, H_i\right)$$

$$h_i = h\left(O_{i-1}, H_{i+1}\right), 1 \le i \le n$$

$$H_i \to H_{i+1} : \left\{O_k \middle| 0 \le k \le i\right\}$$

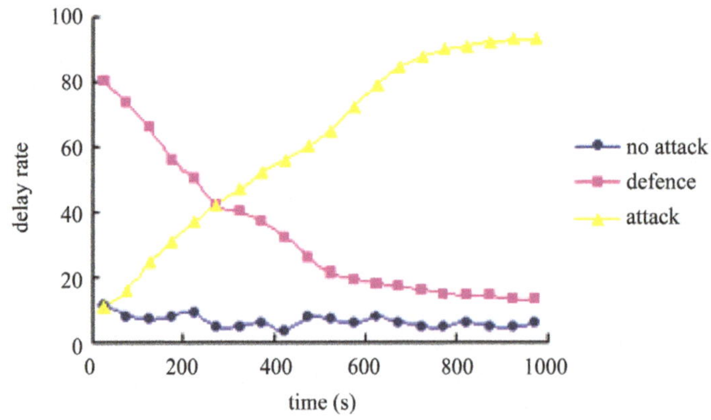

Figure 1. Results contrast figure.

When the antibody begins to roam, H_0 produces a random number r_0 and a o_0, and then calculate h_0 and O_0, send O_0 to H_i. When the antibody reaches H_{i+1}, H_{i+1} checks the chain relationship $h_k, 1 \le k \le i$.

When the antibody moves to the position near the goal node, it clones, mutates and compresses [7] and produces a large number of antibodies that can kill the "virus". These antibodies will isolate the communication between misbehavior nodes and other nodes in network and achieve the purpose of eliminating harmful nodes. When misbehavior nodes disappear, blocking agent will vanish by itself.

The mathematical description of antibody clone is as follows:

For binary coding, antibody $a \in B^l$. Among them, $B^l = \{0,1\}^l$ represents the collection formed by all the binary strings with length of *l*. Antibody cluster $A = \{a_1, a_2, \cdots, a_n\}$ is *n* tuple of antibody *a*.

Clone operator is defined as follows: $\Theta(A) = \{\Theta(a_1), \Theta(a_2), \cdots, \Theta(a_n)\}$. Among it, $\Theta(a_i) = I_i \times a_i, (i = 1, 2, \cdots, n)$. I_i is q_i dimensional row vector $q_i = g(N_c, f(a_i))$. Usually take $q_i = Int\left[N_c \times \frac{f(a_i)}{\sum_{j=1}^{n} f(a_j)} \right], i = 1, 2, \cdots, n$. Among it, $N_c > n$ is concerned set value of clone size and $Int(*)$ is top integral function. After the clone operation, it generates intermediate population $A' = \{A, A'_1, A'_2, \cdots, A'_n\}$ and $A'_i = \{a_{i1}, a_{i2}, \cdots, a_{iq-1}\}$.

The mathematical description of the variation is as follows:

$$p(a_i \to a'_i) = \begin{cases} 0 & a_i \in A \\ p_m^{H(a_i,a')} (1-p_m)^{l-H(a_i,a')} & a_i \in A'_j \end{cases},$$

H is Hamming distance. After variation, it generates intermediate population $A'' = \{A, A''_1, A''_2, \cdots, A''_n\}$ and $A''_i = \{a'_{i1}, a'_{i2}, \cdots, a'_{i,q-1}\}$.

The main effect of compression choice is to choose *n* antibodies with the highest affinity in intermediate population A'' to form a new generation of antibody group $A(t+1)$.

Evaluate the immune system in the form of weighted sum, and the evaluation function is $Sl = Mat + K_1 * Sim - K_2 * Sup$. In this formula, K_1 and K_2 are weighted system of stimulating effect and inhabitant activity.

4.2. Result Analysis

It is can be seen from **Figure 1** that when there is no attack in network, the message delay maintains at a relatively low level and the network communication is normal shown in the figure as "●"; when the network appears attack, the delay rate of message will rise rapidly. With the extension of time, network performance becomes worse and worse and the worst result is network paralysis, shown as "▲"; after applying the method of immune agent in network, the number of message delay of the system will reduce sharply, shown as "■". Although it is higher than the message delay rate when there is no attack, it is still in a lower level and acceptable.

References

[1] Zheng, S.R., Wang, H.T. and Zhao, Z.F. (2005) Ad Hoc Network Technology. Posts and Telecom Press, Beijing.

[2] Cohen, F. (1999) Simulating Cyber Attacks, Defenses, and Consequences [EB/OL]. http://all.net/journal/ntb/simulate/simulate.html

[3] Steve, C. and Uwe, A. (2002) A Recommender System Based on the Immune Network. *Proceedings CEC* 2002, Honolulu, 807-813.

[4] Wang, L. (2001) Intelligent Optimization Algorithms with Applications. Tsinghua University Press, Beijing.

[5] Magic, G. (2012) Telescript/Odyssey [EB/OL]. http://www.genmagic.com/agents

[6] Agent Transport Protocol Specification. IBM Aglet Work Bench [EB/OL]. http://www.trl.ibm.com.jp/aglets

[7] Zhou, G.Y. (2000) Principles of Immunology. Shanghai Scientific and Technological Literature Press, Shanghai.

7

Analysis of Modern Design Approach for Anti-Air Radar Screen

Jungwan Hong[1], Suhwan Kim[2], Yongjin (James) Kwon[1*]

[1]Department of Industrial Engineering, Ajou University, Suwon, South Korea
[2]1st Division, 3rd Department, Agency of Defense Development, Daejeon, South Korea
Email: *yk73@ajou.ac.kr

Abstract

This study is a preparation phase for visualization of utilized information using ergonomic user interface and standardization of elements for anti-air radar screen. Therefore, we investigated the instances of anti-air radar screen for air defense weapon system for defense advanced country. Based on the collected data, we compared and analyzed the air defense weapon system radar screen design. In addition, we carry out a research for layout, configuration, standardization and design of the radar screen's elements. Ultimately, it is essential to share a variety of battle field conditions such as enemy threat, enemy/friendly information, terrain information that can be effectively recognized. In this paper, we conduct case study for ergonomically development of automated/standardized radar screen. It is expected that this research improves the situational awareness and reduces the user's task load.

Keywords

Anti-Air Radar Screen, Air Defense Weapon System, Human Interface, Ergronomics, Situational Awareness, Task Load

1. Introduction

It will be expected to lead to major changes of mission operation environment for air defense mission in the future shown as **Table 1**. According to dramatically improvement of missile target detection range with the development of radar detection technology and missile propellants, ballistic missile's flight speed and accuracy are also improved. Thus, it is necessary to quickly and effectively exhibit the enemy missiles that fly at rapid speed because ballistic missile will fly at Mach 3 to 10 in the future. Future integrated battlefield environment is required exhibition of various elements such as atmosphere, climate, terrain, enemy/friendly information factor [1]. Therefore, the study of the radar screen configuration suitable for mission operation environment condition is required. **Table 2** shows direction of design for anti-air radar screen. As a result, effective handle is possible, grasp the trajectory of ballistic missile three-dimensionally [2]. Applying a touch-screen with a zoom-in function that can be observed a particular area of the wide range of attacks and it must be able to obtain detailed informa-

*Corresponding author.

Table 1. Major changes of mission operation environment.

Changed Situation	Requirements
Increase in detection range and missile propellant range	• Development of radar detection and missile propulsion technology • Target detection range is dramatically increased • Missile range, flying speed, accuracy are improved • Expansion of scope for battle field
Short time of the battle by the fast flight speed of ballistic missile	• The speed of the ballistic missile is faster than normal aircraft speed • The high-speed combat will be occurred • Quick and liquidity of certain capability will be required
Increased probability to engage with an number of target	• Clear situation awareness of commanders is required • Fast and precise command decision is required

Table 2. Direction of design for radar screen.

Direction of Design in Changed Situation	Description
Zoom-in/out function through the touch-screen	• Ability to see wide range of attacks quickly using zoom-function • Display detailed information that does not degrade the overall situational awareness when zooming
Switch function/ Simultaneous display function 2D, 2.5D, 3D (dimension)	• Determine the trajectory of the ballistic missile in 3D as well as 2D • Determine efficiencies for various radar screen's component layout
Rapid and accurate situational awareness	• Utilization of agronomical symbol for component of radar screen • Representation of friendly/enemy/missile/asset with high readability • Quick and accurate threat analysis through rapid situational awareness
Expression of intensive information through the large screen	• Apply high-resolution LCD touch-screen • Minimize the time to load information • Prevention of essential missing information

tion to that area. In addition, represent assets of enemy and friendly using the symbol based on ergonomics of the radar screen's elements and then it should be designed to allow rapid situation awareness. Through a large screen like a 30-inch LCD touch-screen allows broad sense representations of information, further it must be minimize the time to import battlefield information [3].

2. Case Study

Air Defense Weapon System Radar Screen

- Case study for 12 kinds of radar screen for air defense weapon systems.
- Attack range/altitude, mission, intercept altitude, operating country etc.
- Threat information, friendship information, tactical situation information, the operation of the tactical map equipment, the type of air defense weapon system components etc.
- Air defense weapon system radar screen's configuration and layout.
- Major Air defense weapon systems are shown as **Table 3**.
- **Tables 4-7** show results of analysis for radar screen.

3. Results

This study analyzes the common characteristics and flow of air defense weapon system's radar screen. Also features of each radar screen elements are analyzed as a graph.

3.1. Radar Screen Shape/Location/Background/Color

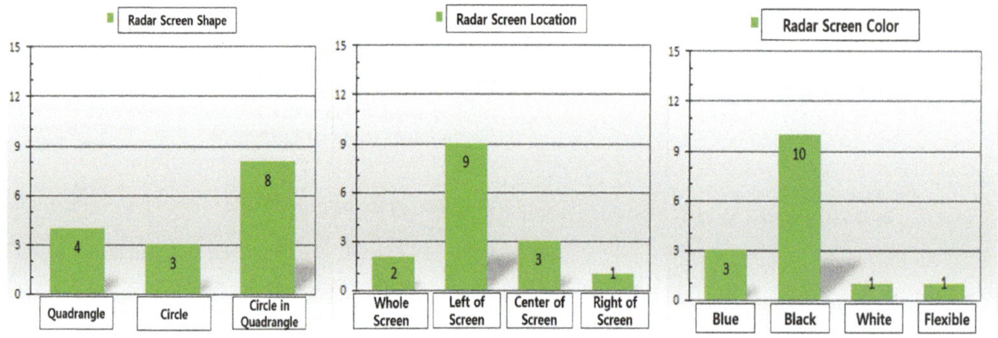

Figure 1. Analysis results of shape, location, color for radar screen.

3.2. Radar Screen Text Information and Tactical Map

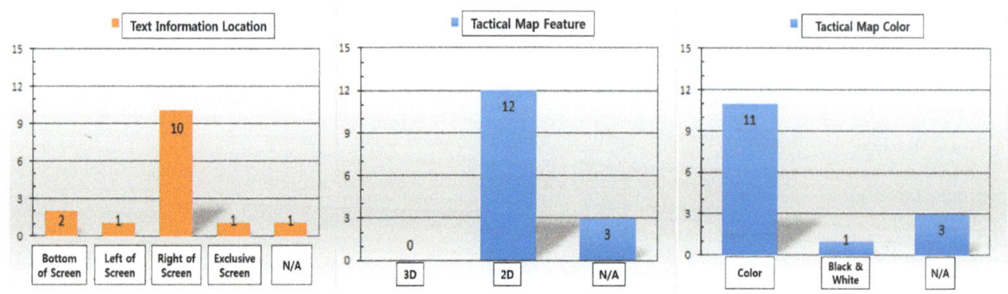

Figure 2. Analysis results of text information and tactical map for radar screen.

Table 3. Major air defense weapon systems.

Air Defense Weapon System	Nation	Attack Range/Altitude
THAAD	USA (Lockheed)	200/100
Patriot (PAC-1, 2, 3)	USA (Raytheon)	160/24
Arrow-2, 3	USA + Israel	100/50
MEADS	USA + Germany + Italy	30 (Detection Range)
PANTSIR-S1	Russia	32/15
Tor M-1	Russia	25/12
OSA	Russia	15/12
S-300/400	Russia	25/70
Polyana d4m1	Russia	2400 (Detection Range)
SAMP/T	France	40/15
MCDS	Italy	30/15

4. Conclusion

Radar screen which form is circular shape onto a quadrangle screen taking the largest proportion. Position of radar screen mostly is displayed on the left side of its screen. Its background's color is usually dark color such as black, blue, indigo. Text information of radar screen is normally on its right side. Recent radar screen system has evolved into free-form with adapting its components to user's convenience. Recent trend of radar screen is including Tactical map into its components. Previous trend is using 2D color Tactical map. In the future warfare, radar screen will need 2.5D, 3D Tactical map. Most of radar screens have 4 monitor up down left right side. The number of user will be 2 or 3, and their assignments are detecting radar screen, management of arm and total control. **Table 8** shows results.

Table 4. Radar screen layout analysis (1).

Name		THAAD	Patriot PAC-1, 2 (old type)	Patriot PAC-3 (the newest)
Radar Screen Layout				
Radar Screen	Shape	Quadrangle	Circle	Circle in Quadrangle
	Location	Whole screen	Center of screen	Center of screen
	Color (Background/Symbol)	Blue/White	Black/Green	Black/Various color
Text Info.	Shape	N/A	Text type	Text type
	Location	N/A	Bottom of screen	Bottom of screen
	Color	White	Green	White
Tactical Map (Available/Resolution/Color)		Available/2D/Color	N/A	Available/2D/Color
Number of Operator		2	3	3
Role		Radar detection, Targeting and Fire operation	Radar detection, Targeting and Fire operation, Whole situational awareness	Radar detection, Targeting and Fire operation, Whole situational awareness
Total Display		2	2	4

Table 5. Radar screen layout analysis (2).

Name		Arrow-3	MEADS	Pantsir-S1 MD
Radar Screen Layout			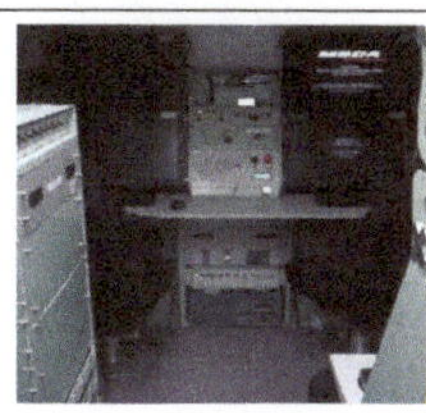	
Radar Screen	Shape	Circle in Quadrangle	Quadrangle	Circle
	Location	Left of screen	Left of screen	Left of screen
	Color (Background/Symbol)	Blue/White	N/A	Black/Red, Yellow
Text Info.	Shape	Window Box type	Window Box type	Window Box type
	Location	Right of screen	Right of screen	Right of screen
	Color	N/A	N/A	Various Color
Tactical Map (Available/Resolution/Color)		Available/2D/Color	Available/2D/Color	Available/2D/Color
Number of Operator		2	2	2
Role		Radar detection, Targeting and Fire operation	Radar detection, Targeting and Fire operation	Radar detection, Targeting and Fire operation
Total Display		4	4	2

Table 6. Radar screen layout analysis (3).

Name		Tor M-1 (old type)	Tor M-1 (the newest)	OSA-2
Radar Screen Layout				
Radar Screen	Shape	Quadrangle	Circle in Quadrangle	Circle
	Location	Whole screen	Left of screen	Left of screen
	Color (Background/Symbol)	Black/Green	Black/Blue	Black/Red, Yellow
Text Info.	Shape	N/A	Window Box type	Window Box type
	Location	Exclusive Display	Right of screen	Right of screen (Flexible layout)
	Color	Red	Various Color	Various Color
Tactical Map (Available/Resolution/Color)		N/A	Available/2D/Color	Available/2D/Color
Number of Operator		2	3	3
Role		Radar detection, Targeting and Fire operation	Radar detection, Targeting and Fire operation, Whole situational awareness	Radar detection, Targeting and Fire operation, Whole situational awareness
Total Display		4	4	3

Table 7. Radar screen layout analysis (4).

Name		S-300	Polyana D4M1	SAMP/T
Radar Screen Layout				
Radar Screen	Shape	Circle in Quadrangle	Quadrangle	Circle in Quadrangle
	Location	Left of screen	Left of screen	Left of screen
	Color (Background/Symbol)	Black/Various Color	White/Black, Red, Blue	Black/Various Color
Text Info.	Shape	Window Box type	Window Box type	Window Box type
	Location	Right of screen (Flexible layout)	Right of screen	Right of screen
	Color	Various Color	Black	Various Color
Tactical Map (Available/Resolution/Color)		N/A	Available/2D/White & Black	Available/2D/Color
Number of Operator		2	7	3
Role		Radar detection, Targeting and Fire operation	Radar detection, Targeting and Fire operation, Whole situational awareness	Radar detection, Targeting and Fire operation, Whole situational awareness
Total Display		4	7	5

Table 8. Radar screen layout analysis results.

Division	Analysis Results	Case of Air Defense Weapon System
Radar Screen Feature	Circle Radar Screen	Patriot series, S-300, Tor M-1, Arrow-3
Radar Screen Location	Left of Display	Arrow-3, Pantsir-S1 etc.
Radar Screen Color	Black	SAMP/T, MCDS, OSA etc.
Text Information Location	Right side of Radar Screen	Pantsir-S1, Arrow-3 etc.
Tactical Map Feature	2D Tactical Map	THHAD, the Newest Patriot etc.
Tactical Map Color	Color Type	THHAD, the Newest Patriot etc.

Acknowledgements

This work was supported by the Agency for Defense Development (ADD) under the Contract No. UD140066CD. The authors wish to express sincere gratitude for the financial support.

References

[1] Olmos, O., Wickens, C.D. and Chudy, A. (2014) Tactical Displays for Combat Awareness: An Examination of Dimensionality and Frame of Reference Concepts and the Application of Cognitive Engineering. *The International Journal of Aviation Psychology*, **10**, 247-271. http://dx.doi.org/10.1207/S15327108IJAP1003_03

[2] Brown, C., Fagan, P., Hepplewhite, A., Irving, B., Lane, D. and Squire, E. (2014) Real-Time Decision Support for the Anti-Air Warfare Commander. *Proc. 6th Int. Command Control Res. Technol. Symp.* (*ICCRTS*), 2001.

[3] Filippidis, A., Blandford, S., Foster, K. and Moran, G. (2006) Simulation Activities Using Gateway and Tactical Digital Information Links. Defence Science and Technology Organisation, 1-13.

An Improved Histogram Equalization Method in the Traffic Monitoring Image Processing Field

Jia Shi, Kejian Yang

Department of Computer Science and Technology, Wuhan University of Technology, Wuhan, China
Email: 838357982@qq.com, ykjcgs@whut.edu.cn

Abstract

To process the traffic monitoring image, a local Histogram Equalization method based on fuzzy mathematics was proposed in this paper. In this paper, firstly, we define a function to measure the similarity degree of two images. Then, a suitable Gaussian fuzzy distribution function was chose to generate a 3 × 3 matrix of influential factors. In order to reduce the artificial boundaries, we combined the 3 × 3 influential matrix with a 3 × 3 smooth filter matrix to get the final smooth-influence matrix. Finally, the smooth-influence matrix was used to process the center block image. The simulation results demonstrated that the proposed method can reduce time consumption while improving the image contrast and can get satisfactory results.

Keywords

Image Process, Histogram Equalization, Fuzzy Mathematics, Traffic Monitoring Image

1. Introduction

With the rapid development of IoT (Internet of thing), traffic monitoring technology has been more and more widely used. But due to various reasons, the traffic monitoring image cannot be applied directly. Among them, the device, environments, and the intensity of light are all the factors that influence the image quality. In order to make secondary use, we have to process the image and enhance the image contrast. As an important step in image processing, contrast enhancement has becoming more widely used.

Currently, there are many methods of contrast enhancement; among them, histogram equalization is a well-known method because of its less complexity and high efficiency [1] [2]. Generally, histogram equalization can be divided into two categories: one is global histogram equalization; the other is local histogram equalization. In global histogram equalization, the transformation function is the cumulative distribution function of the whole input image. However, in the global histogram, because of the transformation process of the image information application in each pixel, the processed image may lose a lot of local details. Especially when the gray level of the input image gathers in the left part, the gray level of the output image will gather in the right part. And the processed image shows a blenching effect and has an unnatural look. In contrast, the local histogram equalization used a sliding window method to produce an enhanced contrast effect but the process is complex and time-consuming.

In recent years, many researchers have tried different ways to improve the traditional histogram equalization algorithm, and have acquired some great achievements. Among them, BHE (bi-histogram equalization) method firstly divided the input image gray level into two parts by the average gray value [3], and then use HE algorithm in each part independently. The results showed that BHE method substantially eliminates the excessive bleaching phenomenon of the image, and retains an average luminance; however some minor details still have not been enhanced. To compensate for this defect, the local histogram equalization came into being. AHE (adaptive histogram equalization) method divides the image into a series of sub-blocks, then HE method is implemented in each sub-block [1], meanwhile the bi-linear interpolation is used to eliminate the artificial boundaries between sub-blocks. Like AHE method [4], BOHE (over-lapped sub-block histogram equalization) method divides the image into a series of rectangular sub-blocks, and then moves the central sub-block one pixel by one pixel. This method can acquire a perfect processed image, but it is too complex to be applied in reality. Therefore, POSHE (partially over-lapped sub-block histogram equalization) method was proposed to improve the BOHE method [4], this method uses a low pass filter to generate the transformation function of central block, the simulation results show that it has increased the time efficiency by 100 times than that of the BOHE method.

To process the traffic monitoring image [5], there are three aspects need to be taken into consideration. Firstly, most parts of the image are black because of the asphalt pavement. Therefore, it is necessary to prevent the bleaching phenomenon of the processed image. Secondly, in order to ensure the time efficiency, the proposed method should not be too complex. Thirdly, in order to ensure the method to be processed automatically, it should not have too many arguments. From the above three aspects, we proposed our method. Firstly, the image was divided into a series of sub-blocks, and then according to the similarity function and Gaussian fuzzy distribution function, a 3 × 3 influential factors matrix was generated to show the influence degree of the peripheral sub-blocks on the central block. Moreover, the 3 × 3 influential factors matrix was combined with a 3 × 3 smoothing filter matrix to reduce the artificial boundaries. Finally, the combined smooth-influence matrix was used to process the center sub-block.

The rest of this paper is organized as follows: Section 2 is to describe our method, Section 3 is to show the simulation and comparison results, Section 4 is to analyze our method, and Section 5 is to conclude this whole paper.

2. Proposed Methodology

2.1. Histogram Equalization

During the process of Histogram Equalization, the discrete transformation function is as follows [1]:

$$s_k = T(r_k) = \sum_{j=0}^{k} p_r(r_j) \qquad (k = 0,1,2,\cdots,L-1) \tag{1}$$

where,

$p_r(r_j) = \frac{n_j}{n}$; n is the total number of pixels of the input image;

n_j is the number of pixels of r_j gray level; L is the total gray levels of input image;

r_k is the gray level of input image; s_k is the gray level of output image.

In order to simplify the analysis the continuous form of formula (1) is figured out as follows [1]:

$$s = T(r) = \int_0^r p_r(w)dw \tag{2}$$

As the Formula (2) shows, there are differential relationships between $T(r)$ and $P(r)$. When the value of $P(r)$ is large, the growth rate of $T(r)$ is high, so the intensive gray level will be stretched. When the value of $P(r)$ is small, the growth rate of $T(r)$ is low, so the sparse gray level will be compressed. In this way, the gray level of the whole image will be distributed uniformly, and the image will be enhanced.

2.2. Similarity Function

We use the difference of the pixels amount of the same gray level between two sub-block images to measure the similarity degree of them. In order to simplify the process of the measurement, firstly, the normalization preprocessing of each gray level was implemented. The similarity degree varies between 0 and 1, especially, when the two sub-block images are the same, the similarity degree is 1. Therefore, the similarity function is defined as follows:

$$S(image_a, image_b) = 1 - \sum_{i=0}^{L-1} \left| \frac{n_{ai}}{n_a} - \frac{n_{bi}}{n_b} \right| = 1 - \sum_{i=0}^{L-1} \left| p_{ar}(r_i) - p_{br}(r_i) \right| \tag{3}$$

where,

$image_a, image_b$ represent the two sub-block images;

n_a, n_b are the total pixels number of the two sub-block images;

L is the total gray level of the two sub-block images;

n_{ai}, n_{bi} represent the ith gray level pixel of sub-block $image_a$ and $image_b$ separately;

$P_{ar}(r_i), P_{br}(r_i)$ represent the ratio of gray level r_i of sub-block $image_a$ and $image_b$ separately after the normalization preprocessing.

Defined in this way, the similarity function not only can decrease the computational complexity but also is suitable for sub-blocks of different sizes [6].

2.3. The Calculation of Influential Factor Matrix

The domain of the similarity degree is defined as $X = [0,1]$. Fuzzy set A represents "low-influence degree" and the influential factor is 1/16; Fuzzy set B represents "mid-influence degree" and the influential factor is 1/8; Fuzzy set C represents "high-influence degree" and the influential factor is 1/4.Because only when the similarity degree is suitable, the value of the influence degree can be maximum. Therefore, Fuzzy distribution function of the intermediate type should be assigned as the membership function. In this paper, we choose Gaussian function as the membership function of Fuzzy set A, B and C. The formulas of these three functions are listed as follows [5] [7]:

$$A(x) = \exp\left\{ -\left(\frac{x - a_1}{\sigma_1} \right)^2 \right\} \quad x \in [0,1] \tag{4}$$

$$B(x) = \exp\left\{ -\left(\frac{x - a_2}{\sigma_2} \right)^2 \right\} \quad x \in [0,1] \tag{5}$$

$$C(x) = \exp\left\{ -\left(\frac{x - a_3}{\sigma_3} \right)^2 \right\} \quad x \in [0,1] \tag{6}$$

where,

x is the similarity of the two sub-block images;

a_1, σ_1 are the arguments of membership function $A(x)$;

a_2, σ_2 are the arguments of membership function $B(x)$;

a_3, σ_3 are the arguments of membership function $C(x)$.

After several tests, the arguments of the three membership functions are determined, on the condition that $a_1 = 0.15$, $a_2 = 0.45$, $a_3 = 0.75$, and $\sigma_1 = \sigma_2 = \sigma_3 = 1$, the processed image is the best.

In order to simplify the description, we number the sub-blocks of the split image, which can be shown in **Figure 1**. According to the three membership functions and the principle of maximizing membership, the influential factors from the around sub-blocks to the central block is calculated (that is to calculate the influential factors form sub-blocks A, B, C, D, E, F, G, H to the center block), and then the influential factors matrix $M_{\text{influence}}$ can be acquired. Because the similarity degree of the center block is 1, the center value of $M_{\text{influence}}$ is always 1/4 of "high-influence degree".

2.4. The Combination of Smooth-Influence Matrix

Because we have divided the input image into a series of sub-blocks before the processing, there are some in

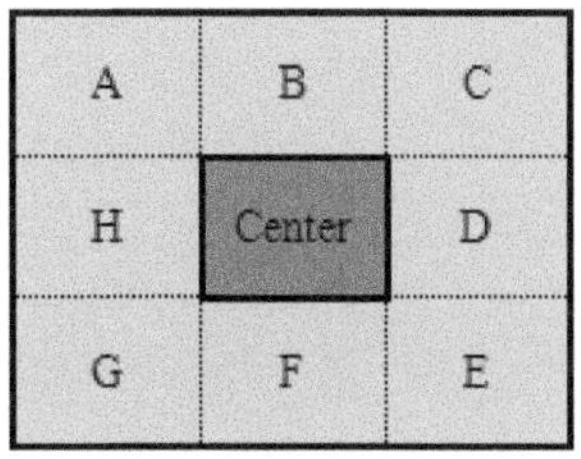

Figure 1. The graph of the split image.

evitable artificial boundaries in the processed image, In order to decrease the computational complexity and reduce the quantity of artificial boundaries, the 3 × 3 smoothing filter matrix is combined with $M_{\text{influence}}$ to acquire the final smooth-influence matrix to process the image. In this way, the artificial boundaries will not be processed independently.

Define operator $\oplus$ to represent the combination of matrix, which can be shown as follows:

$$M_A \oplus M_B = \begin{bmatrix} a_{11} & a_{12} & a_{13} \\ a_{21} & a_{22} & a_{23} \\ a_{31} & a_{32} & a_{33} \end{bmatrix} \oplus \begin{bmatrix} b_{11} & b_{12} & b_{13} \\ b_{21} & b_{22} & b_{23} \\ b_{31} & b_{32} & b_{33} \end{bmatrix} = \begin{bmatrix} \frac{a_{11}+b_{11}}{sum} & \frac{a_{12}+b_{12}}{sum} & \frac{a_{13}+b_{13}}{sum} \\ \frac{a_{21}+b_{21}}{sum} & \frac{a_{22}+b_{22}}{sum} & \frac{a_{23}+b_{23}}{sum} \\ \frac{a_{31}+b_{31}}{sum} & \frac{a_{32}+b_{32}}{sum} & \frac{a_{33}+b_{33}}{sum} \end{bmatrix} \tag{7}$$

where,

$sum = \sum_{i=1}^{3}\sum_{j=1}^{3}(a_{ij} + b_{ij})$;

$a_{11} \cdots a_{33}$ are the elements of M_A;

$b_{11} \cdots b_{33}$ are the elements of M_B.

According to the operator defined in Formula (7), the 3 × 3 smoothing filter matrix is combined $M_{\text{influence}}$ with to acquire the final smooth-influence matrix.

$$M_{\text{smooth-influence}} = M_{\text{influence}} \oplus M_{\text{smooth-fliter}} \tag{8}$$

where,

$M_{\text{smooth-fliter}} = \frac{1}{16} \times \begin{bmatrix} 1 & 2 & 1 \\ 2 & 4 & 2 \\ 1 & 2 & 1 \end{bmatrix}$ is the 3 × 3 smoothing filter matrix;

$M_{\text{influence}}$ is the influential factor matrix;

$M_{\text{smooth-influence}}$ is the smooth-influence matrix.

According to the smooth-influence matrix, we compute the histogram of the center block and get the transformation function to process the center block image. In this way, the process can retain the information of the center block and peripheral sub-blocks, meanwhile reducing the artificial boundaries.

3. Simulation and Comparison

The proposed method can be divided into the following steps [8] [9]:

Step 1: Color transformation (RGB- > HSV);

Step 2: Extract the V component, and divide it into a series of sub-blocks;

Step 3: Compute $M_{\text{smooth-influence}}$, and get the transformation function of the center block;

Step 4: Implement the equalization process of the center block;

Step 5: Revert the V component and color transformation (HSV- > RGB), then output the image.

In the proposed method, the processing of each sub-block needs the auxiliary processing of the 8 peripheral blocks. So we have to do some additional processing on the edge blocks. The results of additional processing is showed in **Figure 2**.

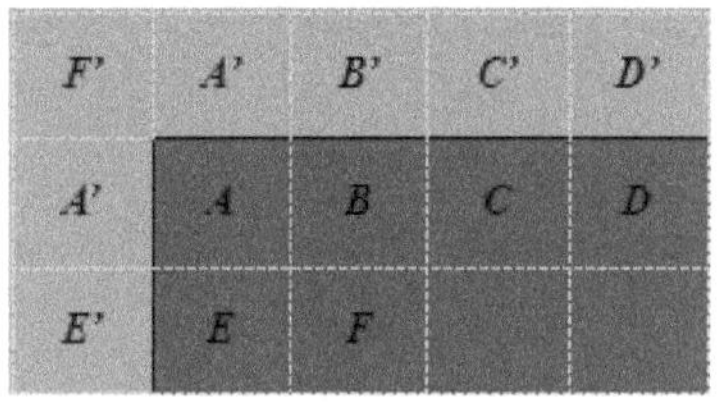

Figure 2. The graph of additional process.

For the edge blocks, we copy their reflectional symmetrical blocks as the filled blocks (as the **Figure 3** shows, block A and block A', block B and block B' and so on); for the corner blocks, we copy their rotational symmetrical blocks as the filled blocks (As the **Figure 3** shows, block F and block F').

Then two traffic monitoring images of low contrast were chose (such as a dawn image and a fog image) to do experiments:

We have tested HE, BHE, AHE and POSHE methods respectively. The results of the experiment are showed in **Figure 3**. And the corresponding histograms are showed in **Figure 4**. Form **Figure 3**, we can know that the results of local histogram equalization get by AHE and POSHE methods are better than the results of global histogram equalization get by HE and BHE methods. As the **Figure 3** shows, the gray level of the original image gathers in the left part, so the result of HE method shows a bleaching effect and the histogram of the processed image gathers in the right part. The BHE method doesn't do enough contrast enhancements on the image. The processed images of AHE and POSHE methods have a great enhancement, and the gray level of the whole image distributes uniformly, meanwhile the details are retained (e.g. the wire in the top right corner of image, and the lamp of vehicle etc.). Our proposed method produced satisfactory results. On the one hand, the contrast of the processed image have been enhanced (e.g. road, vehicle and the roadside trees); on the other hand, as the histogram shows, our proposed method have stretched the contrast, and the gray level of the whole image distributes uniformly.

In order to have a more objective understanding of the proposed method, we compared HE, BHE, AHE, POSHE and the proposed method from following three aspect: pre-processing, artificial boundaries process and average time. The results of the comparison are shown in **Table 1**.

As **Table 1** shows, the local methods such as AHE and POSHE need to eliminate artificial boundaries because they divide the image before processing .So the local methods are slower than the global methods such as HE and BHE. The execution time of the proposed method is between BHE and POSHE, which is about 0.2705 seconds. Its execution speed is very fast. That is to say it can be applied to embed device and real-time processing.

4. Performance Evaluation

4.1. Analysis of Time Efficiency

In the process of our method, the peripheral sub-blocks A, B, C, D, E, F, G and H affect the histogram equalization of the center block. The main time consumption concentrates on the computation of $M_{\text{smooth-influence}}$ because the computation of the three membership functions includes exponential operation and square operation. For each block, the transformation function can be written as:

$$s_k = T_{center}(r_k) = \sum_{i=1}^{3}\sum_{j=1}^{3} M_{\text{smooth-influence}}(i,j) \times T(r_k)_{i,j} \tag{9}$$

where,

$M_{\text{smooth-influence}}(i,j)$ is the corresponding element of $M_{\text{smooth-influence}}$

$T(r_k)_{i,j}$ is the corresponding transformation function of each peripheral sub-block.

If the image is divided into $n \times m$ sub-blocks and each block have $size$ pixels, we can suppose the execution time of similarity function as $T_a(size)$, the execution time of membership function as T_b and the execution time of HE for each block as $T_c(size)$. According to Formula (9), for each block, the similarity function needs to compute 8 times, the membership function needs to compute 24 times, and the transformation function needs to compute 1 time.

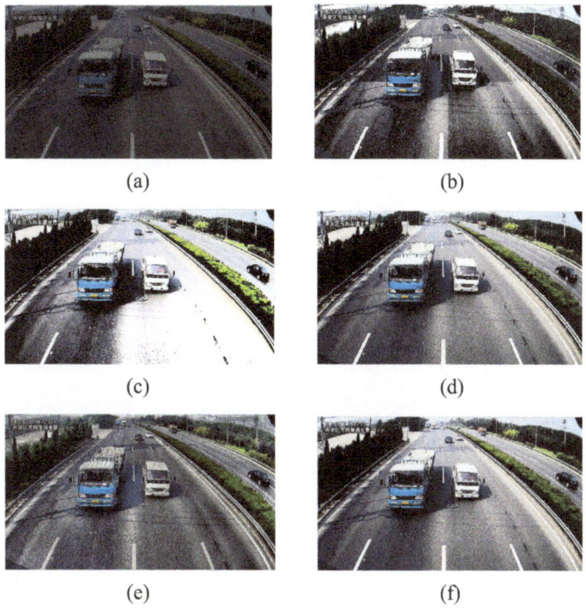

Figure 3. Simulation results. (a) Original; (b) Proposed; (c) Heed; (d) BHEed; (e) AHEed; (f) POSHEed.

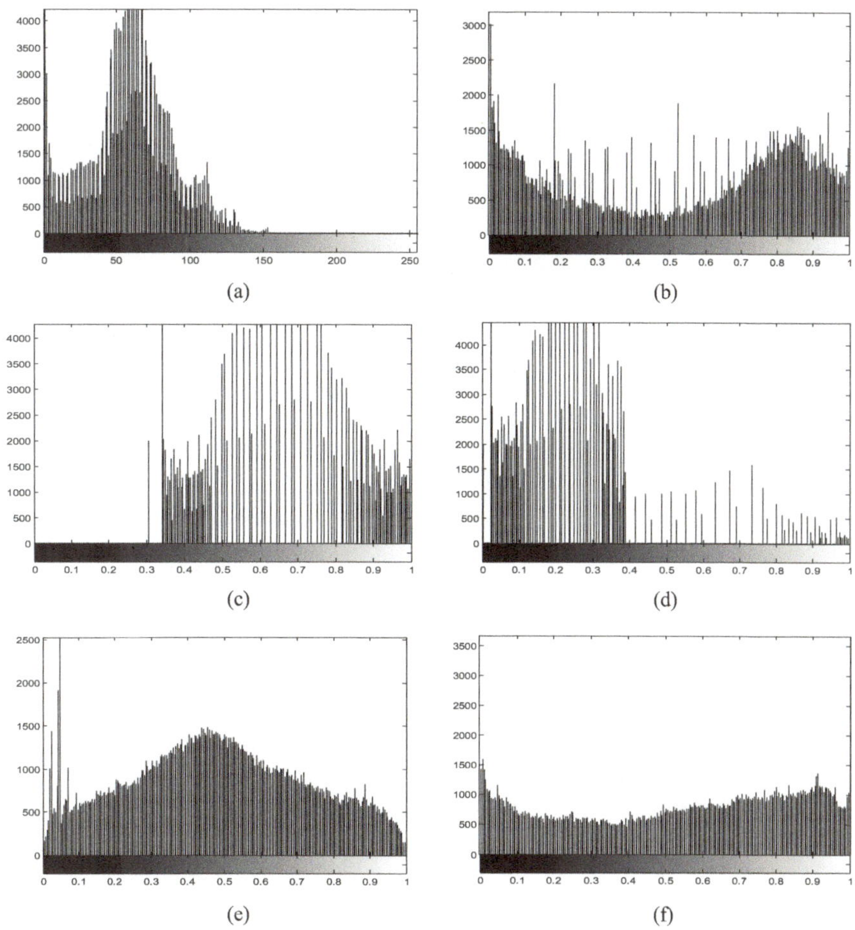

Figure 4. Histogram of **Figure 3**. (a) Original; (b) Proposed; (c) Heed; (d) BHEed; (e) AHEed; (f) POSHEed.

Table 1. The result of comparison.

Method	Pre-processing	Artificial boundaries process	Average time
Proposed	Divide	Using smooth filter	0.270552
HE	NULL	NULL	0.256914
BHE	NULL	NULL	0.256914
AHE	Divide	Bi-linear interpolation	0.293260
POSHE	Divide	Detect and eliminate	0.289349

a. Experimental environment: Matlab 7.06, Intel (R) Celeron (R) CPU, 4.00 GB RAM; b. Experimental data: 720 P traffic monitoring images.

So the total execution time of processing an image can be written as:

$$T = n \times m \times \left(8 \times T_a(size) + 24 \times T_b + T_c(size)\right) \tag{10}$$

From the Formula (10), we know the total execution time is proportional to the number of sub-blocks, that is, when the value of $n \times m$ is increased, the total execution time is also increased; meanwhile the total execution time is also proportional to the size of sub-blocks, that is, when the size of a sub-block is increased, the total execution time is also increased; However, the number of sub-blocks is inversely proportional to the size of sub-blocks, because when a fixed size image is divided into $n \times m$ sub-blocks, the more sub-blocks you divided, the less size of sub-blocks you will get. So in order to reduce the total execution time, the main problem is to balance the number of sub-blocks and the size of sub-blocks. Firstly, an image was divided into different sizes, then our method was tested and the total execution time was recorded.

As the **Figure 5** shows, we know when the number of sub-blocks is between 4 × 6 and 8 × 8, the total execution time is relative short; when the number of sub-blocks is small, the total execution time is a little shorter than that of HE method; when the number of sub-blocks is large, the total execution time is very long, because the computation of membership function is very time-consumed. For the monitoring image of 1280 × 720 pixel image, the most suitable segmentation approach is dividing the image into 4 × 6 or 6 × 6 sub-blocks.

4.2. Analysis of Membership Function

In this paper, there are two reasons why do we choose Gaussian function as the membership function. On one hand, the Gaussian function is suitable for describing the phenomenon of nature, human society, psychology and education; On the other hand, the Gaussian function's arguments are relatively simple compared to the other intermediate types of distribution fuzzy function, it only needs two arguments to control the position and shape of the function.

Form the subject point, on the condition that the shape controlling parameter σ is static, when a_1, a_2, a_3 are small and concentrated, the influences of the peripheral sub-blocks on the central sub-block are great, and the processing results of the central sub-block will loss too much details. When a_1, a_2, a_3 are large and concentrated, the influences of the peripheral sub-blocks on the central sub-block are small, and the image processing results will show obvious artificial boundaries. Therefore, arguments a_1, a_2, a_3 should distribute on the coordinate axis uniformly one by one, then the influences of the around sub-blocks on the central sub-block can be better measured.

As the **Figure 6** shows, three different groups of arguments was chose to experiment. The arguments of (b) distribute on the axis uniformly, the result looks better. The arguments of (c) are small and concentrated, the results lost a lot of details (e.g. the goods of truck). On contrast, the arguments of (d) are large and concentrated, there are obvious artificial boundaries in the results.

From the experimental results, we know the more suitable arguments are $a_1 = 0.15, a_2 = 0.45, a_3 = 0.75$, and $\sigma_1 = \sigma_2 = \sigma_3 = 1$.

5. Conclusion

With the rapid development of traffic monitoring technology, the conventional image processing method is no longer suitable for the traffic monitoring images. In this paper, three features of traffic monitoring images have

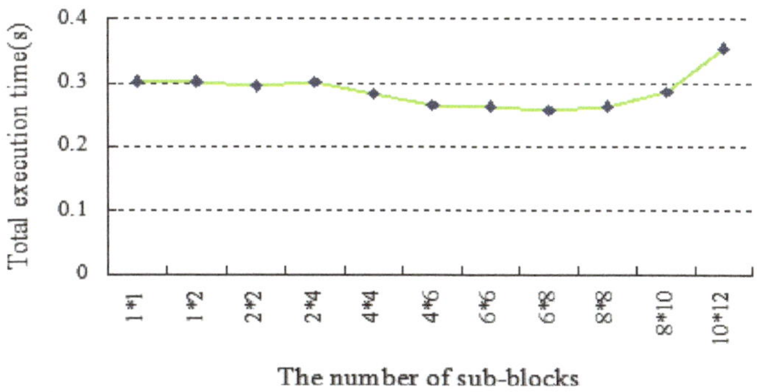

Figure 5. Execution time graph.

(a) (b) (c) (d)

Figure 6. Experimental results with different arguments. (a) Original image; (b) $a_1 = 0.15, a_2 = 0.45, a_3 = 0.75$; (c) $a_1 = 0.75, a_2 = 0.85, a_3 = 0.95$; (d) $a_1 = 0.15, a_2 = 0.20, a_3 = 0.25$.

been analyzed. On the basis of these features, a local histogram equalization method based on fuzzy mathematics was proposed. This method is suitable for contrast enhancement of traffic monitoring images. The contrast of the processed images is stretched, meanwhile the time consumption of our method is relatively low. In order to reduce artificial boundaries, we use a 3 × 3 smooth filter. But when observed carefully, some slight artificial boundaries will be found. Because the artificial boundaries and the image content are closely related, the method of reducing artificial boundaries completely and quickly is more challenging. So one of the focuses of future research is to eliminate the artificial boundaries.

References

[1] Conzalez, R.C. and Woods, R.E. (2001) Digial Image Processing. Prentice-Hall, Inc., New Jersey.

[2] Gonzalez, R.C., Woods, R.E. and Eddins, S.L. (2004) Digital Image Processing Using MATLAB. Prentice Hall.

[3] Kim, Y. (1997) Contrast Enhancement Using Brightness Preserving Bi-Histogram Equalization. *IEEE Transactions on Consumer Electronics*, **43**, 1-8. http://dx.doi.org/10.1109/30.580378

[4] Youn, J., Sup, K.L. and Hwang, K.S. (2001) An Advanced Contrast Enhancement Using Partially Overlapped Sub-Block Histogram Equalization. *IEEE Transactions Circuits and Systems for Video Technology*, **11**, 475-484. http://dx.doi.org/10.1109/76.915354

[5] Zadeh, L.A. (2005) Fuzzy Set and Fuzzy Information-Granulation Theory. Beijing Normal University.

[6] Tan, K.K. and Oakley, J.P. (2005) Physics-Based Approach to Color Image Enhancement in Poor Visibility Conditions. *Journal of the Optical Society of America A*, **18**, 2460-2467. http://dx.doi.org/10.1364/JOSAA.18.002460

[7] Wang, J.-H. and Wang, X.-Z. (2004) An Initial Analysis of Parametric Sensitivity to Fuzzy Extension Matrix. *Proceedings of* 2004 *International Conference on Machine Learning and Cybernetics*, **3**, 1884-1888. http://dx.doi.org/10.1109/ICMLC.2004.1382085

[8] Meylan, L. and Süsstrunk, S. (2004) Color Image Enhancement Using a Retinex-Based Adaptive Filter. *Proceedings of IS&T Second European Conference on Color in Graphics, Image, and Vision* (*CGIV* 2004), **2**, 359-363.

[9] Terai, Y., Goto, T., Hirano, S. and Sakurai, M. (2009) Color Image Contrast Enhancement by Retinex Model. *The* 13*th IEEE International Symposium on Consumer Electronics* (*ISCE*), 392-393.

9

Segmentation of Hyper-Acute Ischemic Infarcts from Diffusion Weighted Imaging Based on Support Vector Machine

Yuqing Peng[1,2], Xiaodong Zhang[1], Qingmao Hu[1*]

[1]Research Lab for Medical Imaging and Digital Surgery, Shenzhen Institutes of Advanced Technology, Chinese Academy of Sciences, Shenzhen, China
[2]Shenzhen College of Advanced Technology, University of Chinese Academy of Sciences, Shenzhen, China
Email: *qm.hu@siat.ac.cn

Abstract

Accurate and automatic segmentation of hyper-acute ischemic infarct from magnetic resonance imaging is of great importance in clinical trials. Manual delineation is labor intensive, exhibits great variability due to unclear infarct boundaries, and most importantly, is not practical due to urgent time requirement for prompt therapy. In this paper, segmentation of hyper-acute ischemic infarcts from diffusion weighted imaging based on Support Vector Machine (SVM) is explored. Experiments showed that SVM could provide good agreement with manual delineations by experienced experts to achieve an average Dice coefficient of 0.763 ± 0.121. The proposed method could achieve significantly higher segmentation accuracy and could be a potential tool to assist clinicians for quantifying hyper-acute infarction and decision making especially for thrombolytic therapy.

Keywords

İschemic Stroke, Infarct Segmentation, Feature Selection, SVM

1. Introduction

Ischemic stroke remains the second cause of death worldwide with social costs increasing year by year [1]. Hyper-acute ischemic stroke is an emergency as brain tissues would be deprived of oxygen and blood supply after stroke. At hyper-acute stage diffusion weighted (DW) imaging is preferred due to its high sensitivity in detecting stroke lesions [2]. Ischemic infarcts are hard to be delineated from DW imaging on account of diverse locations, variable shapes and unclear boundaries even for experienced experts, which is especially the case for infarcts at hyper-acute stage as they are still undergoing progression and exhibiting most variations [3]. Therefore an automated and effective method is necessary to segment infarcts to aid the decision making for therapy.

Prakash *et al.* [3] segmented infarcts from DW images (DWIs) by first identifying axial slices with ischemic

*Corresponding author.

lesions followed by binarization with a global DWI intensity threshold derived from histogram divergence. Tsai and co-workers segmented acute cerebral infarcts from DWIs and apparent diffusion coefficient (ADC) maps base on fuzzy C-means clustering (FCM) determining the histographic distribution and defining self-adjusted intensity thresholds [4].

We investigate support vector machine (SVM) on hyper-acute ischemic infarcts segmentation from DW imaging. Features are derived from DW imaging, and then carefully chosen by several feature selection methods. The infarcts are initially determined from our previous work based on hypo-intense ADC map [5], followed by refinement through training and classifying based on optimized features and SVM. To evaluate the contribution of each image, we conduct experiments through training one image and one image left out.

2. Material and Methods

2.1. Datasets

In this study, 98 ischemic stroke patients (31 women and 66 men, with age range 24 - 78 years) within 6 hours from the symptom onset were used. They were collected from Tiantan Hospital and Tianjin Huanhua Hospital since 2005. The protocol of the research has been approved by the Institutional Review Board of both hospitals. All patients gave written consent and provided permission for scientific and educational purpose. The baseline DW imaging was carried out with two 3.0 Tesla scanners (Trio-Tim, Erlagen, Germany) with a spin-echo, multislice, and single shot echo-planar of b = 0 (B0 image) and 1000 or 1500 s/mm^2 (DWI) and the corresponding ADC map. The imaging covered the whole brain, with 19 - 24 axial slices of a 5 mm slice thickness and 1 - 1.5 mm gap, most matrixes being 128 × 128 with few being 156 × 156 or 384 × 384 to have in-plane resolution of 0.60 mm to 1.80 mm.

2.2. Overview of the Study

Our study consist of the following components: extracting relevant features, employing feature selection algorithms to choose more robust features, and applying SVM to segment ischemic infarcts. The algorithm was implemented with Matlab version 2013a running on Microsoft windows with Core i5 3.2 GHz processor and 8GB memory. The schematic diagram is shown in **Figure 1**.

2.3. Preprocessing

All images were normalized to the range of [0, 255]. Asymmetry map (ASM) was calculated to reflect the characteristics of infarcts in DWI and ADC: being hyper-intense on DWI, hypo-intense on ADC, and being generally asymmetrical with respect to the midsagittal plane (MSP) extracted from B0 [6].

$$DWI\_ADC = \begin{cases} DWI - ADC, \text{if } DWI - ADC > 0 \\ 0, \qquad \text{otherwise} \end{cases} \tag{1}$$

$$DIF(x,y,z) = DWI\_ADC(x,y,z) - \max_{(u,v)\in N_s(x_0,y_0)} DWI\_ADC(u,v,z) \tag{2}$$

$$ASM(x,y,z) = \begin{cases} DIF(x,y,z), if\ DIF(x,y,z) > 0 \\ 0, \qquad \text{otherwise} \end{cases} \tag{3}$$

where (x_0, y_0) and (x, y) are symmetrical with respect to the midsagittal line (which is the intersection between the MSP and the axial slice) on the axial slice z, $N_s(x_0, y_0)$ is the neighborhood of (x_0, y_0) and the window size of window is 5 × 5. Axial slices of B0, DWI, ADC and ASM of a patient are shown in **Figure 2**. It is clear that the infarct region on the right hemisphere has been enhanced to be more easily discernable than both the DWI and ADC.

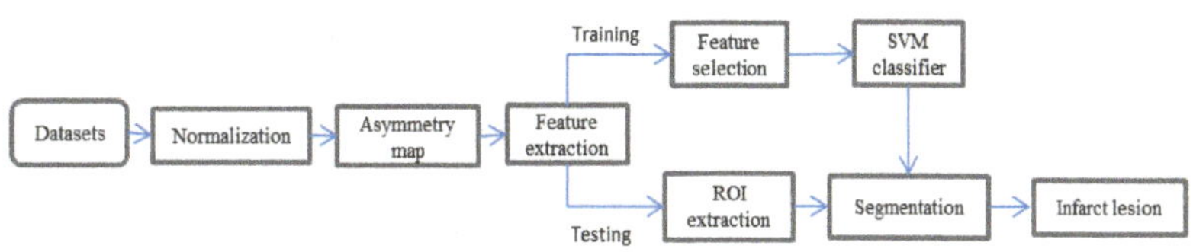

Figure 1. Flowchart of the proposed method.

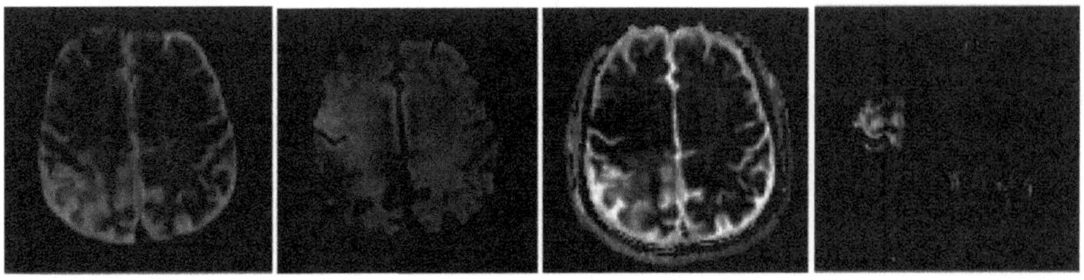

Figure 2. From left to right, axial slices of B0, DWI, ADC and ASM.

2.4. Region of Interest

To alleviate the interference of normal brain tissues and speed up the calculation, candidate infarcts in the form of regions of interest (ROIs) are extracted from thresholding ADC maps subjected to DWI [5]. Specifically, denote the most frequent ADC value for all voxels within the brain as ADC_{ref}, any voxels with ADC not greater than $0.75*ADC_{ref}$ are to formulate connected components. Any connected components with average DWI values not smaller than the intensity average plus the intensity standard deviation of brain voxels on DWI at the corresponding axial slice are kept as part of region R_1. Voxels with ADC value within $(0.75*ADC_{ref}, 0.85*ADC_{ref}]$ are checked to formulate connected components and added to R_1 if the components has at least one neighboring voxels in R_1. R_1 is then increased by morphological dilation with a structuring element of radius R_d (being 2 voxels). Suppose the z coordinates of R_1 are in the range of z_0 to z_n with $z_n >= z_0$, then R_1 region on the axial slice z_0 and z_n are respectively pasted to axial slice $z_0 - 1$ and $z_n + 1$ to attain the eventual ROIs. **Figure 3** shows an axial slice of DWI, ADC and the corresponding ROIs (in color), where regions in red are initial ischemic infarcts and those in yellow are added voxels through the described morphological dilation.

2.5. Feature Extraction

For training, the voxels in infarct regions are the positive samples and drawn by a specialist. While the negative samples are those voxels close to the infarct regions obtained through iterative dilation of the infarct regions such that the number of negative samples is just equal to or greater than that of the positive samples. The features for each voxel are: intensity value, mean value, standard deviation, the weighted mean intensity value with Gaussian kernel (mean being 0 and standard deviation being 1), minimum value, maximum value, median value, skewness, kurtosis. All of these features are calculated from a local window of size 3×3 except the intensity value. The feature vector is formulated form the 9 quantities of B0, DWI, ADC and ASM to be 36 dimensional.

2.6. Training

Segmentation of ischemic infarcts can be considered as a classification of two classes: infarcts and non-infarcts. Two-fold cross-validation is applied in this research. The classifier is SVM while features are optimized through the feature selection procedure detailed below.

SVM was first proposed by Vapnik [7] as a supervised learning method to transform training samples into a multidimensional hyperspace and then computes decision boundary in hyperplane by solving the convex optimization problem. The margin between the positive and negative samples from the separating hyperplane is to be maximal. To obtain the maximal margin, we employ LIBSVM tool box to obtain optimal parameters [8], using radial basis kernel functions (RBFs). Two important parameters of the RBF, penalty factor C and the parameter γ, are respectively set to 32 and 0.5 using grid search method to obtain a good performance for training samples.

2.7. Feature Selection

Redundancy and irrelevant features may lead to non-effective classification model. It is desirable to rank the importance of the features and select the distinguishing features. Five feature selection methods are employed and compared in our experiments.

Fisher score (FS) [9] identifies features with small within-class distance and large between-class distance. Threshold value is used to differentiate effective features.

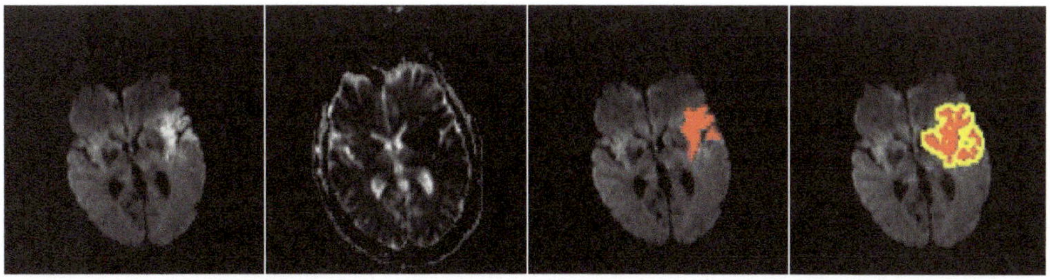

Figure 3. From left to right, an axial slice of DWI, ADC, the segmented ischemic regions form [5] on DWI and the corresponding ROIs for detecting infarction overlaid in color on DWI, where the region in red are initial ROIs from [5] while those in yellow are from morphological dilation.

Gini index (Gini) [10] quantifies the ability of feature to differentiate the classes. The features are relevant with smaller Gini Index. The top k features with the smallest Gini index are selected.

Mutual-information (MI) [11] describes the relationship between the features and categories. It calculates information between the feature and the category with a larger MI for larger information.

Minimum redundancy maximum relevance (mRMR) [12] is a filtering method based on space search. It searches for the feature subsets by using information gap and information entropy to obtain the feature subsets.

ReliefF algorithm [13] computes feature weight to rank features with linear time according to the number of features extracted from datasets.

3. Results

The proposed segmentation method has been evaluated on 2-fold cross validation. All datas are divided into two groups randomly with equal numbers.

3.1. Performance Evaluation Metrics

To assess the performance of the segmentation, three popular evaluation metrics are used, the Dice coefficient (Dice), sensitivity and specificity as calculated below: Dice = 2TP/(FP + FN), sensitivity = TP/(TP + FN), and specificity = TN/(TN + FP), where TP is for the number of correctly segmented infarct voxels, TN is for the number of correctly segmented non-infarct voxels, FP is for the number of wrongly segmented infarct voxels, and FN is for the number of wrongly segmented non-infarct voxels.

3.2. Classifiers with Different Image Combinations

To see the relative importance of different images in the classifier, we designed classifiers with each singular image and classifiers with one image being excluded. For instance, we designed a classifier from ASM image to obtain an average Dice of 0.749 ± 0.122. The results are shown in **Figure 4**.

3.3. Performance of Different Feature Selection Methods

We compared the suitability of a number of feature selection methods. Each selector is applied to the training data and returns a ranking of the feature vector. Different sizes of features are tested on each selector. The mean Dice over all testing images on SVM are shown in **Figure 5**.

The highest Dice is achieved from feature selection using on MI algorithm with 15 features on SVM. The mean Dice, sensitivity and specificity are 0.763 ± 0.121, $76.08\% \pm 14.8\%$, $99.95\% \pm 0.098\%$ respectively. Two slices with good and bad segmentation are shown in **Figure 6**.

3.4. Comparison with Existing Methods

The proposed method has been compared with divergence measure based method (DM method) [3] and FCM based method [4]. For a fair and relevant comparison, we have tried our best to implement them to attain the best performance on the 98 datasets within 6 hours from symptom onset. The Dice, sensitivity and specificity for [3] ([4]) are respectively 0.215 ± 0.213 (0.597 ± 0.204), 0.565 ± 0.346 (0.585 ± 0.221) and 0.983 ± 0.024 (0.999 ± 0.001).

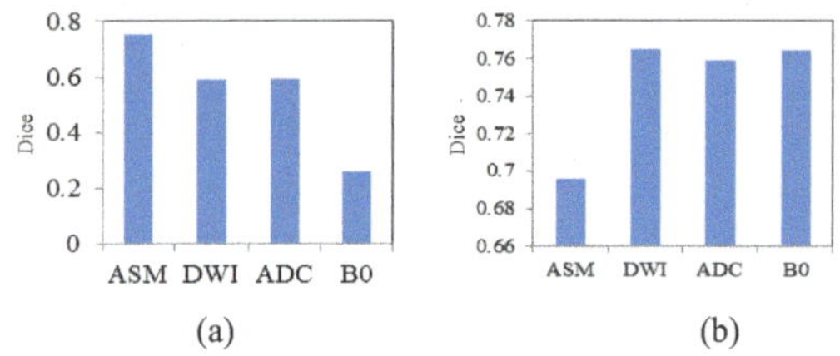

Figure 4. Contributions of different images: (a) Dice obtained from single image; (b) Dice obtained with one left out.

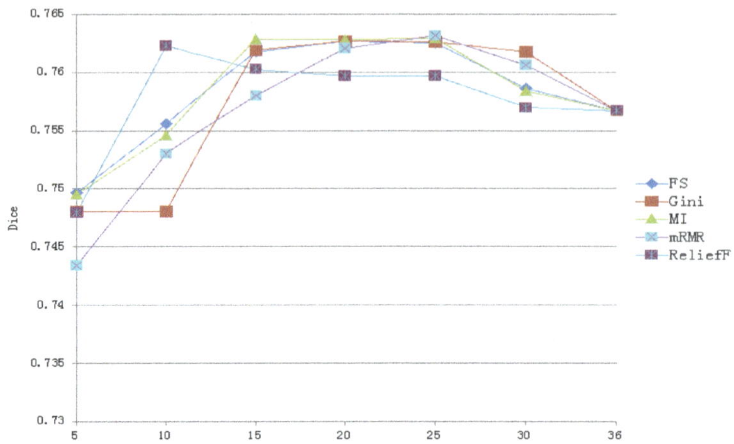

Figure 5. The Dice obtained from different feature selection algorithms: FS, Gini, MI, mRMR and ReliefF.

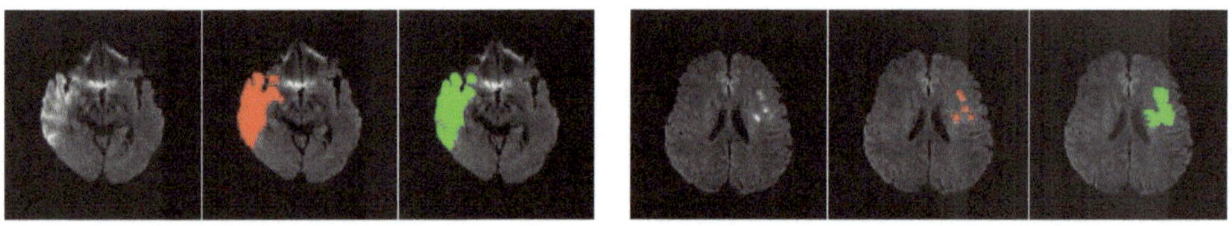

(a) (b)

Figure 6. Segmentation of two DW axial slices with high (a) and low (b) Dice coefficient, red for ground truth infarcts and green for segmented infarcts. (a) Good segmentation; (b) Bad segmentation.

4. Discussion

In this paper, we have explored automated segmentation of hyper-acute infarcts from DW imaging based on carefully chosen features and SVM. As the DM method [3] is basically a global thresholding method to determine the DWI threshold based on divergence measure, it is not appropriate for processing data imaged within 6 hours due to the substantial variability in intensities. For the FCM based method [4], it is dependent on the prominent edge on DWI for confirmation of infarcts, which could fail because the DWI might have ill-defined edges. The proposed method could cope with the variability in intensities and ill-defined edges of DWI and ADC data imaged within 6 hours through learning the pattern from training samples. As such, the proposed method attain significantly higher accuracy in terms of Dice and sensitivity all with $p < 0.001$.

From **Figure 4**, it can be seen that the classifier from ASM could yield the best Dice (0.749 ± 0.122) which is close to the Dice based the four images (0.763 ± 0.121). This may suggest that ASM has integrated the most relevant information from DWI, ADC and B0. On the other hand, the classifiers from DWI or ADC achieve a similar Dice (0.588 ± 0.244 and 0.592 ± 0.179 respectively), which may suggest that both DWI and ADC alone could not achieve the optimum segmentation and they could provide complementary information to enhance infarct segmentation. Classifier based on B0 could only obtain a very low Dice (0.256 ± 0.213), which is consistent to the consensus that B0 carries little information on hyper-acute infarcts. From **Figure 5**, it can be seen that the best Dice is not achieved with the full features, indicating there are feature redundancy in the original 36 features. The best Dice is achieved with 15 features chosen from MI feature selection method.

To conclude, segmentation of hyper-acute ischemic infarcts based on SVM has achieved good results and features chosen from MI could yield an average Dice of 0.763 ± 0.121, which is significantly higher than existing methods. The proposed method could be a potential tool for clinicians to quantify the hyper-acute ischemic stroke and assist the decision making especially for thrombolytic therapy.

Acknowledgements

This work has been supported by: National Program on Key Basic Research Project (Nos. 2013CB733800 and 2013CB733803), National Science and Technology Pillar Program during the Twelfth Five-year Plan Period (No. 2011BAI08B09), Shenzhen Key Technical Development Grant (No. CXZZ20140610151856719), and Shenzhen Basic Research Grant (No. JCYJ20140414170821262). Authors would like to thank Dr. Yiqun Zhang for her valuable discussion on manually delineating infarct regions without clear boundaries.

References

[1] Du Plooy, J.N., Buys, A., Duim, W. and Pretorius, E. (2013) Comparison of Platelet Ultrastructure and Elastic Properties in Thrombo-Embolic Ischemic Stroke and Smoking Using Atomic Force and Scanning Electron Microscopy. *Plos One*, **8**. http://dx.doi.org/10.1371/journal.pone.0069774

[2] Muir, K.W., Buchan, A., von Kummer, R., Rother, J. and Baron, J.C. (2006) Imaging of Acute Stroke. *Lancet Neurol*, **5**, 755-768. http://dx.doi.org/10.1016/S1474-4422(06)70545-2

[3] Prakash, K.N.B., Gupta, V., Hu, J.B. and Nowinski, W.L. (2008) Automatic Processing of Diffusion-Weighted Ischemic Stroke Images Based on Divergence Measures: Slice and Hemisphere Identification, and Stroke Region Segmentation. *Int J Comput Ass Rad*, **3**, 559-570.

[4] Tsai, J.Z., Peng, S.J., Chen, Y.W., Wang, K.W., Wu, H.K., Lin, Y.Y., Lee, Y.Y., Chen, C.J., Lin, H.J., Smith, E.E., Yeh, P.S. and Hsin, Y.L. (2014) Automatic Detection and Quantification of Acute Cerebral Infarct by Fuzzy Clustering and Histographic Characterization on Diffusion Weighted Mr Imaging and Apparent Diffusion Coefficient Map. *Biomed Res Int*, **13**. http://dx.doi.org/10.1155/2014/963032

[5] Ma, L., Gao, P.Y., Hu, Q.M., Lin, Y., Jing, L.N., Xue, J., Wang, X.C., Chen, Z.J., Wang, Y.L., Liao, X.L., Liu, M.L. and Chen, W.J. (2010) Prediction of Infarct Core and Salvageable Ischemic Tissue Volumes by Analyzing Apparent Diffusion Coefficient without Intravenous Contrast Material. *Acad Radiol*, **17**, 1506-1517. http://dx.doi.org/10.1016/j.acra.2010.07.010

[6] Hu, Q.M. and Nowinski, W.L. (2003) A Rapid Algorithm for Robust and Automatic Extraction of the Midsagittal Plane of the Human Cerebrum from Neuroimages Based on Local Symmetry and Outlier Removal. *NeuroImage*, **20**, 2153-2165. http://dx.doi.org/10.1016/j.neuroimage.2003.08.009

[7] Cortes, C. and Vapnik, V. (1995) Support-Vector Networks. *Mach. Learn.*, **20**, 273-297. http://dx.doi.org/10.1007/BF00994018

[8] Chang, C.C. and Lin, C.J. (2011) LIBSVM: A Library for Support Vector Machines. *Acm T Intel Syst Tec*, **2**.

[9] Cao, L., Li, J., Sun, Y., Zhu, H. and Yan, C. (2010) EEG-Based Vigilance Analysis by Using Fisher Score and PCA Algorithm. 2010 *IEEE International Conference on Progress in Informatics and Computing* (*PIC*), 175-179.

[10] Zhu, W. and Lin, Y. (2013) Using Gini-Index for Feature Weighting in Text Categorization. *J Comput Inf Sys*, **9**, 5819-5826.

[11] Tourassi, G.D., Frederick, E.D., Markey, M.K. and Floyd, C.E. (2001) Application of the Mutual Information Criterion for Feature Selection in Computer-Aided Diagnosis. *Med Phys*, **28**, 2394-2402. http://dx.doi.org/10.1118/1.1418724

[12] Peng, H., Long, F. and Ding, C. (2005) Feature Selection Based on Mutual Information: Criteria of Max-Dependency, Max-Relevance, and Min-Redundancy. *IEEE Trans Pattern Anal Mach Intell*, **27**, 1226-1238. http://dx.doi.org/10.1109/TPAMI.2005.159

[13] Robnik-Sikonja, M. and Kononenko, I. (2003) Theoretical and Empirical Analysis of ReliefF and RReliefF. *Mach. Learn.*, **53**, 23-69. http://dx.doi.org/10.1023/A:1025667309714

10

Research on Destriping Based on Rough Set

Xiangyang Chen, Liping Chen, Sile Wang, Wenzhu Yang, Sukui Lu

School of Computer Science and Technology, Hebei University, Baoding, China
Email: againcome@sina.com

Abstract

In order to remove the stripe noises in cotton foreign fiber images by line scanning camera collected, in multi threshold segmentation of rough set, every region's color is instead of the statistics color of the region. This method can retain the detail information of original image as far as possible, and do well in the stripe noise removal. The roughness of rough set was calculated respectively using directional diagram, Canny operator and Sobel operator. Comparing the three methods, the results indicate that the Canny operator keeps the more details of image, and directional diagram and Sobel operator have the better effects on denoising.

Keywords

Stripe Noises, Multi Threshold Segmentation, Rough Set, Denosing

1. Introduction

In using line scan camera to acquire the cotton fiber image, it is often to produce stripe noises, which affects the extraction and recognition of the foreign fiber [1]. Today there are many methods to remove the stripe noises. The traditional method of denoising can be divided into the frequency domain and the spatial domain. Frequency domain denoising depends that the noise frequency is different larger with the image frequency, and therefore with the noise of the image is converted from the spatial domain to the frequency domain, to change the transforming coefficient of the noise, as far as possible to retain the transforming coefficients of the image information, and then inversing transform image to reducing image noises. But the denoising process tends to lose the part detail information of the image, resulting in blurred images. Commonly used methods are Wavelet transform [2] and Fourier transform [3]. Spatial denoising is mainly through calculating the image gray values, and using the normal pixel gray value as a reference to correct the noise pixel gray value, so as to reduce the influence of noise pixels. Typical methods are median filter [4], histogram equalization [5] [6] and the moment matching [7].

Using rough set theory [8] in multi threshold image segmentation has a good effect. If using statistical color replacement the color of segmentation region, it will realize the target image denoising. The distribution and intensity of stripe noise in the cotton fiber image is not uniform, in order to ensure the clarity of the image, the multi threshold segmentation of rough set is used to remove the noise.

2. Rough Set Theory

Since 1982 Poland scholar Z. Pawlak proposed the rough set theory, it has been a hot research topic in the field of artificial intelligence, and has been widely applied machine learning, data mining, process control and image processing fields.

Definition 1: information systems can be represented as a four tuple, the formalized definition is: $S=\{U,A,V,f\}$, in which U : finite set of objects; A : finite set off properties; $V:V=\bigcup_{p\in A}V_p$, V_p is p attribute domain; $f:U\times A\rightarrow V$ is general function, makes $f(x_i,q)\in V_q$ for each $x_i\in U,q\in A$.

Definition 2: $X\subset U$ is a subset of individuals on the global, $P\subseteq A$. The lower approximation and the upper approximation of the X are respectively

$$\underline{P}X=\{Y\in U\mid P:Y\subseteq X\} \tag{1}$$

$$\overline{P}X=\{Y\in U\mid P:Y\cap X\neq\varnothing\} \tag{2}$$

$\underline{P}X$ is the confirmed largest defined set contained within the X, $\overline{P}X$ is the smallest defined set contained X.

Therefore, the boundary region can't be clearly defined as:

$$Bnd_p(X)=\overline{P}X-\underline{P}X \tag{3}$$

The larger the boundary area is, the more the set X can't express. The uncertainty of the set X can be expressed by the roughness of rough sets, the formula is as follows:

$$\rho_P(X)=1-\frac{\underline{P}X}{\overline{P}X} \tag{4}$$

The $\rho_P(x)$ is [0, 1], the greater the value is, the more uncertainty the set X is.

3. Rough Set Denoising Process

The core idea of rough set denoising is that the color of a segmentation region can be replaced by a same color, so as to achieve the purpose of noise removal. Therefore, how to determine the region is a core problem. The linear scanning image of the cotton fiber is a color image, and the information contained in the RGB color image is often more than the gray image, and it is more consistent with the human visual sense. If only gray level histogram is used to distinguish different gray areas, the approximate relationship between pixels in color image cannot be reflected.

The histogram can be used to express the number of pixels in each gray level, so the basic histogram can be used as the lower approximation of the region. The histon histogram considered the color difference between the pixels in the neighborhood, and the uncertain information is described. So the histon histogram is used as the upper approximation.

Let f for a RGB color image of $M\times N$, which is decomposed into f_r, f_g and f_b in R, G, B three color channels. The basic histogram and histon histogram are defined respectively:

Definition 3: basic histogram

$$h_i(g)=\sum_{m=1}^{M}\sum_{n=1}^{N}\sigma\left(I(m,n,f_i)-g\right);\ 0\le g\le L-1, i=\{R,G,B\} \tag{5}$$

The σ is an impulse function; the value L is 256, representing the number of gray level of images; f_i represents a gray image; $I(m,n,f_i)$ is the gray value of pixel $I(m,n)$.

Definition 4: histon histogram

$$H_i(g)=\sum_{m=1}^{M}\sum_{n=1}^{N}\left(1+X(m,n)\right)\sigma\left(I(m,n,f_i)-g\right);\ 0\le g\le L-1, i=\{R,G,B\} \tag{6}$$

$X(m,n)$ is a matrix of $M\times N$, which indicates that the difference of the neighborhood pixel color, can be obtained by the following Equation (7):

$$X(m,n)=\begin{cases}1, & D(m,n)<exp\\ 0, & other\end{cases} \tag{7}$$

where $D(m,n)$ represents the distance sum of between each pixel and the center pixel point in the neighborhood, and exp represents a threshold constant.

The process of removal of the stripe noise using rough sets is shown in **Figure 1**. Since the histogram is calculated by the gray scale, it is necessary to decompose the RGB image from R, G and B three channels respectively.

Roughness calculation: the calculating the basic image histogram and histon histogram, then get the roughness images by Equation (4).

Bands calculation: due to many bands in the original image, in order to remove the noise information and need to unite the less different band. Firstly, the position vector and the number N of all the peaks in the image are obtained according to the roughness, Height threshold Th of the peaks can be calculate by Equation (8):

$$Th = Mp - \sqrt{\sum_{i=1}^{N}(p_i - Mp)^2 / N} \tag{8}$$

Which Mp is mean of the maximum and minimum peak value, P_i is the peak point. The experimental results show that the best of the width threshold Tw of the peak is 10. Then the peaks of less than Th are abandoned, and the width of the peaks of less than Tw is united.

The minimum value between two peaks is the valley value, the gray level between the two valleys is a band.

Band's colors calculation: a band using a replace gray color, the color of band $[k_1, k_2]$ can be calculated by Equation (9):

$$BandColor = \sum_{i=k1}^{k2} i * \frac{h(i)}{SH} \tag{9}$$

The $h(i)$ is the number of pixels in the gray level i of the basic histogram, SH is the sum of the number of pixels between $[k_1, k_2]$, can be calculate by the Equation (10):

$$SH = \sum_{i=k1}^{k2} h(i) \tag{10}$$

A new composed gray color can be used to represent the color of the band.

In the R, G and B channels, the band information and band colors are calculated respectively, the color of each band is replaced by the calculating the gray color, and then to synthesize a RGB color image.

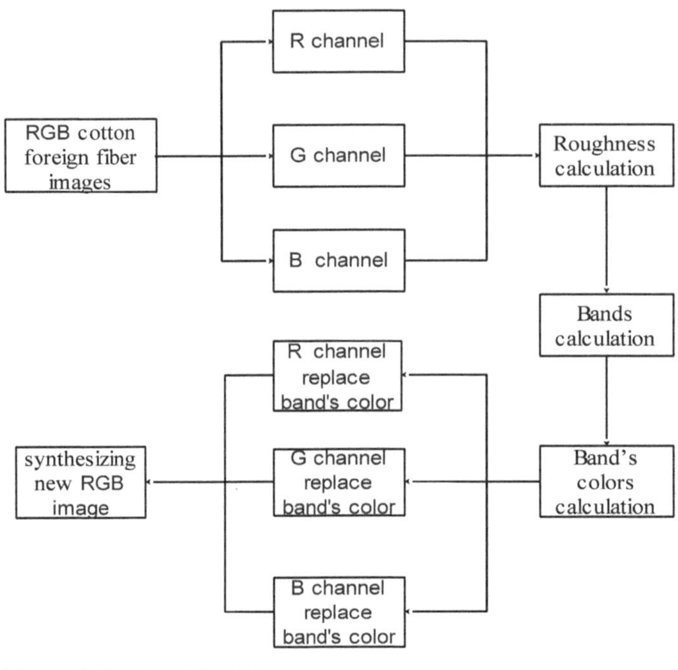

Figure 1. Rough set denoising process.

4. Experimental Results and Analysis

Experimental picture is the image obtained from the color line scan camera. The key of rough set denosing is to determine the boundary of the color region, namely, the difference between the upper approximate histon his to gram $H(g)$ and the lower approximate basic histogram $h(g)$, so as to determine the roughness of the image. If the difference between the upper approximation and the lower approximation is obtained, that is the edge of the region, the roughness of the image can be calculated. Taking into account the image edge is the biggest difference between the images, the use of edge operator and the direction of the image should be able to get the approximate effect, so compared respectively with the three methods of denoising effect.

Method 1: directional diagram. $f(m, n)$ using the Equation (11) to calculate the distance between this point and $f(m-1, n-1)$ for the image f of $M \times N$.

$$ColorDis_{(m,n)} = abs\left(\arctan\left(f(m,n) - f(m-1,n-1)\right)\right) \tag{11}$$

and $1 \le m \le M, 1 \le n \le N$;

Method 2: using canny operator.

Method 3: using Sobel operator.

The experimental results are as follows:

Figure 2 is a cotton color image with a foreign fiber, among **Figures 3-5**, the (a) and (b) is the edge difference image and the denoising image by using the directional diagram, Canny operator and Sobel operator respectively. The experimental results show that the three methods do well in the strip noise removal. Compared with directional diagram and Sobel operator method, Canny operator method for denoising retains more image details, but some noise was also retained, so the denoising effects of the other two methods are better than its. These three methods can enhance the distinction between foreign fiber and others in the image. The rough set denosing not only realizes the removal of stripe noise, but also realizes the enhancement of the extraction region. It can provide higher quality image for the next step of image segmentation, and is helpful to the extraction and recognition of foreign fiber.

Figure 2. Color image of cotton.

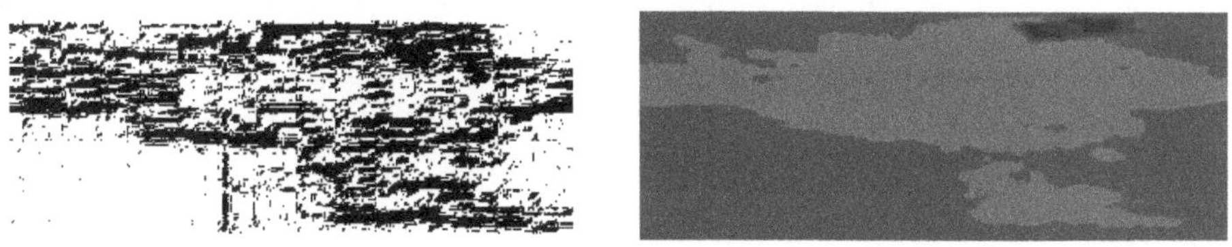

(a) (b)

Figure 3. Directional diagram method.

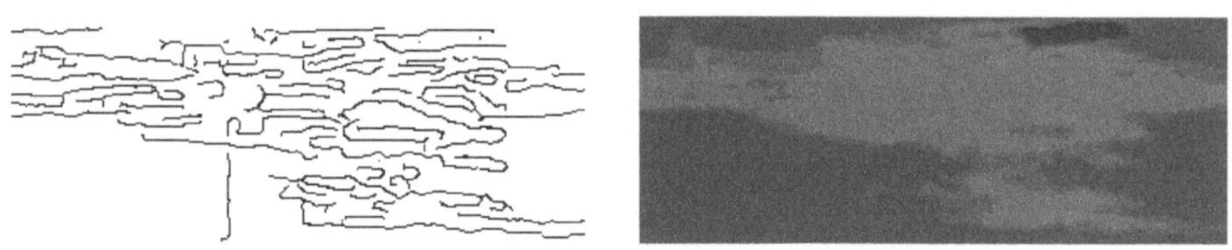

(a) (b)

Figure 4. Canny operator method.

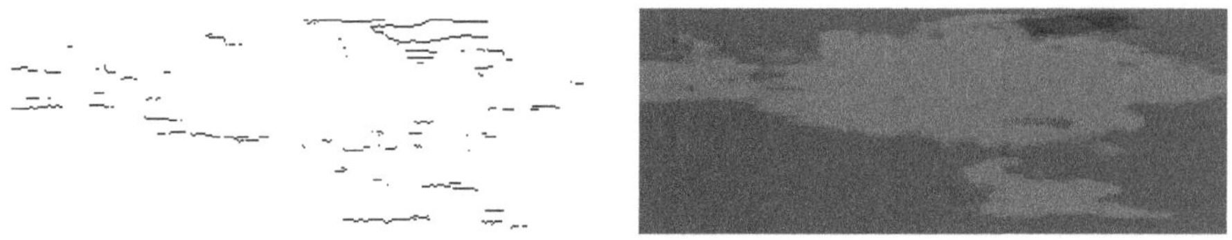

(a) (b)

Figure 5. Sobel operator method.

5. Conclusion

The basic histogram and his ton histogram of the RGB color image is used as the lower approximation and the upper approximation of the image, and the roughness of the image is obtained by them. The image is divided into regions according to the roughness, and replaced the original color with statistical color within the region. The experimental results show that the rough set algorithm can achieve the removal of stripe noise very well. In the edge calculation of the image, the influences of the three methods are compared. Canny operator method has retained the more details, and the results of the direction and Sobel operator method are much better. The foreign fiber region is enhanced, which makes the foreground and background more obvious, and provides a high quality image for the following image segmentation.

Acknowledgements

The authors thank Hebei Natural Science Foundation (F2015201033), The Ministry of Science and Technology of the People's Republic of China (2013DFA11320), for their financial support.

References

[1] Wang, S.L., Fan, S.Y. and Lu, S.K. (2013) Visual Saliency Map Based Color Image Segmentation for Foreign Fiber Detection. *Computer Engineering and Design*, **34**, 2783-2787.

[2] Lagha, M., Tikhemirine, M. and Bergheul, S. (2013) De-Noised Estimation of the Weather Doppler Spectrum by the Wavelet Method. *Digital Signal Processing*, **23**, 322-328. http://dx.doi.org/10.1016/j.dsp.2012.08.001

[3] Houda, B., Rachid, S. and Ludovic, J. (2015) Comparison of Leaf Surface Roughness Analysis Methods by sensitivity to Noise Analysis. *Biosystems Engineering*, **136**, 77-86. http://dx.doi.org/10.1016/j.biosystemseng.2015.04.012

[4] Huang, Y., Lei, T. and Fan, Y.Y. (2015) Adaptive Decision-Based Unsymmetric Trimmed Median Filter. *Computer Science*, **42**, 303-307.

[5] Kang, C.Q., Zhang, Q.L. and Zheng, Y. (2013) Non Uniformity Correction Algorithm for IR Images Based on Midway Equalization Histogram. *Laser & Infrared*, **43**, 1240-1242.

[6] Anis, F.M.R., Hishammuddin, A. and Rohayanti, H. (2015) A Low Lighting or Contrast Ratio Visible Iris Recognition Using Iso-Contrast Limited Adaptive Histogram Equalization. *Knowledge-Based Systems*, **74**, 40-48. http://dx.doi.org/10.1016/j.knosys.2014.11.002

[7] Jiang, W., Shen, H.F. and Zeng, Ch. (2014) Destriping Method for Band 28 of Terra MODIS Images. *Geomatics and Information Science of Wuhan University*, **39**, 526-530.

[8] Wang, S., Zhu, Q.X. and Zhu, W. (2014) Graph and Matrix Approaches to Rough Sets through Matroids. *Information Sciences*, **288**, 1-11. http://dx.doi.org/10.1016/j.ins.2014.07.023

11

Image Denoising Combining the P-M Model and the LLT Model

Qian Yang

School of Mathematical Science, University of Electronic Science & Technology of China, Chengdu, China
Email: dianzikedashuxue@126.com

Abstract

In this paper, we present a noise removal technique by combining the P-M model with the LLT model. The combined technique takes full use of the advantage of both filters which is able to preserve edges and simultaneously overcomes the staircase effect. We use a weighting function in our model, and compare this model with the P-M model as well as other fourth-order functional both in theory and numerical experiment.

Keywords

P-M Model, LLT Model, Fourth-Order PDEs, Combination, Image Denoising

1. Introduction

In image processing, the methods based on variation and PDEs play an important role. In recent years, many scholars have done a lot of research in this respect, and many good results have been achieved on image such as enhancing, sharping, and denoising images. See for instance [1]-[7]. The total variation minimization introduced by Rudin, Osher, and Fatemi [6] is the most significant of such methods. But now, one will consider the following more general functional

$$\min E(u)=\int_{\Omega}\left\{f\left(|\nabla u|\right)+\frac{\lambda}{2}\left(u-u_0\right)^2\right\}\mathrm{d}x\mathrm{d}y. \tag{1}$$

By using the functional variation principle to figure out the corresponding Euler-Lagrange equation, we can find the solution of Equation (1) to approach the original image. These second order partial differential equations (PDEs) models [8]-[10] are shown to be very useful in image denoising, restoration, inpainting, etc. Meanwhile, most of the second-order models have a drawback; the drawback is that these models can produce the blocky effects. The authors in [11]-[15] also put forward a variety of fourth-order PDEs models. In the domain with a smooth changed in the intensity value, most of second-order models have a tendency to converge to a piecewise constant image, while the fourth-order PDEs models converge to a piecewise planar image. In this regions, the gradient of the fourth-order PDEs maintains a constant (not 0); thus it reduces that image would not have the extra edge. However, many fourth-order PDEs models have some common shortcomings: blurring the image feature to some extent and producing speckle. Our principal contribution of this paper is to produce a new solution by taking the best from the second-order and the fourth-order PDEs models by a convex combination. A weighting function is used in an iterative way to generate a new model. This weighing function is used to balance the relative weights of the two methods. Our intention is to improve the quality of the recovered image, such as suppressing the noise, overcoming the blocky effect and preserving the sharp edges. This paper is organized as follows: we give a brief description of the classic second-order PDEs model (the P-M model) and fourth-order PDEs model (the LLT model), and will give a method to find out the solution of the scale parameter in Section 2. In Section 3, we consider the convex combination of the above two models. Implementation details are given in Section 4. In Section 5, we give the numerical experiments along with comparisons among the P-M model and various fourth-order models. Finally, we conclude with a brief discussion in Section 6.

2. Prior Work

2.1. A Typical Second-Order Model: The P-M Model

The P-M model has a good performance in many details and it is the origin of the other models. It is well known that the model represented by Perona and Malik [10] is derived from the following functional,

$$\min E(u) = \int_{\Omega} \left\{ f(|\nabla u|) + \frac{\lambda_1}{2}(u-u_0)^2 \right\} \mathrm{d}x\mathrm{d}y \tag{2}$$

$$J_1(u) := \int_{\Omega} f(|\nabla u|) \mathrm{d}x\mathrm{d}y \tag{3}$$

where $\Omega \subset R^2$ is an open, bounded and connected domain with Lipschitz boundary, $\lambda_1 > 0$ is a parameter that balances the relative weights of the two terms, and some times, it can also be interpreted as a Lagrange multiplier. u is taken in an appropriate Sobolev space $W^{1,p}$ on the space $BV(\Omega)$ of function with bounded variation on Ω and f is a function of the magnitude of u, we know that $f(\cdot)$ is a non negative monotonically increasing function. $f(s) = \frac{k^2}{2}\ln^{(k^2+s^2)}$ defined by Perona and Malik. Under these conditions, Perona and Malik minimized the energy functional (2), thus and obtained the following equation

$$\begin{cases} \dfrac{\partial u}{\partial t} = \nabla \cdot (g(|\nabla u|)) - \lambda_1 (u-u_0) \quad \text{in } (0,T)\times\Omega \\ u(0,x,y) = u_0(x,y) \\ g(|\nabla u|) = \dfrac{1}{1+\left(\dfrac{|\nabla u|}{k}\right)^2} \end{cases} \tag{4}$$

here g is the diffusion coefficient, and $g(\cdot)$ is a non negative monotonically decreasing function with $g(0) = 1$ as well as $\lim_{s\to\infty} g(s) = 0$, $\nabla\cdot$ and ∇ denote the divergence and the gradient. one can get a result that the intensity of the level image is the global minima of the energy functional from (2) and (4). One should notice that there are many fast methods to minimize (2) such as [14].

2.2. A Fourth-Order PDEs Model

The P-M model can remove the noise, at the same time it will produce the staircase effect, while the fourth-order model can remove the noise and will not generate the blocky effect. Since the Laplacian of the image at a pixel is zero only if the image is planar in its neighborhood, there are many fourth-order PDEs models attempt to remove noise by approaching piecewise planar image. Piecewise planar image looks more natural than step images which anisotropic diffusion, in other word, the fourth-order model will not produce the staircasing effect. It also is a strength of the LLT model. This model is put forward by Lysaker and Tai *et al.* In [4]:

$$\min E_2(u) = \int_\Omega \left(u_{xx}^2 + u_{xy}^2 + u_{yx}^2 + u_{yy}^2\right)^{\frac{1}{2}} \mathrm{d}x\mathrm{d}y + \frac{\lambda_2}{2}\int_\Omega (u-u_0)^2 \,\mathrm{d}x\mathrm{d}y \tag{5}$$

where the parameter $\lambda_2 > 0$ which balances the relative weights of the two terms. This model is called the LLT model. For simplicity we introduce the notation $|D^2u| = \left(u_{xx}^2 + u_{xy}^2 + u_{yx}^2 + u_{yy}^2\right)^{\frac{1}{2}}$ and denote (5) by

$$\min E_2(u) = \int_\Omega \left\{|D^2u| + \frac{\lambda_2}{2}(u-u_0)^2\right\} \mathrm{d}x\mathrm{d}y \tag{6}$$

$$J_2(u) := \int_\Omega |D^2u| \mathrm{d}x\mathrm{d}y. \tag{7}$$

Through the calculation of this variation, the gradient descent PDE of the minimization is

$$\frac{\partial u}{\partial t} = -\left(\frac{u_{xx}}{|D^2u|}\right)_{xx} - \left(\frac{u_{xy}}{|D^2u|}\right)_{yx} - \left(\frac{u_{yx}}{|D^2u|}\right)_{xy} - \left(\frac{u_{yy}}{|D^2u|}\right)_{yy} - \lambda_2(u-u_0). \tag{8}$$

To avoid singularities in the above systems, we replace $|D^2u|$ by $|D^2u|_\varepsilon = \sqrt{|D^2u|^2 + \varepsilon}$, where ε is a small positive constant. So, from (8), it implies that

$$\frac{\partial u}{\partial t} = -\left(\frac{u_{xx}}{|D^2u|_\varepsilon}\right)_{xx} - \left(\frac{u_{xy}}{|D^2u|_\varepsilon}\right)_{yx} - \left(\frac{u_{yx}}{|D^2u|_\varepsilon}\right)_{xy} - \left(\frac{u_{yy}}{|D^2u|_\varepsilon}\right)_{yy} - \lambda_2(u-u_0). \tag{9}$$

Since the problem is convex, the steady state solution of the gradient descent PDE (9) is the minimizer of the energy functional. Numerical experiments show that the LLT model is able to greatly supress the staircase effect, but blur image and produce the speckle effect.

2.3. Solving Scale Parameters

In order to find an appropriate scale parameter in our approach, firstly, we will give the specific calculation method [4] [16]. As commonly done, we can also view (2), (6) as a constrained problems:

$$u := \arg\min J_1(u) \qquad \text{such that } \|u-f\|_2^2 = |\Omega|_\sigma^2 \tag{10}$$

$$u := \arg\min J_2(u) \qquad \text{such that } \|u-f\|_2^2 = |\Omega|_\sigma^2 \tag{11}$$

where σ^2 is the variance of an additive noise added in a noisy image *f* and $|\Omega|$ is the area of Ω. We merely multiply (4) and (9) by $(u-u_0)$ and integrate by parts over Ω. If steady state has been reached, the left side of (4) and (9) vanishes. Then we can get

$$\lambda_1 = -\frac{1}{|\Omega|_\sigma^2}\int_\Omega \left(\frac{\nabla u}{1+\frac{|\nabla u|^2}{k}}\right)\cdot\nabla(u-u_0)\mathrm{d}x\mathrm{d}y \tag{12}$$

$$\lambda_2 = -\frac{1}{|\Omega|\sigma^2}\int_\Omega \left(\frac{u_{xx}}{|D^2u|_\varepsilon}(u-u_0)_{xx} + \frac{u_{xy}}{|D^2u|_\varepsilon}(u-u_0)_{yx} + \frac{u_{yx}}{|D^2u|_\varepsilon}(u-u_0)_{xy} + \frac{u_{yy}}{|D^2u|_\varepsilon}(u-u_0)_{yy}\right)\mathrm{d}x\mathrm{d}y. \tag{13}$$

These equations give us some dynamic values, which appear to converge as $t \to \infty$. In our approach, we can get the dynamic parameter by this calculation method.

3. Convex Combination-Based Second and Fourth Order Diffusion

In Section 2, the P-M model and the LLT model are presented in front of us, and in this section, we will combine these models with a weighting function. The image enhancement techniques for image deboising is a good approach to combine the advantages of the second-order PDEs method and the fourth-order PDEs method, thus has been popular, such as [17] [18] method. Now, we will give a new combination model. Firstly, we use the notation v and u to distinguish the solution of (4) and (8), respectively. One can apply variational principle to minimize the energy functional which is put forward by Perona *et al.*, and get the steady-state solution of the nonlinear diffusion process

$$\begin{cases} \frac{\partial v}{\partial t} = \nabla \cdot \left(g\left(|\nabla v| \nabla v \right) \right) - \lambda_1 \left(v - v_0 \right), \\ v\left(0, x, y\right) = u_0\left(x, y\right). \end{cases} \tag{14}$$

And the LLT model as follows

$$\begin{cases} \frac{\partial u}{\partial t} = -\left(\frac{u_{xx}}{|D^2 u|_\varepsilon} \right)_{xx} - \left(\frac{u_{xy}}{|D^2 u|_\varepsilon} \right)_{yx} - \left(\frac{u_{yx}}{|D^2 u|_\varepsilon} \right)_{xy} - \left(\frac{u_{yy}}{|D^2 u|_\varepsilon} \right)_{yy} - \lambda_2 \left(u - u_0 \right) \\ u\left(0, x, y\right) = u_0\left(x, y\right). \end{cases} \tag{15}$$

Set Ω is a rectangular domain, $\boldsymbol{n} = \left(n_1, n_2\right)$ denotes the outward normal vector of $\partial\Omega$, with the boundary conditions. When $\boldsymbol{n}$ is orthogonal to the y-axe

$$\begin{cases} u_{xx} = 0, \\ \left(\frac{u_{xx}}{|D^2 u|_\varepsilon} \right)_x + \left(\frac{u_{xy}}{|D^2 u|_\varepsilon} \right)_y = 0 \end{cases} \tag{16}$$

where $\boldsymbol{n}$ is orthogonal to the x-axe

$$\begin{cases} u_{yy} = 0, \\ \left(\frac{u_{yx}}{|D^2 u|_\varepsilon} \right)_x + \left(\frac{u_{yy}}{|D^2 u|_\varepsilon} \right)_y = 0. \end{cases} \tag{17}$$

And the initial condition is:

$$u\left(0, x, y\right) = u_0\left(x, y\right) \qquad \text{such that} \quad \left(x, y\right) \in \Omega.$$

To some extent, these two models are able to suppress the noise, but it is well known that the first model is prone to massive effect, while the second model is tend to blur images and produce the speckles. Both methods have their pros and cons depending on the character of the image interest. We should find out a way for demonstrating the merits of these two models, at the same time, the shortcoming is suppressed by adjusting the parameter. Considering (14) and (15), we try to create a new model by a convex combination $w = \alpha u + \left(1 - \alpha\right) v$ for $\alpha \in \left[0, 1\right]$. We prefer that the weighting function α can be found adaptively. When $\alpha = 1$, it is the LLT model that makes full use of piecewise planarity conditions (PPCs) with which unconstrained fourth-order variational model in continuum converge to a piecewise planar image. The LLT model looks more natural than the P-M model. When $\alpha = 0$, it is the P-M model. But we expect to have a new model, which has the advantages of the two models, so α should satisfy $0 < \alpha < 1$, thus we will calculate the above weight function through a number of different ways. one can find that when $\frac{1}{2} < \alpha < 1$ the effect of image denoising is more outstanding. In this paper, we find out a $\alpha = 0.625$ by repeated experiments, it works fine. As for more detailed algorithms, we will give them in the following.

4. Difference Scheme

As in [19], one can minimize (4) and (8), respectively, by using evolution Equations (14) and (15), and one can discrete (14) with difference scheme, the expression as follows: first, let Δx and Δy be the square lattice sizes for the x and y variables, $\Delta t > 0$ be the artificial time step in the descent direction. And we denoted $u(i\Delta x, j\Delta y, k\Delta z)$ and $v(i\Delta x, j\Delta y, k\Delta z)$ by u_{ij}^k and v_{ij}^k, respectively, where $i\Delta x, j\Delta y$ are the grid points and $i = 0,1,2,\cdots,N$; $j = 0,1,2,\cdots,M$. By applying forward difference scheme in time, we can get

$$\frac{\partial v_{ij}}{\partial t} \approx \frac{v_{ij}^{k+1} - v_{ij}^k}{\Delta t}$$

where $k \geq 0$ is the iteration number. Finally, we iteratively compute the discrete minimizer by

$$v_{ij}^{k+1} = v_{ij}^k + \Delta t \cdot \left(c_N \cdot \sigma_N v + c_S \cdot \sigma_S v + c_E \cdot \sigma_E v + c_W \cdot \sigma_W v\right)_{ij}^k - \lambda_1 \left(v_{ij}^k - v_0^k\right).$$

Note that N, S, E, W represent the North, the South, the East and the West, respectively. And the symbol σ indicates 4-nearest-neighbors differences:

$$\sigma_N v_{ij}^k = v_{i-1,j}^k - v_{ij}^k,$$
$$\sigma_S v_{ij}^k = v_{i+1,j}^k - v_{ij}^k,$$
$$\sigma_E v_{ij}^k = v_{i,j-1}^k - v_{ij}^k,$$
$$\sigma_W v_{ij}^k = v_{i,j+1}^k - v_{ij}^k.$$

And the diffusion coefficient $g\left(|\nabla v|\right) = \dfrac{1}{1 + \left(\dfrac{|\nabla v|}{k}\right)_\varepsilon^2}$ is continually updated with the increase of the number of iterations.

$$c_{Nij}^k = g\left(\left|\nabla_N v_{ij}^k\right|\right), \qquad c_{Sij}^k = g\left(\left|\nabla_S v_{ij}^k\right|\right)$$
$$c_{Eij}^k = g\left(\left|\nabla_E v_{ij}^k\right|\right), \qquad c_{Wij}^k = g\left(\left|\nabla_W v_{ij}^k\right|\right)$$

Similarly, we can obtain the LLT discrete form: firstly, see [4], we can get

$$\left|D^2\right|_\varepsilon^2 = \left(D_{xx} u_{ij}^l\right)^2 + \left(D_{xy}^+ u_{ij}^l\right)^2 + \left(D_{yx}^+ u_{ij}^l\right)^2 + \left(D_{yy} u_{ij}^l\right)^2 + \varepsilon$$

$$u_{ij}^{l+1} = u_{ij}^l - \Delta t \left[D_{xx}\left(\frac{D_{xx} u_{ij}^l}{\left|D^2 u_{ij}^l\right|_\varepsilon}\right) + D_{xx}\left(\frac{D_{xx} u_{ij}^l}{\left|D^2 u_{ij}^l\right|_\varepsilon}\right) + D_{xx}\left(\frac{D_{xx} u_{ij}^l}{\left|D^2 u_{ij}^l\right|_\varepsilon}\right) + D_{xx}\left(\frac{D_{xx} u_{ij}^l}{\left|D^2 u_{ij}^l\right|_\varepsilon}\right)\right] - \lambda_2 \left(u_{ij}^l - u_0^l\right)$$

$$D_{xx} u_{i,j}^l = \frac{D_x^+ u_{i,j}^l - D_x^+ u_{i-1,j}^l}{\Delta x}$$

$$D_{xy}^+ u_{i,j}^l = \frac{D_x^+ u_{i,j+1}^l - D_x^+ u_{i,j}^l}{\Delta y}$$

$$D_{yx}^- u_{i,j}^l = \frac{D_y^- u_{i,j}^l - D_y^- u_{i-1,j}^l}{\Delta x}$$

$$D_{yy} u_{i,j}^l = \frac{D_y^+ u_{i,j}^l - D_y^+ u_{i,j-1}^l}{\Delta y}$$

These are discrete format of the P-M model and the LLT, but our focus points are the following:

$$w = \alpha \cdot u_{ij}^{l+1} + (1-\alpha) \cdot v_{ij}^{k}$$

Here, the value of α is 0.625, it is a preferred value of α. When α is different, the effect of image denoising is distinct. In the following numerical experiments, it will prove that a combination model is effective and accurate.

5. Numerical Experiments

In this section, we will present numerical results obtained by applying our proposed model to image denoising. For comparison, we also present some results from application of the P-M model (14) and the LLT model (15) to the same images. From the result of experiment, we will utilize the pictures to verify these models, especially demonstrate the advantages of the convex combining model: the new model can reduce the blocky effects and avoid leaving the speckle artifacts. The following two classical single PDE schemes are:

(a) The model proposed by Perona and Malik [10]

$$\frac{\partial v}{\partial t} = div\left(\frac{\nabla v}{1+\left(\frac{\nabla v}{k}\right)^2}\right).$$

(b) Lysaker and Tai *et al.* raised such model [12]

$$\frac{\partial u}{\partial t} = -\left(\frac{u_{xx}}{|D^2 u|_\varepsilon}\right)_{xx} - \left(\frac{u_{xy}}{|D^2 u|_\varepsilon}\right)_{yx} - \left(\frac{u_{yx}}{|D^2 u|_\varepsilon}\right)_{xy} - \left(\frac{u_{yy}}{|D^2 u|_\varepsilon}\right)_{yy}.$$

Note that, to avoid singular in the LLT model, we utilize the above expression. The contrast parameter $k > 0$ can be chosen in a variety of ways. Further, we proposed our model to compare with the above models.

(c) Convex combination of the above two models.

$$w = \alpha \cdot u + (1-\alpha) \cdot v$$

From the expression, we know that our model is built on the basis of two classical models. Our model is more natural, and the effect is good in comparison with the model which is obtained by an energy functional that contains gradient functions and Laplace functions. Our experimental pictures are 256 × 256 sized gray-scale image Lena. We consider the noisy image $f = u + n$ where u is the original image, and the n is a white Gaussian noise with the expectation zero and standard deviation σ. In **Figure 1**, we show some results on Lena image. Top left is the original image, Top medium is a noisy Lena gray scale image with the white Gaussian noise with standard deviation $\sigma = 15$. Top right is the denoised image u by our new model with $\lambda 1 = 1$, $\lambda 2 = 1$, and bottom left is the denoised images u by the convex combination model with $\lambda 1 = 0.3$, $\lambda 2 = 0.5$, another two sets of data are $\lambda 1 = 0.1$, $\lambda 2 = 0.2$ and $\lambda 1 = 0.05$, $\lambda 2 = 0.1$, respectively. From these data, we can know that when the other parameters are unchanged, the near the parameter $\lambda 1$ and $\lambda 2$ approximate to 0.5, the better the result. The SNR of the top right picture is 20.0640 and the PSNR value is 26.6707, similarly, from bottom left picture to bottom right picture, we get that the SNR values are 23.0891, 22.7871, 20.9781, respectively and the PSNR values are 29.6958, 29.3938 and 27.5848, but finally we will found that the near the parameter $\lambda 1$ and $\lambda 2$ approximate to 0.5, the speckles of the image is more obvious. On the whole, one can find out that $\lambda 1 = 0.1$ and $\lambda 2 = 0.2$ is suitable.

Figure 2 shows a comparative result for the noisy Lena gray scale image ($\sigma = 15$) with the classical P-M model and the LLT model. In these pictures, we can see that the convex combination model presents a more distinct image than the other models. The new model ($\alpha = 0.625$) avoids the staircasing artifacts associate with the P-M model, at the same time, it prevents that the edges are blurred. **Figure 3** shows a comparative result for a small part of the Lena image ($\sigma = 15$). In this group of images, we will see that the P-M model appears obvious blocky effect and the LLT model leaves the speckle artifacts. Our proposed model can avoid the staircase and the speckle effect while removing noise. Note that we use signal to noise (SNR) [5] [17] [20] as means of judging performance:

Figure 1. Top left: The original image u, top medium: a noisy image f with a white Gaussian noise (σ = 15), top right: the denoised Lena image by the convex combination model with $\lambda_1 = 1$, $\lambda_2 = 1$, bottom left, bottom center, bottom; right: the denoised image u by the convex combination model with $\lambda_1 = 0.3$, $\lambda_2 = 0.5$; $\lambda_1 = 0.1$, $\lambda_2 = 0.2$; $\lambda_1 = 0.05$, $\lambda_2 = 0.1$; respectively.

(a) (b) (c) (d)

Figure 2. Comparison of denoising results on noisy Lena image. Top left: the noisy image with a white Gaussian noise σ = 15, top right: the denoised image u by The LLT model, with PSNR = 29:2871, bottom left: the denoised image u by the PM model with PSNR = 27:9767, bottom right: the denoised image u by the convex combination model with PSNR = 29:4037.

(1) $$SNR(u,f) = 20 \cdot \log_{10}^{\left\{\frac{\|f-\bar{f}\|_{L^2}}{\|f-u\|_{L^2}}\right\}}$$

where u is the original image, f is the recovery and f is its mean value.

(2) Signal to noise ratio (PSNR) is defined by:

$$PSNR = 10 \cdot \log_{10}^{\left(\frac{255^2}{MSE}\right)}$$

where $u_{\max}$ is the maximum value of the image u, in 8-bit images $u_{\max} = 25.5$, and MSE is the mean squared error defined by.

(3) $$MSE = \frac{1}{MN} \sum_{i=0}^{M-1} \sum_{j=0}^{N-1} \left\| f(i,j) - u(i,j) \right\|^2$$

where f denotes the denoised image and u denotes the original image.

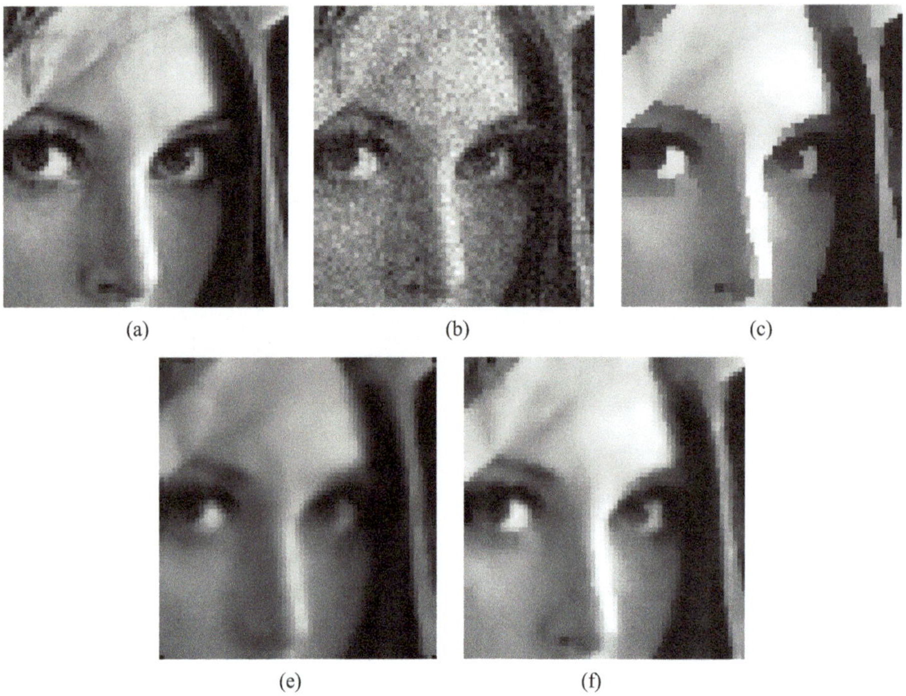

Figure 3. A small portion of the Lena image is shown to better emphasize disparties in the smooth surface.

Table 1. SNR and PSNR comparison of various models for the Lena image.

Image	The PM model	The LLT model	My model
SNR (dB)	21.3700	22.6804	22.7970
PSNR (dB)	27.9767	29.2871	29.4037

In **Table 1**, we give the comparison results of the SNR and the PSNR using these three models. As can be seen, the convex combination model performs well for a variety of images (Lena, the small part of the Lena).

6. Conclusion

This paper proposes a convex combination model by the P-M model and the LLT model, and $\alpha = 0.625$ is found to balance these two models. Numerical results represent the competitive performance of the new model, for noise removal while maintaining the jump discontinues better, and suppressing the speckles.

References

[1] Catt, F., Lions, P., Morel, J. and Coll, T. (1992) Image Selective Smoothing and Edge Detection by Nonlinear Diffusion. *SIAM Journal on Numerical Analysis*, **29**, 182-193. http://dx.doi.org/10.1137/0729012

[2] Duran, J., Buades, A., Coll, B., *et al.* (2014) A Nonlocal Variational Model for Pansharpening Image Fusion. *SIAM Journal on Imaging Sciences*, **7**, 761-796. http://dx.doi.org/10.1137/130928625

[3] Jin, Y., Jost, J. and Wang, G. (2004) A New Nonlocal Model for Image Denoising. *Journal of Mathematical Imaging and Vision*, **48**, 93-105. http://dx.doi.org/10.1007/s10851-012-0395-2

[4] Lysaker, M., Lundervold, A. and Tai, X.C. (2003) Noise Removal Using Fourth-Order Partial Differential Equation with Applications to Medical Magnetic Resonance Images in Space and Time. *IEEE Transactions on Image Processing*, **12**, 1579-1590. http://dx.doi.org/10.1109/TIP.2003.819229

[5] Lieu, L.H. and Vese, L.A. (2008) Image Restoration and Decomposition via Bounded Total Variation and Negative Hilbert-Sobolev Spaces. *Applied Mathematics & Optimization*, **58**, 167-193. http://dx.doi.org/10.1007/s00245-008-9047-8

[6] Rudin, L., Osher, S. and Fatemi, E. (1992) Nonlinear Total Variation Based Noise Removal Algorithms. *Physica D*, **60**, 259-268. http://dx.doi.org/10.1016/0167-2789(92)90242-F

[7] Surya Prasath, V.B. and Vorotnikov, D. (2004) On a System of Adaptive Coupled PDEs for Image Restoration. *Journal of Mathematical Imaging and Vision*, **48**, 35-52. http://dx.doi.org/10.1007/s10851-012-0386-3

[8] Alvarez, L., Lions, P. and Morel, J. (1992) Image Selective Smoothing and Edge Detection by Nonlinear Diffusion. II. *SIAM Journal on Numerical Analysis*, **29**, 845-866. http://dx.doi.org/10.1137/0729052

[9] Jin, Y., Jost, J. and Wang, G. (2012) A Nonlocal Version of the Osher-Sol-Vese Model. *Journal of Mathematical Imaging and Vision*, **44**, 99-113. http://dx.doi.org/10.1007/s10851-011-0313-z

[10] Perona, P. and Malik, J. (1990) Scale-Space and Edge Detection Using Anisotropic Diffusion. *IEEE Transaction on Pattern Analysis and Machine Intelligence*, **12**, 629-639. http://dx.doi.org/10.1109/34.56205

[11] Kim, S. and Lim, H. (2009) Fourth-Order Partial Differential Equation for Effective Image Denoisng. 7th Mississippi State-UAB Conference on Differential Equation and Computational Simulation. *Electronic Journal of Differential Equations*, **17**, 107-121.

[12] Liu, X.W., Huang, L.H. and Guo, Z.Y. (2011) Adaptive Fourth-Order Partial Differential Equation Filter for Image Denoising. *Applied Mathematics Letters*, **24**, 1282-1288. http://dx.doi.org/10.1016/j.aml.2011.01.028

[13] Wang, L.Q., Xiao, L., Huang, L.L. and Wei, Z.H. (2011) Nonlocal Total Variation Based Speckle Noise Removal Method for Ultrasound Image. 2011 *4th International Congress on Image and Signal Processing* (*CISP*), **2**, 709-713. http://dx.doi.org/10.1109/cisp.2011.6100293

[14] You, Y.L. and Kaveh, M. (2000) Fourth-Order Partial Differential Equation for Noise Removal. *IEEE Transaction on Image Processing*, **9**, 1723-1731. http://dx.doi.org/10.1109/83.869184

[15] Zhou, G.X., Zhang, Q.S. and Tan, X.H. (2011) Image Enhancement and Denoising by Forward and Backward Fourth Order Partial Differential Equations. In: Lin, S. and Huang, X., Eds., *Advances in Computer Science*, *Environment*, *Ecoinformatics*, *and Education*, Springer-Verlag, Berlin, 514-519. http://dx.doi.org/10.1007/978-3-642-23339-5_94

[16] Gilboa, G. and Osher, S. (2007) Nonlocal Linear Image Regularization and Supervised Segmentation. *Multiscale Modeling & Simulation*, **6**, 595-630. http://dx.doi.org/10.1137/060669358

[17] Liu, T.H. and Xiang, Z.Y. (2013) Imaging Restoration Combing the Second-Order and Fourth-Order PDEs. *Mathematical Problems in Engineering*, **2013**, Article ID: 743891.

[18] Wang, H.B., Wang, Y.Q. and Ren, W.Q. (2012) Image Denoising Using Anisotropic Second and Fourth Order Diffusions Based on Gradient Vector Convolution. *ComSIS*, **9**, 1493-1511. http://dx.doi.org/10.2298/CSIS120219060W

[19] Li, X.P. and Chen, T.W. (1994) Nonlinear Diffusion with Multiple Edginess Thresholds. *Pattern Recognition*, **27**, 1029-1037. http://dx.doi.org/10.1016/0031-3203(94)90142-2

[20] Lou, Y.F., Zhang, X.Q., Osher, S. and Bertozzi, A. (2010) Image Recovery via Nonlocal Operators. *Journal of Scientific Computing*, **42**, 185-197. http://dx.doi.org/10.1007/s10915-009-9320-2

12

Image Classification Based on Histogram Intersection Kernel

Hanbin Xi, Tiantian Chang

School of Science, Xi'an University of Posts and Telecommunications, Xi'an, China
Email: changtiantian@xupt.edu.cn

Abstract

Histogram Intersection Kernel Support Vector Machines (SVM) was used for the image classification problem. Specifically, each image was split into blocks, and each block was represented by the Scale Invariant Feature Transform (SIFT) descriptors; secondly, k-means cluster method was applied to separate the SIFT descriptors into groups, each group represented a visual keywords; thirdly, count the number of the SIFT descriptors in each image, and histogram of each image should be constructed; finally, Histogram Intersection Kernel should be built based on these histograms. In our experimental study, we use Corel-low images to test our method. Compared with typical RBF kernel SVM, the Histogram Intersection kernel SVM performs better than RBF kernel SVM.

Keywords

Classification, Bag of Word, Support Vector Machine, Kernel Function, Visual Keywords

1. Introduction

With the rapid increase of digital images, it is impossible to label them manually. How to classify these images quickly and accurately becomes an important research topic currently.

Support vector machine (SVM) is a kernel based supervised learning classifier, based on Statistical Learning Theory (SLT), which is a widely used method in classification problems, e.g., natural language processing [1], information retrieval [2] [3] and data mining [4] [5]. However, how to choose a right kernel is a challenge work in the SVM. Wu proposed an interpolation kernel, which has less subjectivity and more predominance of generalization than most of traditional kernel functions while needing vast calculation [6]. Neumann use wavelet kernel SVM as a classifier which solves difficult signal classification problems in cases where solely using a large margin classifier may fail [7]. Radial Basis kernel Function was proved its effective learning performance, while its extrapolation capability decreases with the parameters' increasing. Polynomial kernel function has a good performance in global, while its performance is poor in local. Different kernels result in different results; therefore, it's essential to choose a right kernel for specific classification task.

In this paper, Histogram Intersection Kernel SVM was used in the image classification. Firstly, the original images were split into blocks by the regular grid with equal size of B × B. Secondly the Scale Invariant Feature Transform (SIFT) descriptors were extracted from those blocks. In order to get the dictionary of Bag Of Word

(BOW) model, we use k-means clustering algorithm to cluster those descriptors into k groups, each of them was regarded as a visual keywords which making up the dictionary, and we can get the histogram of each images, that gives the frequency of each visual keywords contained in the dictionary. Finally, Histogram Intersection Kernel should be constructed with these histograms.

2. Image Preprocessing and Feature Extraction

2.1. Image Preprocessing

We use the Corel-low image to test our method, and the original images were split into blocks which have equal size of B × B with the regular grid. **Figure 1** shows the schematic diagram of how to split the original images.

2.2. Extracting Feature

Scale Invariant Feature Transform (SIFT) was first proposed by David G.L. in 1999 [8], and was improved in 2004 [9]. SIFT descriptor is an invariant image local feature description operator of invariant image that is based on scale space, image rotation and even affine transformation. The core issue of object recognition is that to match a target to its images no matter if they are at different times, different light, and different poses. However, since the state of target and the impact of the environment, the image in the same type differs from each other greatly in different scenes, and therefore people can identify the same object by the local common character. The so-called local feature descriptor is used to describe the local feature of the image. The ideal local feature should be robust to translation, scaling, rotation, the lighting changes, affine and projector impact. SIFT descriptor is a typical local feature.

Then we extract SIFT descriptor from each block after preprocessing of the original images, **Figure 2** shows the diagram of the extraction of SIFT descriptor, and the green marks are the SIFT descriptor. A SIFT descriptor is denoted by a circle, representing its support region, and one of its radii, representing its orientation.

2.3. Bag-of-Words Model

Bag-of-Words (BOW) model was used to express a document (or sentence) as a histogram of frequency of words, ignoring the order of these words in natural language processing firstly [10]. It was used to perform the classification of computer images by Li [11].This model can learn characteristic labels of scenes without human intervention in the training database. But its performance for the images of indoor scenes is lackluster [11].

Since SIFT descriptors were extracted, the local feature of each image was obtained. Then these descriptors were clustered into k groups via k-means clustering algorithm. The k groups are regarded as visual keywords making up the dictionary. In order to let the images represented by the visual keywords of the dictionary, every descriptor of each image is replaced by a nearest visual keyword in Euclidean distance. All the descriptors of the

Figure 1. Each image was partitioned into blocks with regular grid.

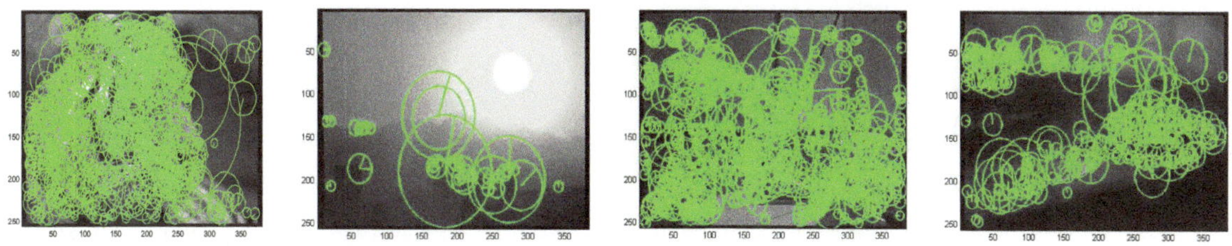

Figure 2. Example of extracting SIFT descriptor.

images can be represented as the visual keywords. After that, the images were expressed with a histogram of the frequencies of the visual keywords. Assume that each image is expressed by a histogram $h_i \in R_+^d\left(i=1,2,\ldots,l\right)$, where l is the number of all the images.

3. Support Vector Machine

Support Vector Machines (SVM) was proposed by Vapnik [12] [13], which is commonly used in classification, regression, and other learning tasks [14]. SVM is built on the theory of max-margin classification hyper-plane.

Given $\left(x_i,y_i\right)\left(i=1,2,\ldots,l\right)$, where x_i is training data and y_i is the corresponding label. Maximizing the margin is to solve the optimization problem:

$$\begin{aligned} &\min_w \frac{1}{2}\|w\|^2 + C\sum_{i=1}^{l}\xi_i \\ &s.t.\quad y_i\left(w\cdot x_i+b\right)\geq 1-\xi_i,(i=1,\cdots,N) \\ &\xi_i \geq 0 \end{aligned} \tag{1}$$

where C is the penalty factor, ξ is the slack factor [15].

The dual problem of the Equation (1) is defined as follow.

$$\begin{aligned} &\min_\alpha \frac{1}{2}\sum_{i=1}^{l}\sum_{j=1}^{l}\alpha_i\alpha_j y_i y_j K(x_i,x_j)-\sum_{i=1}^{l}\alpha_i \\ &s.t\sum_{i=1}^{l}\alpha_i y_i = 0 \\ &0\leq \alpha_i \leq C(i=1,\cdots,l) \end{aligned} \tag{2}$$

where α^* is the Lagrange multiplier. By solving the Equation (2), α^* should be obtained, and we have:

$$w=\sum_{i=1}^{l}\alpha_i y_i x_i \tag{3}$$

Then we get the decision function:

$$y_j=\sum_{i=1}^{l}\alpha_i y_i K(x_i,x_j)+b \tag{4}$$

where b is a bias value.

However, in the practical applications, the problems are nonlinear separable. The solution to this problem is to induce a kernel function into the optimization problem, which maps the data into a higher dimensional space. To avoid complicated calculation in the higher dimension, the kernel function that satisfies Mercer's theorem has calculated the data in the original dimension beforehand. Given $x_i,x_j \in R_+^d\left(i,j=1,2,\ldots,l\right)$, the kernel function is defined as: $K\left(x_i,x_j\right)=\left\langle\phi\left(x_i\right)\cdot\phi\left(x_j\right)\right\rangle$, and $\phi\left(x_i\right)$ is a mapping [16].

Here are some typical kernel functions [17]:

- Linear kernel: $K\left(x_i,x_j\right)=x_i^T x_j$
- Polynomial kernel: $K\left(x_i,x_j\right)=\left(\gamma x_i^T x_j+r\right)^d,\gamma>0$
- Radial Basis Function (RBF) kernel: $K\left(x_i,x_j\right)=\exp\left(-\gamma\left\|x_i-x_j\right\|^2\right),\gamma>0$

Here γ,r and d are kernel parameters.

4. Histogram Intersection Kernel Construction

Each image is divided into blocks with the equal size B × B on a regular grid, the spatial position can be characterized for each block in an image, and the SIFT descriptors are extracted from each block. The frequency of these visual keywords in an image can be represented as a histogram $h_i,\left(i=1,\cdots,l\right)$, where l is the number of the images. The histogram intersection kernel is defined as:

$$K_{HIS}(x,z)=\sum_{m=1}^{d}\min(x_m,z_m) \tag{5}$$

This function is a positive definite kernel, so it can be used to quantify the similarity of two images, which also satisfies the Mercer's theorem.

Histogram intersection kernel function, which contains the spatial location of objects, is robust to many transformations. Such as, distractions in the background of the object, view point changing, occlusion [18].

5. Experiment and Analysis

Our PC configuration is: windows 8.1 system, Intel Core i7 2.3 GHz, with 6 GB memory and the programming software is Matlab 2012a. We use the Corel-low images as our experimental data, which contains 59 classes and each class has 100 pictures. The number of classes has an influence on the result of classification. In order to make sure that whither the classification result of Histogram Intersection kernel SVM is robust to different numbers of classes Thus, two groups of experiment were designed. The first group contains 12 classes selected from the 59 classes randomly, and 70 pictures of each class are used for training and the rest for testing randomly too, we name it the group one. The second is to use all the 59 classes and take 80 pictures of each class for training and the rest for testing, we name it the group two.

The size of the regular grids is given 16 × 16, split the original images into blocks. Then SIFT descriptors were extracted from every block. The size of dictionary has a great influence on the classification accuracy. At the same time the larger size cost more in the calculation. So we get the dictionary size by seeking a tradeoff between the classification accuracy and the calculation cost. Finally the size of the dictionary is 1000 for the group one and 3000 for the group two. Note that RBF kernel is a typical kernel for SVM [14], so RBF kernel function is taken as ground-truth to compare with histogram intersection kernel SVM. RBF kernel function has one kernel parameter γ. In C-Support Vector Machine (C-SVM), there is a penalty parameter C which is the cost of C-SVM. We find the parameter by grid searching method.

Table 1 shows the size of dictionary, best parameters of SVM and average classification accuracy of the two kernels for both groups of experiments. The average classification accuracy of RBF kernel is about 5.8% lower than Histogram Intersection kernel in group one and 0.8% lower than Histogram Intersection kernel in group two. The RBF kernel costs more in the calculation and needs lots of time to find the best parameters since it requests two parameters.

And **Figure 3** is the confusion matrix of these two kernels for Group One. From the confusion matrix, we can see that except class Flower and Africa, the accuracy of Histogram Intersection kernel is higher than the RBF kernel.

Figure 4 is the confusion matrix of these two kernels for Group Two. There are 37 classes whose prediction accuracy of Histogram Intersection kernel is higher than or equal to the RBF kernel in the 59 classes.

Comparing histogram intersection kernel with RBF kernel, we made a conclusion that histogram intersection kernel has higher accuracy and requests less computation time than RBF kernel for image classification.

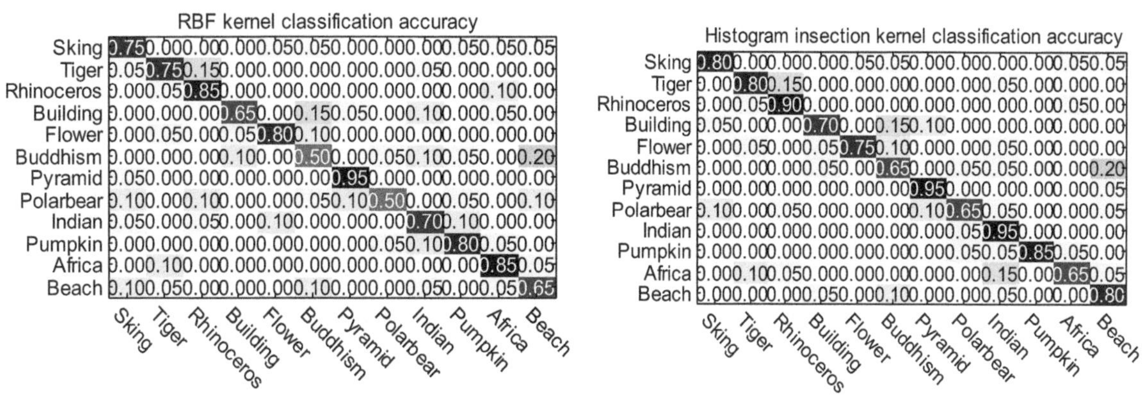

Figure 3. Confusion matrix of group one.

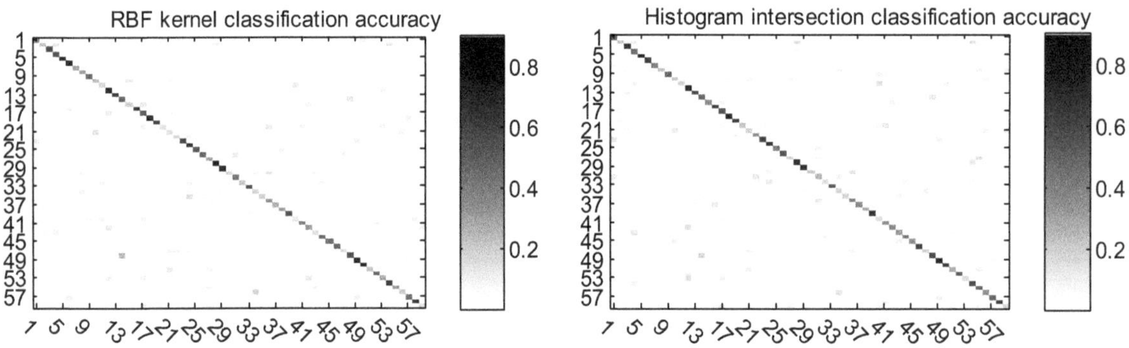

Figure 4. Confusion matrix of group two.

Table 1. The parameters and average classification of the two groups.

	RBF Kernel		Histogram Intersection Kernel	
	Group One	Group Two	Group One	Group Two
Size of Dictionary	1000	3000	1000	3000
Best γ	2.8284	0	-	-
Best C	64	1024	2	8
Average Classification Accuracy (%)	72.9167	46.7232	78.7500	47.5141

6. Conclusions

Image classification is an important research field in pattern recognition; SVM is a good classifier for non-linear classification problem. And the kernel function selection is the core for the classification. To find a right kernel is important to the classification problem. The experimental results show that, the performance of the Histogram Intersection Kernel is better than RBF kernel SVM. In our future work, there are several problems we should consider. 1) The size of dictionary influences on both the classification accuracy and the computation time greatly, therefore a new method that is used to represent an image, needs to be proposed. 2) A kernel whose effect is robust to the increase of the number of classes has to be proposed in the future.

Acknowledgements

This work was jointly supported by the Natural Science Fund Project of Education Department of Shaanxi Provincial Government (Grant No. 14JK1658), the Natural Science Foundation of China (Grant No. 61302050), and the College students innovation and entrepreneurship training program of Xi'an University of Posts and Telecommunications.

References

[1] Kudo, T.K. and Matsumoto, Y.J. (2001) Chunking with Support Vector Machines. *Proceedings of the Second Meeting of the North American Chapter of the Association for Computational Linguistics on Language Technologies*, **816**. http://dx.doi.org/10.3115/1073336.1073361

[2] Zhang, D. and Lee, W.S. (2003) Question Classification Using Support Vector Machines. *SIGIR*, **3**, 26-32. http://dx.doi.org/10.1145/860435.860443

[3] Yang, C.H., Lin, K.H.Y. and Chen, H.H. (2007) Emotion Classification Using Web Blog Corpora. *IEEE/WIC/ACM International Conference on Web Intelligence*, 275-278. http://dx.doi.org/10.1109/WI.2007.51

[4] Yoshikawa, Y.Y., Iwata, T.H. and Sawada, H.S. (2014) Latent Support Measure Machines for Bag-of-Words Data Classification. *Advances in Neural Information Processing Systems*, **27**.

[5] Kolari, P., Finin, F. and Joshi, A. (2006) SVMs for the Blogosphere: Blog Identification and Splog Detection. *AAAI Spring Symposium on Computational Approaches to Analysing Weblogs*, University of Maryland, Baltimore County.

[6] Wu, T., He, H.G. and He, M.K. (2003) Interpolation Based Kernel Function's Construction. *Chinese Journal of Computers*, **26**, 990-996.

[7] Neumann, J., Schnörr, C. and Steidl, G. (2005) Efficient Wavelet Adaptation for Hybrid Wavelet-Large Marge Classifiers. *Pattern Recognition*, **38**, 1815-1830. http://dx.doi.org/10.1016/j.patcog.2005.01.024

[8] Lowe, D.G. (1999) Object Recognition from Local Scale-Invariant Features. *IEEE International Conference on Computer Vision*. http://dx.doi.org/10.1109/iccv.1999.790410

[9] Lowe, D.G. (2004) Distinctive Image Features from Scale Invariant Keypoints. *International Journal of Computer Vision*, **60**, 91-110. http://dx.doi.org/10.1023/B:VISI.0000029664.99615.94

[10] Jiu, M.Y., Wolf, C., Garcia, C. and Baskurt, A. (2012) Supervised Learning and Codebook Optimization for Bag-of-Words Models. *Cognitive Computation*, **4**, 409-419. http://dx.doi.org/10.1007/s12559-012-9137-4

[11] Li, F.F. and Pietro, P. (2005) A Bayesian Hierarchical Model for Learning Natural Scene Categories. *Computer Vision and Pattern Recognition*, **2**, 524-531.

[12] Cortes, C. and Vapnik, V. (1995) Support-Vector Networks. *Machine Learning*, **20**, 273-397. http://dx.doi.org/10.1007/BF00994018

[13] Vapnik, V. (2000) The Nature of Statistical Learning Theory. 2nd Edition, Springer-Verlag, New York. http://dx.doi.org/10.1007/978-1-4757-3264-1

[14] Chang, C.C. and Lin, C.J. (2011) LIBSVM: A Library for Support Vector Machines. *ACM Transactions on Intelligent Systems and Technology*, **2**, 27. http://dx.doi.org/10.1145/1961189.1961199

[15] Chang, T.T., Horace, H.S.Ip. and Feng, J. (2010) Large Scale Image Clustering with Support Vector Machine Based on Visual Keywords. *MDMKDD* 2010. http://dx.doi.org/10.1145/1814245.1814250

[16] Keerthi, S.S. and Lin, C.J. (2003) Asymptotic Behaviors of Support Vector Machines with Gaussian Kernel. *Neural Computation*, **15**, 1667-1689. http://dx.doi.org/10.1162/089976603321891855

[17] Aneja, K., Saroj and Jyoti (2013) A Genetic Algorithm Approach to Kernel Functions Parameters Selection for SVM. *International Journal of Current Engineering and Technology*, **2**, 713-716.

[18] Boughorbel, S., Tarel, J.-P. and Nozha, B. (2005) Generalized Histogram Intersection Kernel for Image Recognition. *ICIP* 2005. http://dx.doi.org/10.1109/icip.2005.1530353

13

Analysis of Abnormal Vehicle Behavior Based on Trajectory Fitting

Enyuan Jiang, Xuejun Wang

College of Communication Engineering, Jilin University, Changchun, China
Email: jeyenna@163.com, xjwang@jlu.edu.cn

Abstract

In order to analysis the abnormal vehicle behavior by trajectory fitting effectively, the whole process is divided into three steps: target detection and tracking, vehicle trajectory analysis, vehicle behavior detection. Firstly, a three-frame-differencing method is used to achieve initial target location and an improved tracking algorithm based on Kalman predictor is proposed; then, an adaptive segmented linear fitting algorithm is proposed to achieve vehicle trajectory fitting; finally, two parameters including the rate of velocity variation and the rate of direction variation are used to establish vehicle abnormal behavior detection model. Experiment result shows that the three high dangerous vehicle behaviors in road surveillance videos can be detected effectively: sharp brake, sharp turn, and sharp turn brake.

Keywords

Target Detection, Target Tracking, Trajectory Fitting, Vehicle Behavior Detection

1. Introduction

Nowadays traffic surveillance systems are widely used on highways and city roads. And the traditional manual surveillance watching cannot meet the needs of analyzing the huge amount of video data. Using the "intelligent" computer to discover the abnormal behavior in surveillance video becomes more important and gets more attention [1]-[3].

Recently, many scholars have done their own research works in this field. Zhao, Y.T., Li, X.Y. and Luo, D.H. [4] used the vehicle's tracking trajectory to separate the vehicles behavior into four elements: forward, backward, stop, left or right and then a vehicle behavior model is built based on them. Wang, W.G. and Ma, R. G. [5] proposed a weighed support vector machine (SVM) method based on the importance of the samples to solve the low detection problem caused by imbalanced samples. Liu, Q.C., Lu, J. and Chen, S.Y. [6] proposed a traffic incident detection method based on the random forest. Amin, M.S., Reaz, M. B. I. and Nasir, S. S. [7] proposed an incident detection and location method by fusing vehicle acceleration and GPS data.

To analysis abnormal vehicles behavior effectively, the whole process is divided into three steps in this paper: Firstly, a three-frame-differencing method and the improved tracking algorithm based on Kalman predictor are proposed to track vehicle. Then, an adaptive segmented linear fitting algorithm based on the least square method is proposed to achieve vehicle trajectory fitting which can reduce the amount of computation. Finally, two parameters including the rate of velocity variation and the rate of direction variation are used to establish vehicle abnormal behavior detection model. So the three highly dangerous vehicle behavior including sharp brake, sharp turn and sharp turn brake in surveillance video can be detected effectively.

2. Target Detection and Tracking

Vehicle motion parameters will affect the detection of the abnormal vehicle behavior directly. The Kalman and the Camshift algorithm [8] can improve the tracking result, but it needs to initialize the search window manually and the computation is complex. Aiming at this shortage, we make a further improvement as below.

2.1. Three-Frame-Differencing

In traditional Camshift algorithm [9], we must manually locate the initial position of a target, this artificial selection may introduce extra background color information and affect the tracking results. The three-frame-differencing method can locate the initial position automatically, and let this position as the initial window in the later Camshift algorithm.

This method is as follows: suppose $f_{k-1}(x,y)$, $f_k(x,y)$ and $f_{k+1}(x,y)$ are three consecutive frames, and respectively calculate the difference between $f_{k-1}(x,y)$ and $f_k(x,y)$ and between $f_k(x,y)$ and $f_{k+1}(x,y)$ to get the difference images $D_{(k-1,k)}$ and $D_{(k,k+1)}$. Then the binary images $R_{(k-1,k)}$ and $R_{(k,k+1)}$ is obtained through a proper threshold T and use logic "and" operation to get a binary image R_k. From the binary image R_k the contour can be extracted, and the initial window in Camshift algorithm is obtained. The process of three-frame-differencing is shown in **Figure 1**.

2.2. Improved Tracking Algorithm Based on Kalman Predictor

Kalman predictor [10] is an optimal estimation based on least error covariance, and the actual motion parameters are modified by the estimated value of future motion state. It can improve the tracking result in Camshift algorithm effectively when the target moves fast or it is partially occluded. The formula of occlusion judging is as follows.

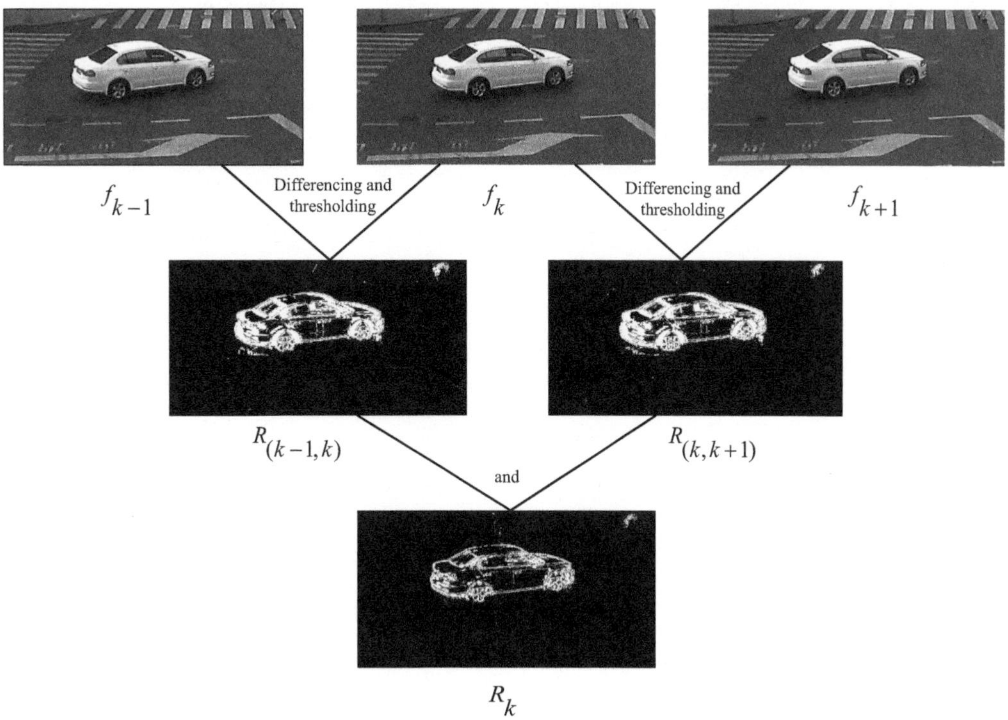

Figure 1. Diagram of three-frame-differencing.

$$\begin{cases} d(y) = \sqrt{1-\rho(p(y),q)} < T_b & \text{no occlusion} \\ else & \text{occlusion} \end{cases} \quad (1)$$

$d(y)$ is Bhattacharyya distance, $\rho(y) = \rho(p(y),q) = \sum_{u=1}^{m} \sqrt{p_u(y) q_u}$ is Bhattacharyya coefficient, $p_u(y)$ is the color histogram of initial window obtained by three-frame-differencing, $q_u(y)$ is the color histogram of tracking window, T_b is the threshold of occlusion judging.

When the occlusion occurs, the tracking result in Camshift algorithm is not credible, so the result of kalman predictor are outputted. When the occlusion ends, the result of camshaft are outputted. The improved algorithm using the Kalman predictor improves the tracking accuracy effectively. It is shown in **Figure 2**.

3. Analysis of Tracking Trajectory

After target detection and tracking, analysis of tracking trajectory is necessary. The traditional fitting trajectory is curve-fitting, but it requires a large number of complex operations and the following process is more complex. Here, an adaptive segmented linear fitting method based on least square is adopted. According to the principle of overall minimum fitting error, the best segmentation fitting points are automatically determined, so the speed and accuracy can be improved. When the trajectory appears curvilinear, the linear fitting method can not be closely matched the trajectory. Therefore, the tracking points are grouped and fitted in each group, that is the idea of the segmented linear fitting.

The function of a line is $y = a + bx$, b is the slope of the line, a is ordinate at the origin.

Define error:

$$Q = \sum_{i=1}^{N} (y_i - y)^2 - \sum_{i=1}^{N} (y_i - a - bx_i)^2 \quad (2)$$

The ordered track point is divided into *k* group:

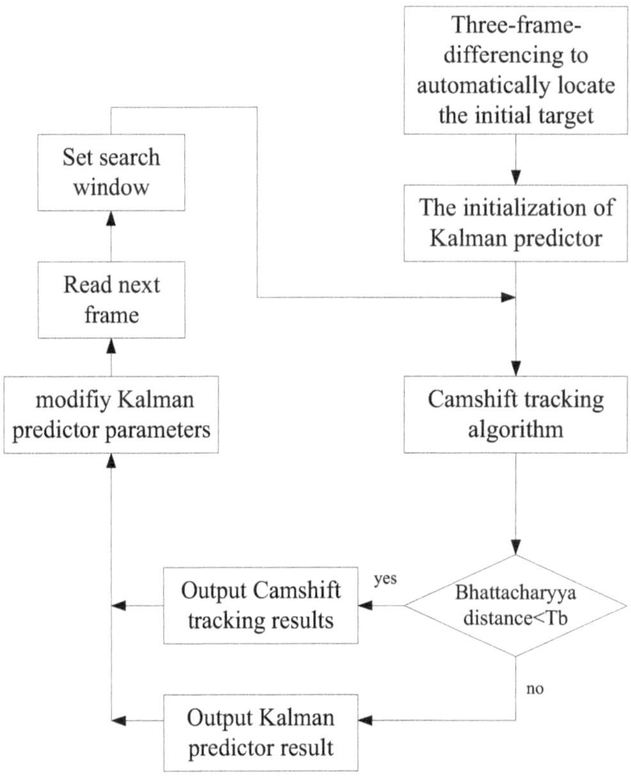

Figure 2. Process of improved tracking algorithm.

$$\begin{aligned} N_1 &= (x_{11}, y_{11}), (x_{12}, y_{12}), \ldots, (x_{1m_1}, y_{1m_1}) \\ N_2 &= (x_{21}, y_{21}), (x_{22}, y_{22}), \ldots, (x_{2m_1}, y_{2m_1}) \\ &\ldots \\ N_k &= (x_{k1}, y_{k1}), (x_{k2}, y_{k2}), \ldots, (x_{km_k}, y_{km_k}) \end{aligned} \tag{3}$$

The overall fitting error is related to the number of groups k and the value of segmentation point $(m_1, m_2, \ldots, m_{k-1})$. Therefore, according to the principle of overall minimum fitting error, the value of k and $(m_1, m_2, \ldots, m_{k-1})$ are automatically determined and the adaptive segmentation of the trajectory is realized.

Define overall fitting error:

$$Q^k = \sum_{i=1}^{m_1} (y_i - a_1 - b_1 x_i)^2 + \ldots + \sum_{i=m_{k-1}+1}^{N} (y_i - a_k - b_k x_i)^2 \tag{4}$$

when $Q^k = Q^k_{\min}$, $(m_1, m_2, \ldots, m_{k-1})$ are the best segmentation points for track points divided into k groups.

Define difference of errors:

$$q_k = Q^{k+1}_{\min} - Q^k_{\min} \tag{5}$$

q_k is the difference between the overall minimum fitting error of $k+1$ groups linear fitting and the overall minimum fitting error of k groups linear fitting, if the difference q_k is less than the setting error-threshold T_{q_k}, This shows that the overall fitting error is relatively small when the track points divided into k groups, and k is the number of groups.

4. Detection of Abnormal Vehicle Behavior

Usually, sharp brake, sharp turn, and sharp turn brake are dangerous events in driving [11]. The detection of abnormal vehicle behavior can control the transportation system effectively. Here, we pay more attention to the variation of velocity and direction in a short time, so these two parameters are used to establish abnormal vehicle behavior detection model.

Define the intersection point of two adjacent fitting lines as the inflection point of trajectory:

$$T_i = \left(\frac{a_i - a_{i+1}}{b_{i+1} - b_i}, \frac{(a_i - a_{i+1}) a_i}{b_{i+1} - b_i} + b_i \right) \tag{6}$$

a_i and b_i are the parameters of linear fitting equation.

Define the rate of direction variation:

$$D = \frac{\pi - \arccos\left(\dfrac{|T_i - T_{i-1}|^2 + |T_{i+1} - T_i|^2 - |T_{i+1} - T_{i-1}|^2}{2|T_i - T_{i-1}| * |T_{i+1} - T_i|} \right)}{\pi * (\Delta t_{i-1\sim i+1})} \tag{7}$$

Define the rate of velocity variation:

$$A = \frac{\dfrac{|T_i - T_{i+1}|}{\Delta t_{i\sim i+1}} - \dfrac{|T_i - T_{i-1}|}{\Delta t_{i-1\sim i}}}{\Delta t_{i-1\sim i+1}} \tag{8}$$

T_i is the inflection point of trajectory, $|T_i - T_j|$ is the Pixel distance of two inflection points, $\Delta t_{i\sim j}$ is the time interval of two inflection points, D is the rate of direction variation at the inflection point T_i, A is the rate of velocity variation at the inflection point T_i.

When a vehicle brakes, the rate of velocity variation increases rapidly. When a vehicle takes a sharp turn, the rate of direction variation increases rapidly. when a vehicle takes a sharp turn brake, the rate of velocity variation and direction variation increase rapidly. Specifically as follows **Table 1**.

5. Results and Analyses

Test results are shown in **Figures 3-5**, the result of target tracking is (a), the result of linear fitting is (b), **Figure 3** is two scenes of sharp brake, n is the number of inflection points that satisfy $|A| > T_a(A < 0)$ and $|D| \le T_d$ which is bigger than the half of inflection points' total number N, **Figure 4** is two scenes of sharp turn, n is the number of inflection points that satisfy $|D| > T_d$ and $|A| \le T_a$ which is bigger than the half of inflection points' total number N, the **Figure 5** is two scenes of sharp turn brake, n is the number of inflection points that satisfy $|A| > T_a(A < 0)$ and $|D| > T_d$ which is bigger than the half of inflection points' total number N. Based on many experiments, we set $T_a = 0.35$, $T_d = 0.85$.

Experiment result shows that the three high dangerous vehicle behaviors in the road surveillance videos can be detected effectively: sharp brake, sharp turn, and sharp turn brake.

6. Conclusion

In this paper, an abnormal vehicle behavior detection method based on trajectory fitting is proposed. Firstly, a three-frame-differencing method and the improved tracking algorithm based on Kalman predictor are proposed. Then, an adaptive segmented linear fitting algorithm based on the least square is proposed. Finally, two parameters including the rate of velocity variation and the rate of direction variation are adopted to establish abnormal

(a) (b) (c) (d)

Figure 3. Sharp brake.

(a) (b) (c) (d)

Figure 4. Sharp turn.

(a) (b) (c) (d)

Figure 5. Sharp turn brake.

Table 1. Vehicle abnormal behavior analysis.

Three abnormal behaviors	Equations
Sharp brake	$|A| > T_a(A < 0)$ & $|D| \le T_d$
Sharp turn	$|D| > T_d$ & $|A| \le T_a$
Sharp turn brake	$|A| > T_a(A < 0)$ & $|D| > T_d$

vehicle behavior detection model. Experiment result shows that the three high dangerous vehicle behaviors can be detected effectively: sharp brake, sharp turn, and sharp turn brake.

References

[1] Versavel, J. (1999) Road Safety through Video Detection. 1999 *IEEE/IEEJ/JSAI International Conference on Intelligent Transportation Systems*, 753-757. http://dx.doi.org/10.1109/itsc.1999.821155

[2] Qimei, L.B.C. (2006) Vehicle Activity Analysis from Freeway Traffic Video. *Chinese Journal of Scientific Instrument*, S3. (In Chinese) http://en.cnki.com.cn/Article_en/CJFDTOTAL-YQXB2006S3139.htm

[3] Zhao, X.M., Hui, F., Shi, X., Ma, J.Y. and Yang, L. (2014) Concept, Architecture and Challenging Technologies of Ubiquitous Traffic Information Service System. *Journal of Traffic and Transportation Engineering*, 105-115. (In Chinese) http://www.cnki.com.cn/Article/CJFDTotal-JYGC201404015.htm

[4] Zhao, Y.T., Li, X.Y. and Luo, D.H. (2011) Study on the Methods of Automatic Incident Detection Based on the Video Vehicle Trajectory Model. *Acta Scientiarum Naturalium Universitatis Sunyatseni*, 56-60. (In Chinese) http://www.cnki.com.cn/Article/CJFDTotal-ZSDZ201104012.htm

[5] Wang, W.G. and Ma, R.G. (2013) Weighed Support Vector Machine for Traffic Incident Detection. *Journal of Chang'an University* (*Natural Science Edition*), **6**, 014. (In Chinese) http://www.cnki.com.cn/Article/CJFDTotal-XAGL201306014.htm

[6] Liu, Q.C., Lu, J. and Chen, S.Y. (2014) Design and Analysis of Traffic Incident Detection Based on Random Forest. *Journal of Southeast University* (*English Edition*), **1**, 88-95. http://en.cnki.com.cn/Article_en/CJFDTotal-DNDY201401017.htm

[7] Amin, M.S., Reaz, M.B.I. and Nasir, S.S. (2014) Integrated Vehicle Accident Detection and Location System. *TELKOMNIKA* (*Telecommunication Computing Electronics and Control*), **12**, 73-78. http://jogjapress.com/index.php/TELKOMNIKA/article/view/1787 http://dx.doi.org/10.12928/telkomnika.v12i1.13

[8] Yan, J.H., Chen, S.H. and Ai, S.F. (2014) Target Tracking with Improved CAMShift Based on Kalman Predictor. *Journal of Chinese Inertial Technology*, **4**, 021. (In Chinese) http://www.cnki.com.cn/Article/CJFDTotal-ZGXJ201404021.htm

[9] Bradski, G.R. (1998) Computer Vision Face Tracking for Use in a Perceptual User Interface. http://citeseerx.ist.psu.edu/viewdoc/summary?doi=10.1.1.14.7673

[10] Liu, R.M., Li, X.L., Han, L. and Meng, J. (2013) Track Infrared Point Targets Based on Projection Coefficient Templates and Non-Linear Correlation Combined with Kalman Prediction. *Infrared Physics & Technology*, **57**, 68-75. http://www.sciencedirect.com/science/article/pii/S1350449512001193 http://dx.doi.org/10.1016/j.infrared.2012.12.011

[11] Zhang, X.Y. and Da, Q.D. (2001) Technology of Highway Automatic Incident Detection. *Systems Engineering-Theory & Practice*, **21**, 118-124. (In Chinese) http://www.cnki.com.cn/Article/CJFDTotal-XTLL200106021.htm

14

Researches on Six Lattice-Valued Logic

Hua Li

Department of Information Engineering, Hangzhou Polytechnic College, Hangzhou, China
Email: zjlihua@126.com

Abstract

Based on the direct product of Boolean algebra and Lukasiewicz algebra, six lattice-valued logic is put forward in this paper. The algebraic structure and properties of the lattice are analyzed profoundly and the tautologies of six-valued logic system $L_6P(X)$ are discussed deeply. The researches of this paper can be used in lattice-valued logic systems and can be helpful to automated reasoning systems.

Keywords

Six Lattice-Valued Logic, Lattice Implication Algebra, Filter, Tautology

1. Introduction

Lattice-valued logic is an important case of multi-valued logic, and it plays more and more important roles in artificial intelligence and automated reasoning. Six lattice-valued is a kind of common lattice, which can express logic in real world, such as language values, and evaluation values. It can deal with not only comparable information but also non-comparable information. Therefore, theoretical researches and logic and reasoning systems based on six lattice-valued logic are of great significance.

2. The Structure of Lattice L_6

The set of $L=\{O,a,b,c,d,I\}$ is a lattice, and the order relation of L is shown in **Figure 1**. The complement operator "′"and implication operation "→" are defined in **Table 1** respectively.

L means an lattice implication algebra.

Then set $A=\{O,I\}$, $B=\{O,m,I\}$. As A is the true set of classical binary logic, the operation rules of the complement operation and the implication operation are the same with the classical two-valued logic systems. B is the true-value set of Lukasiewicz system with three-valued logic, and complement operations and implication operations are defined in **Table 2**.

Let $L^* = A\times B$, the order relations, disjunctive, conjunctive, complement operation and implication operation

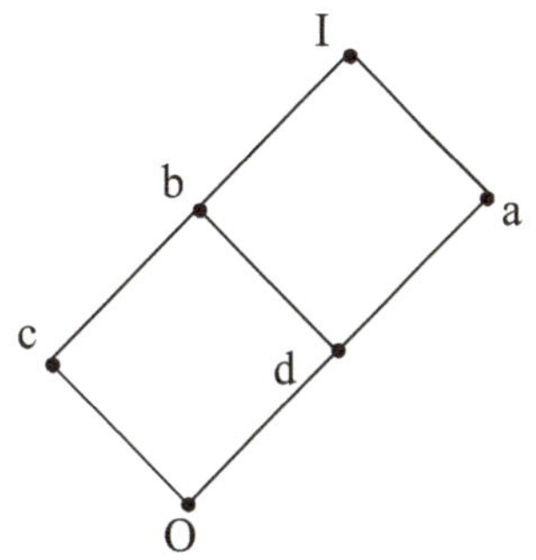

Figure 1. Structure of the six-valued lattice.

Table 1. Computing of the six-valued lattice L_6.

x	x'	$\rightarrow$	O	a	b	c	d	I
O	I	O	I	I	I	I	I	I
a	c	a	c	I	b	c	b	I
b	d	b	d	a	I	b	a	I
c	a	c	a	a	I	I	a	I
d	b	d	b	I	I	b	I	I
I	O	I	O	a	b	c	d	I

Table 2. Computing of L_3.

x	x'	$\rightarrow$	O	m	I
O	I	O	I	I	I
m	m	m	m	I	I
I	O	I	O	m	I

on L are defined as follows:

For any $(a,b)\in L^*$, $(c,d)\in L^*$:

(1) $(x,y)\le(z,r)$, if and only if $x\le z$ and $y\le r$.

(2) $(x,y)=(z,r)$, if and only if $x=z$ and $y=r$.

(3) Under other circumstances, (x, y) cannot be compared with (z, r).

(4) $(x,y)\wedge(z,r)=(x\wedge z,y\wedge r)$, $(x,y)\vee(z,r)=(x\vee z,y\vee r)$.

(5) $(x, y)\rightarrow(z, r)=(x\rightarrow z,y\rightarrow r)$.

(6) $(x,y)'=(x',y')$.

The L^* constitute a six element lattice and its operation diagram is shown in Hasse **Figure 2**.

Theorem 1. L is isomorphic lattice implication of L^*.

Proof:

Obviously, we can construct a upward one-to-one mapping from L to L^*: $f:L\rightarrow L^*$, making

$$f(O)=(O,O),\quad f(a)=(O,I),\quad f(b)=(I,m)$$
$$f(c)=(I,O),\quad f(d)=(O,m),\quad f(I)=(I,I)$$

Clearly f is conjunctive homomorphic mapping and disjunctive homomorphism mapping.

Here is the proof that f is complement homomorphic mapping and implication homomorphism mapping.

According to the definition of implication operations and complement operations, it can be easily obtained in **Table 3**.

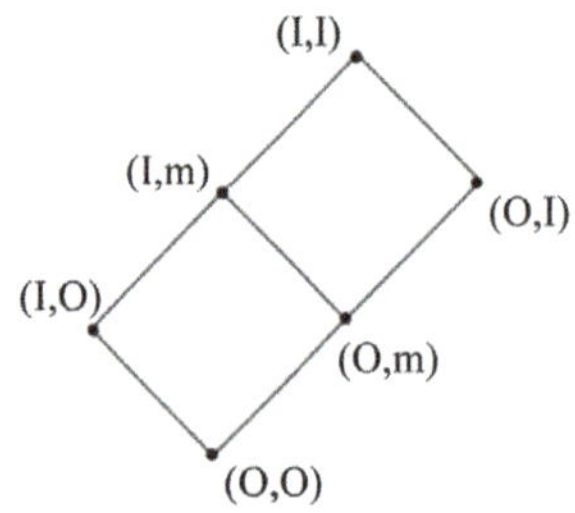

Figure 2. Six-valued lattice generated by the direct product.

Table 3. Six-valued lattice generated by the direct product.

x	x'	$\to$	(O,O)	(O,I)	(I,m)	(I,O)	(O,m)	(I,I)
(O,O)	(I,I)	(O,O)	(I,I)	(I,I)	(I,I)	(I,I)	(I,I)	(I,I)
(O,I)	(I,O)	(O,I)	(I,O)	(I,I)	(I,m)	(I,O)	(I,m)	(I,I)
(I,m)	(O,m)	(I,m)	(O,m)	(O,I)	(I,I)	(I,m)	(O,I)	(I,I)
(I,O)	(O,I)	(I,O)	(O,I)	(O,I)	(I,I)	(I,I)	(O,I)	(I,I)
(O,m)	(I,m)	(O,m)	(I,m)	(I,I)	(I,I)	(I,m)	(I,I)	(I,I)
(I,I)	(O,O)	(I,I)	(O,O)	(O,I)	(I,m)	(I,O)	(O,m)	(I,I)

It can be seen from the **Table 3**, f is the implication operations and the complement operations homomorphic.

In summary, we proofed that:

For any $x,y\in L$, $f(x')=(f(x))'$, $f(x*y)=f(x)*f(y)$, where $*$ is one of disjunctive, conjunctive, complement operation.

Thus L and L^* is isomorphic lattice implication.

3. The Property and Language of Lattice L_6

Due to L_6 is a lattice implication algebra, it not only has all the properties of lattice implication algebra but also properties as follows.

Theorem 2. As shown the six-valued lattice L_6 in **Figure 1**, the implication operation satisfies the following properties: For any $x,y,z\in L_6$:

(1) $z\le y\to x$ iff $y\le z\to x$.

(2) $z\to(y\to x)\ge(z\to y)\to(z\to x)$.

(3) $(x\to y)\vee((x\to y)\to(x'\vee y))=I$.

(4) $y\to z\le(x\to y)\to(x\to z)$.

(5) $x\to(y\to z)=y\to(x\to z)$.

(6) $x\to y\le(x\vee z)\to(y\vee z)$.

Theorem 3. As the true subset of L_6, $L_0=\{O,I,a,c\}$ is a sub lattice implication algebra. What's more, L_0 is a Boolean algebra, and the implication arithmetic of it meets that: for any x, $y\in L_0$, $x\to y=x'\vee y$.

Proof: It is clearly that L_0 is a sub lattice of L_6. For any $x,y\in L_0$, $x'\in L_0$, $x\to y\in L_0$, therefore when regarding L_6, the operation of L_0 is closed, that is to say, L_0 is a sub lattice implication algebras of L_6.

It can be verified easily: for any $x,y\in L_0$, $x\to y=x'\vee y$. Meeting the of Boolean algebra axiom, L_0 is a Boolean algebra.

Any sub-set of power set lattice in a collection is called the set lattice for the collection. The isomorphism from a lattice L to a set lattice $B(X)$ in collection X is named as a isomorphic representation L by $B(X)$, which can be denoted as L for abbreviation. Through establishing the lattice representation, lattice language can be simplified, which is very important for studying the structure and properties of the lattice.

Definition 1 [1]. Let L is a lattice, an element $x\in L$ is called as an join-irreducible element, if

(1) $x \neq O$ (when there is a minimum of O when L);

(2) For any $a,b \in L$, if $x = a \vee b$, then $x = a$ or $x = b$.

Assume L is a finite distributive lattice, $\Im(L)$ denotes the set of all join-irreducible element in the collection, and all the join-irreducible element in L can form under set lattice (*i.e.* ideal Lattice) according to the order relation which can be indicated as $O(\Im(L))$. Then we have the following conclusions:

Theorem 4 [2]. Let L is a finite distributive lattice, and mapping can be constructed as follows:

$$\eta : L \to O(\Im(L))$$

$$\eta(a) = \{x \in \Im(L) | x \leq a\}$$

The η is the lattice isomorphism from L to $O(\Im(L))$.

Theorem 5 [2]. Let L is a finite distributive lattice, then the following equivalent hold:

1) L is a distributive lattice;

2) $L \cong O(\Im(L))$;

3) L is isomorphic to a set lattice;

4) For any $n \geq 0$, L is isomorphic to 2^n sub lattice.

According to Theorem 5, theorem representation of six lattice-valued L_6 can be got easily.

Theorem 6. As shown the six-valued lattice L_6 in **Figure 1**, conclusions as follows can be got:

(1) The set of join-irreducible element in L_6 is $\Im(L_6) = \{a,b,c\}$, and its order relation are shown in **Figure 3**.

(2) The under set lattice (*i.e.* ideal lattice), which is the set of all the join-irreducible element and forms according to its order relation, is $O(\Im(L_6)) = \{\phi, \{c\}, \{a\}, \{b,c\}, \{a,c\}, \{a,b,c\}\}$.

(3) The Hasse diagram $O(\Im(L_6))$ of the ideal lattice of L_6, which forms through inclusion relation, is shown in **Figure 4**. Form the figure, we can see that L_6 is isomorphic of lattice implication to its ideal lattice $O(\Im(L_6))$. Lattice implication isomorphism η is defined as follows:

$$\eta : L \to O(\Im(L))$$

$$h(O) = \phi,\ \eta(a) = \{b,c\},\ \eta(b) = \{a,c\},\ \eta(c) = \{a\}$$

$$\eta(d) = \{c\},\ \eta(I) = \{a,b,c\}$$

4. The Filter of Lattice L_6

Since all Lukasiewicz algebras are lattice implication algebra [1], it can be proved that Lukasiewicz algebra filters are trivial.

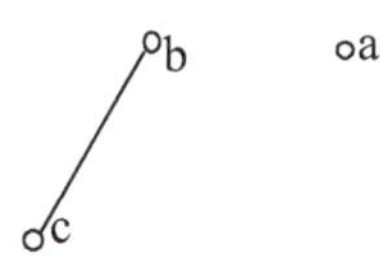

Figure 3. The order of $\Im(L_6)$.

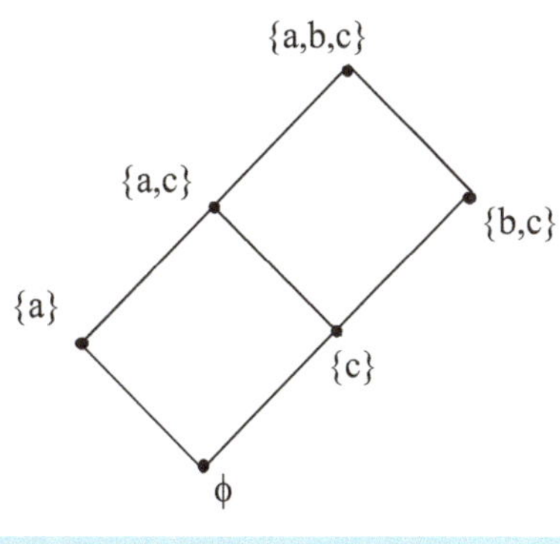

Figure 4. The ideal lattice of $O(\Im(L_6))$.

Theorem 7.

(1) The finite chain of Lukasiewicz only contains trivial filters.

(2) Lukasiewicz algebra [0,1] only contains trivial filters.

Proof: (1) Let's set $L = L_n = \left\{0, \frac{1}{n-1}, \frac{2}{n-1}, \cdots, \frac{n-2}{n-1}, 1\right\} (n = 2,3,\cdots)$. Specific operations are as follows:

For any $x, y \in L$,

$$x \vee y = \max\{x, y\}, \quad x \wedge y = \min\{x, y\}$$

$$x' = 1 - x, \quad x \to y = \min\{1, 1 - x + y\}$$

It is clearly that set {1} and L are trivial filters in L. we can proof that L don't contain any other trivial filters.

From Theorem 6 we can see that filters in L are ideal dual filters of L, and the set of ideal dual filters of L are upper set of L.

If $F = \left\{\frac{k}{n-1}, \frac{k+1}{n-1}, \cdots, 1\right\}$ (where $k \geq 1$) is a filter of L, then

$$\frac{k}{n-1} \to \frac{k-1}{n-1} = \min\left\{1 - \frac{k}{n-1} + \frac{k-1}{n-1}, 1\right\} = \frac{n-2}{n-1} \in F,$$

And $\frac{k}{n-1} \in F$, so it can be seen that the definition of filters: $\frac{k-1}{n-1} \in F$.

This shows that $F = L$, so it demonstrated that L don't contain any other trivial filters.

(2) Let $L = [0,1]$, its upper operation is the same as defined C_2.

It is clearly that set {1} and L are trivial filters in L. we can proof that L don't contain any other trivial filters.

We can see that filters in L are ideal dual filters of L, and the set of ideal dual filters of L are upper set of L. So the filter of L must be an interval containing greatest element 1.

Firstly, we can proof that the filter of L must be a closed interval.

Let us set $F = (u, 1)$ is a filter of L, where $0 < u < 1$, for any x, satisfies $u < x < 1$, then $x \in F$, and $x \to u = \min\{1, 1 - u + x\} = 1 - u + x > u \in F$, conclusion can get $u \in F$.

This shows that F is a closed interval.

Secondly, assume $F = [u, 1]$ is a filter of L, where $0 < u < 1$.

For any x, making $u < x < 1$ and $x + u \geq 1$, then

$$u \to (x + u - 1) = \min\{1 - u + (x + u - 1), 1\} = x \in F$$

thereby $x + u - 1 \in F$, that is contradictory, because $0 \leq x + u - 1 < u$.

So F is an interval.

This proves that Lukasiewicz interval only have trivial filters.

As a special case of Theorem 7, we have the following corollary.

Corollary 1. $C_2 = \{O, I\}$ and $L_3 = \{O, m, I\}$ only contain trivial filters.

Theorem 8. The six element lattice only contains the following four filters:

$\{I\}, L_6, F_a = \{I, a\}, F_{bc} = \{I, b, c\}$.

Proof: According to Theorem 1, L_6 can be seen as the direct product of C_2 and L_3. According to Corollary 1, $C_2 = \{O, I\}$ and $L_3 = \{O, m, I\}$ only contain trivial filters. As followed:

The filters of $C_2 = \{O, I\}$ are $\{I\}$ and $\{O, I\}$.

The filters of $L_3 = \{O, m, I\}$ are $\{I\}$ and $\{O, m, I\}$.

It is easy to know, the filters of L_6 are the direct products of the filters of C_2 and the filters of L_3. So the filters of L_6 are as followed:

$\{(I, I)\}$, $\{(I, I), (O, I)\}$, $\{(I, I), (I, m), (I, O)\}$ and L_6 itself.

In other words: The six element lattice L_6 only contains the following four filters:

$\{I\}, L_6, \{I, a\}, \{I, b, c\}$.

5. The Tautologies of Lattice-Valued Logic System $L_6P(X)$

Here we take the lattice-valued logic system $L_6P(X)$ into consideration, and discuss its tautologies and F-tautologies, the true value domain is L_6.

It is easy to verify:

$$L_6 = C_2 \times L_3$$

where C_2 is a Boolean algebra $\{O, I\}$, L_3 is a Lukasiewicz algebra $\{O, m, I\}$.

Theorem 9. (The definition of tautologies in $L_6P(X)$ [3]) The tautologies in six lattice-valued logic system $L_6P(X)$ process the following relationship:

(1) $T^{L_6} = T^{C_2} \cap T^{L_3} = T^{L_3}$.

(2) $T_a^{L_6} = T^{L_3}$.

(3) $T_b^{L_6} = T^{C_2} \cap T_m^{L_3} = T^{C_2}$.

(4) $T_c^{L_6} = T^{C_2}$.

(5) $T_d^{L_6} = T_m^{L_3} = T^{C_2}$.

Proof: It is noticed that the tautologies in Lukasiewicz three-valued logic system process the following relationship:

$$T^{L_3} \subset T_m^{L_3} = T^{C_2}.$$

Proof of this theorem can be obtained.

From Theorem 7, the six element lattice L_6 only contains four filters as followed:

$\{I\}$, L_6, $F_a = \{I, a\}$, $F_{bc} = \{I, b, c\}$.

Therefore, its non-trivial filters are $F_a = \{I, a\}$, $F_{bc} = \{I, b, c\}$.

We can get the definition of F-tautologies in six lattice-valued logic system $L_6P(X)$ as Theorem 8 similarly.

Theorem 10. (The definition of F-tautologies in $L_6P(X)$ [4]) The F-tautologies in six lattice-valued logic system $L_6P(X)$ process the following relationship:

(1) $T_{F_a}^{L_6} = T^{L_3}$.

(2) $T_{F_{bc}}^{L_6} = T^{C_2}$.

Proof:

Since $T(f)(A) \neq T(g)(A)$, so T is an injection.

Clearly T is a surjection. For any $(U_1, \Psi_L(U_1)), (U_2, \Psi_L(U_2)) \in |\aleph(L)|$,

$\forall \hat{\Re}_f \in H_{S\aleph(L)}((U_1, \Psi_L(U_1)), (U_2, \Psi_L(U_2)))$, G has the inverse image.

Thus G is an isomorphic functor of $\aleph(L)$.

As isomorphic relationship means an equivalence relation, so $S\aleph(L)$ and $\Im(L)$ are isomorphic.

6. Conclusion

In this paper, the six element lattice is built by the direct product of Boolean algebra and Lukasiewicz algebra; the operation of the lattice is defined; the structures, properties and filters are studied; finally the tautologies and F-tautologies of the six lattice-valued logic system are discussed. The results of this paper can be applied to lattice-valued logic systems and automated reasoning applications.

Acknowledgements

The work is supported by the project of Zhejiang province education department of China, Grant No. Y201326675.

References

[1] Yang, X. and Yun, Q.K. (1995) Fuzzy Lattice Implication Algebra. *Southwest Jiaotong University*, **2**, 121-127.

[2] Xu, Y., Ruan, D. and Liu, J. (2004) Progress and Prospect in Lattice-Valued Logic Systems Based on Lattice Implication Algebras. *Proceedings of the 6th International FLINS Conference Applied Computational Intelligence*, 29-34.

http://dx.doi.org/10.1142/9789812702661_0009

[3] Min, H.T. (1996) Georgia, Sequencing Primer Theory and Its Application. Southwest Jiaotong University Press.

[4] Fang, S. and Mei, Z.F. (2010) Six Yuan-Based Language of Logic Attributed True Value Method. *Guangxi Normal University*, **3**, 118-122.

15

Location Characterization in Noisy Range Network Localization

Mingzhu Wei, Ryad Chellali, Yang Yi, Ting Wang, Wen Qin

College of Electric Engineering and Control Science, Nanjing Tech University, Nanjing, China
Email: mingzhu1219@163.com

Abstract

Network localization is a fundamental problem in wireless sensor networks, mainly in location dependent applications. A common family of solutions to this problem is the range-based network localization. The resulting localization algorithms are noise sensitive and thus lacking in terms of robustness. Our contribution provides an algorithm which is robust to measurement errors. We propose an analytical tool to analyze the effect of range errors in the final location and use a distributed method to solve the noisy range localization problem.

Keywords

Network Localization, Noisy Range Measurement, Distributed Computation

1. Introduction

Estimating the positions of agents in networked systems is crucial for many applications ranging from robotics, autonomous vehicles navigation and sensor networks. In network localization, the absolute node positions need to be estimated from partial relative measurements between nodes. For range-only localization problem, the inter-agent distance information is given, and we need to estimate the agents' location. In noiseless case the problem usually has a unique global solution. In practice, the inter-agent distances are often measured with perturbations. Thus, the existence and uniqueness of the global solution are not guaranteed. Indeed, it is well known that in presence of distance noise, a node may have discontinuous deformations (e.g. flip ambiguities and discontinuous flex ambiguities). Then the location estimation error needs to be analyzed. In this paper, we focus on the noisy distance network localization problem.

There are some previous works related with the noisy range network localization. Some authors consider the location error properties with the particular measurement model. [1] derives Cramer-Rao Bound (CRB) error of location for Time-Of-Arrival and Received Signal Strength measurements. [2] derives CRB error from distance and angle measurements model and gives the properties of CRB error with network parameters. Anderson states that when the distance noise is small enough, the noisy minimization problem has a unique solution and the displacements of agent positions have a bound based on the distance noise [3]. Other authors focus on network multi-lateration techniques to search a global solution in noisy range network localization problem. [4] describes a hybrid centralized and distributed algorithm, which provides robust quadrilaterals as subgraph from representative network graph to avoid flip ambiguities. [5] proposes a technique based on Cayley-Menger determinant to formulate the geometric relations among sensors and anchors in sensor networks as quadratic constraints. [6] proposes two computational efficiently algorithms in n-lateration networks to refine the noisy distances based on the Cayley-Menger determinant. [7] proposes a successive refinement method, Semi-Definite Programming (SDP), trying to get a rude solution for the global optimization in the noisy case.

In noisy range-based network localization cases, most of the existing approaches to solve the problem are network multi-lateration based techniques. Typically, successive refinement approaches achieve better statistical performance than multi-lateration approaches in noiseless case. But the existing SDP method is a centralized one. In this paper, we want to investigate the noisy localization performance of a distributed gradient-based algorithm that we propose in [8].

The organization of the paper is as follows. In the Section 2, we formulate the range-based network localization problem in noisy case. In Section 3, we derive the analytical location estimation error characterizations from the Gaussian random distance perturbation model. In Section 4, we will give short overview of the distributed gradient-based localization algorithm. In Section 5, the numerical simulations are performed to prove the robustness of the localization algorithm with the distance noise and accuracies with the network parameters. In Section 6, we summarize the results.

2. Problem Statement

Let $V = \{v_1,\ldots,v_n\}$ be a set of n nodes (representing sensors, agents, robots, vehicles, etc.), and let $P = \{p_1,\ldots,p_n\}$ denote a corresponding set of positions on the Cartesian plane, where $p_i = [x_i, y_i]^{\mathrm{T}} \in \mathrm{R}^2$ are the coordinates of the i-th node. We shall call P a configuration of nodes. Suppose that some pair of nodes, say nodes (i, j), have the possibility of measuring the relative distance between them:

$$\left\|p_i - p_j\right\|^2 = d_{ij}^2, \quad \forall (i,j) \in E. \tag{1}$$

We denote with E the set of unordered node pairs (i, j) such that a distance measurement exists between i and j. Our objective is to determine a node configuration $\{p_1,\ldots,p_n\}$, that minimizes a Least-Squares of fit criterion

$$\sum\nolimits_{(i,j)\in}\left(\left\|p_i - p_j\right\|^2 - d_{ij}^2\right)^2 \tag{2}$$

When the global minimum of f is zero, we say that exact matching is achieved, that means a configuration $\{p_1,\ldots,p_n\}$ is found that exactly matches the given distance measurements. Otherwise, no geometric node configuration can exactly match the given range data, and we say that approximate matching is achieved by the optimal configuration. Actual recovery of the absolute geometric node positions from distance measurements is only possible if the node configuration is generically globally rigid (*ggr*), [9]. In this case, the objective function in (2) has a unique global minimum, if the positions of at least three nodes are known and fixed in advance (anchor nodes, or beacons).

2.1. Problem Setup

In noiseless case, after using anchor and distance information, network localization can be formulated.

$$\begin{aligned} &\min\nolimits_{p\in V_s} \sum\nolimits_{(i,j)\in E_s}\left(\left\|p_i - p_j\right\|^2 - d_{ij}^2\right)^2 \\ &\text{s.t.} \quad p_i = a_i,\ p_i \in V'. \end{aligned} \tag{3}$$

where, V' is the vertex set of anchors, ($a_i \in V', V' \subset V$). V_s is the set of the unknown sensors, $V_s = V \setminus V'$. E' is the set of edges between anchors. E_s is the set of edges between anchors and unknown sensors, $E_s = E \setminus E'$.

In practice, the distance measurements are usually noisy. We assume the distance measurement model, $\hat{d}_{ij}^2 = d_{ij}^2 + \epsilon_{ij}$. Where, $\hat{d}_{ij}$ is the distance measurement, d_{ij} is the true distance and ϵ_{ij} is the measurement noise in the square of the distance.

2.2. Linear Algebra Property of the *ggr* Formation

Let $p \doteq [p_1^{\mathrm{T}}, p_2^{\mathrm{T}} \cdots p_n^{\mathrm{T}}]^{\mathrm{T}}$, denote the vector of node positions. The cost function (2) is rewritten as

$$f(p)=\sum_{(i,j)\in} g_{ij}^2(p),\ g_{ij}(p)\doteq\|p_i-p_j\|^2-d_{ij}^2. \tag{4}$$

The gradient $\nabla g_{ij}(p)$ is a row vectors, $\nabla g_{ij}(p)=2\left[{}_2\cdots{}_2(p_i-p_j)_2\cdots{}_2(p_j-p_i)_2\cdots{}_2\right]$.

Then the rigidity matrix R of the configuration can be written:

$$R=R(p)\doteq\frac{1}{2}\begin{bmatrix}\nabla g_1(p)\\ \vdots\\ \nabla g_{|E|}(p)\end{bmatrix}\in R^{|E|,2|V|}. \tag{5}$$

Let $|C|$ denote the cardinality of set C, and the k-th edge denote edge between nodes i and j, *i.e.*, $\nabla g_k(p)=\nabla g_{ij}(p)$. Notice that in the case when anchor nodes are present, by eliminating the columns of R corresponding to anchors and each row of R corresponding to the edges between the anchors pairs, we have the reduced rigidity matrix $R_r\in R^{|E_s|,2|V_s|}$ as the submatrix obtained from R.

Based on lemma 1 [10], there is $|E_s|\geq 2|V_s|$, which means the number of Equation (1) is larger than the number of the unknown variables, *i.e.* with the assumption of *ggr* formation, the equation set (1) constitutes an overdetermined system of simultaneous equations and then the solution is unique. In the noiseless case, the global optimum can be found with the *ggr* formation assumption. In the noisy case the unique global will not exist, except that the distance bound is typically small [3].

3. Location Estimation Error Properties in Noisy Range Network Localization

We need to know what is the effect of distance errors on the error of network localization. In this section, we propose two analytical ways to derive the error propagation. We assume the distance perturbation, ϵ_{ij} is independent Gaussian random noise, $\epsilon_{ij}\sim N(0,\sigma_{ij}^2)$. We derive the Cramer-Rao Bound (CRB) of position error with independent Gaussian random distance error model. CRB is an independent-algorithm way to express a lower bound on the covariance of unbiased estimator for deterministic parameters. This benchmark implies that if the bound is nearly achieved, then there is no need to improve the accuracy of the algorithm.

For the k-th edge in the global rigid graph, location estimation function becomes: $\hat{d}_k^2=\hat{d}_{ij}^2=\|p_i-p_j\|^2+\epsilon_{ij}$. Let $A_k(p)=A_{ij}(p)=\|p_i-p_j\|^2, k\in|E|$. We stack all the equations and have:

$$D=A(p)+\epsilon$$

where, $D=\left[\hat{d}_1^2,\hat{d}_2^2,\ldots,\hat{d}_{|E_s|}^2\right]^T$, $A(p)=\left[A_1(p),\ldots,A_{|E_s|}(p)\right]^T, \epsilon=\left[\epsilon_1,\epsilon_2,\ldots,\epsilon_{|E_s|}\right]^T$.

Under Gaussian independent random distance assumption, the measurement pdf of the distance $\hat{d}_{ij}$ is:

$$f_{\hat{d}_{ij}|p}(\hat{d}_{ij},p)=\frac{1}{\sqrt{2\pi}\cdot\sigma_{ij}}\cdot\exp\left(-\frac{1}{2\sigma_{ij}^2}\cdot\left(\hat{d}_{ij}^2-\|p_i-p_j\|^2\right)^2\right)$$

Since we assume that the distance measurements are uncorrelated, the covariance matrix Σ is a diagonal matrix and the nonzero element $\Sigma_{k,k}=\sigma_k^2=\sigma_{ij}^2$. Then the measurement pdf is the vector Gaussian pdf:

$$\begin{aligned}f_{D|p}(D,p)&=\prod_{(i,j)\in_s} f_{\hat{d}_{ij}|p}(\hat{d}_{ij},p)\\ &=\frac{1}{(\sqrt{2\pi})^{|s|}\cdot\prod_{(i,j)\in_s}\sigma_{ij}}\cdot\exp\left(\prod_{(i,j)\in_s}\frac{-1}{2\sigma_{ij}^2}\cdot\left(\hat{d}_{ij}^2-\|p_i-p_j\|^2\right)^2\right)\\ &=\frac{1}{(\sqrt{2\pi})^{|s|}\cdot|\Sigma|^{\frac{1}{2}}}\cdot\exp\left(-\frac{1}{2}[D-A(p)]^T\Sigma^{-1}[D-A(p)]\right).\end{aligned} \tag{6}$$

From the measurement pdf (6), we find the Fisher information matrix:

$$\begin{aligned}J(p)&=E\left[(\nabla_p\ln f_{D|p}(p))\cdot(\nabla_p\ln f_{D|p}(p))^T\right]\\ &=\left[G'(p)\right]^T\Sigma^{-1}\left[G(p)\right],\end{aligned} \tag{7}$$

where,

$$[G'(p)]_{k,n} = \frac{\partial A_k(p)}{\partial p_n} = \begin{cases} 2(x_i - x_j), p_n = x_i; \\ 2(y_i - y_j), p_n = y_i; \\ 2(x_j - x_i), p_n = x_j; \\ 2(y_j - y_i), p_n = x_j; \\ 0, \text{otherwise.} \end{cases}$$

According to the definition of the rigidity matrix *R*, by omitting the anchor information we have $G'(p) = 2 \cdot R_r$. We get the new form for the Fisher information matrix as follows: $J(p) = 4R_r^T \cdot \Sigma^{-1} \cdot R_r$.

So the Cramer-Rao Bound matrix is:

$$CRB = J^{-1}(p) = \frac{1}{4} \cdot \left[R_r^T \cdot \Sigma^{-1} \cdot R_r\right]^{-1}. \tag{8}$$

The position estimate error covariance matrix defined as $C = E\left[(\hat{p} - p)(\hat{p} - p)^T\right]$, and it is lower bounded by the Cramer-Rao Bound: $C \geq CRB$.

Distributed Gradient-Based Localization Algorithm

In the previous section, we derived the analytical bounds for the localization error. We will use these bounds for analyzing the performance of the Distributed Gradient-based Method (DGM). This algorithm and its equivalent centralized version (GM) were previously presented in [8] under noise-free distance measurements.

Distributed Gradient-based Method Algorithm: Let *t* denote the number of consensus iterations, τ denote the number of gradient update iterations, $p^{(\tau)}$ denote the configuration by algorithm in update iteration τ. N_i is the neighbors communicated with node *i*. *W* is a symmetric, non-negative with strictly positive diagonal entries, doubly stochastic matrix, compatible with graph *G*.

Input: $t_{\max}$, Maximum number of consensus iterations; $\tau_{\max}$, Maximum number of gradient updates; τ_{wup}, Maximum number of warm-up iterations; η_{abstol}, Absolute update tolerance.

Initialization and Warm-up: Let $\tau = 0$ and set the position estimates p_i^τ, for $i = 1, \ldots, n$, to be the given initial guess for optimization. Mark all nodes as "active". Repeat while $\tau < \tau_{wup}$: Fix $\alpha_\tau = \tilde{\alpha}$ (a small number and equal for all nodes); Update all node position estimates according to,

$$p_i^{(\tau+1)} = p_i^{(\tau)} - \alpha_\tau \nabla f_i(p^{(\tau)}) = p_i^{(\tau)} - \alpha_\tau \sum_{j \in_i} \left(p_i^{(\tau)} - p_j^{(\tau)}\right)^{\mathrm{T}} g_{ij}\left(p^{(\tau)}\right), \quad (i = 1, \ldots, n);$$

Loop until $\tau \geq \tau_{\max}$: (Consensus iterations) Repeat while $t < t_{\max}$: Initialize $\rho_i(\tau(0))$ and $\psi_i(\tau(0))$ as:

$$\rho_i(\tau(0)) \triangleq \left\|p_i^{(\tau)} - p_i^{(\tau-1)}\right\|^2, \quad \psi_i(\tau(0)) \triangleq \left(p_i^{(\tau)} - p_i^{(\tau-1)}\right)\left(\nabla f_i(p^{(\tau)}) - \nabla f_i(p^{(\tau-1)})\right);$$

Update ρ_i and ψ_i according to:

$$\rho_i(\tau(t+1)) \triangleq W_{ii}\rho_i(\tau(t)) + \sum_{j \in_i} W_{ij}\rho_j(\tau(t)), \quad \psi_i(\tau(t+1)) \triangleq W_{ii}\psi_i(\tau(t)) + \sum_{j \in_i} W_{ij}\psi_j(\tau(t)) \tag{9}$$

(Gradient update) At each active node: Compute $\alpha_i(\tau(t))$ according to

$$\alpha_\tau = \alpha_i(\tau(t)) \doteq \frac{\rho_i(\tau(t))}{\psi_i(\tau(t))}$$

Update position estimate according to (9); If $\left\|p_i^{(\tau)} - p_i^{(\tau-1)}\right\| \leq \eta_{abstol}$, then mark node i as "inactive";

Let $\tau = \tau + 1$. If $\tau \geq \tau_{\max}$, or all nodes are inactive, then exit.

End loop.

4. Numerical Experiments

In this section, we analyze the performance of our localization algorithm using the error characterization given the noisy range measurements. We compare the analytical errors with the estimation errors computed by the particular algorithm. We display the root-mean-square (RMS) error associated to the CRB matrix, given by

$$\phi_{CRB} = \sqrt{\frac{\sum_{i=1}^{n-n_{anchors}} \left([CRB]_{2i-1,2i-1} + [CRB]_{2i,2i} \right)}{n - n_{anchors}}} \tag{10}$$

where $n - n_{anchors}$ are non-anchor node positions to be determined. For our localization algorithm under noisy measurements, we display algorithm estimate error ϕ_p defined by

$$\phi_p = \sqrt{\frac{\sum_{i=1}^{n-n_{anchors}} \|\hat{p}_i - p_i\|^2}{n - n_{anchors}}} \tag{11}$$

where $\hat{p}$ are the estimated positions for the non-anchor nodes, and p is the true configuration. In addition, the behavior of the algorithm with parameters such as the number of nodes, the number of anchor-nodes and the noise level are also discussed.

4.1. Network Generation

We have performed several simulations using range limited graphs, which are generically global rigid and have at least three non collinear anchor nodes. Therefore, the global solution of the localization problem is unique. We use random geometric graphs, in which nodes are deployed at random in the square plane, and an edge exists between a pair of nodes if and only if their geometrical distance is smaller than sensing radius, *r*. It has been proved in [9] that if $r > 2\sqrt{2}\sqrt{\log(n)/n}$, the graphs produced by the previous technique are *ggr* with high probability. In the following experiments, the sensing range is $r = 4$. The nodes are distributed in a square $[0,10]\times[0,10]$ with $n = 50$ nodes including 3 anchors.

4.2. Location Error Behavior as Range Error

Firstly, we investigate the error characterization with different standard deviations $\sigma_d = 0.01, 0.05, 0.1, 0.5, 1$ of the distance errors (**Figure 1(a)**). We generate Gaussian noises with zero mean: $\hat{d}_{ij}^2 = |d_{ij}^2 + \sigma_d \cdot N(0,1)|$. The tested noise levels are appropriate for modeling state-of-the-art real measurement techniques [2]. We generate several random geometry graphs with a fixed network size $n = 50$ and with $n_{anchors} = 3$ anchors. For each σ_d, we run 100 Monte-Carlo simulations. The initial guess of configuration is simulated by corrupting the true node positions with Gaussian noises with zero mean and standard deviation $\sigma_p = 5$.

We compare the performance of Distributed Gradient based Method (DGM) to the Semi-Definite Programming (SDP) relaxation-based method, which is a classical successive refinement approach used in range localization scenarios. **Figure 1(a)** shows the localization errors of algorithms DGM and SDP based on Equation (11). When there is no distance error ($\sigma_d = 0$) both algorithms converge to the ground truth. In presence of noises, the results for the DGM are better than the ones associated to the SDP method, for all noise levels. Thus, the distributed gradient-based algorithm is more robust to the range error than the SDP. It can be seen that, as the noise levels in the measurements (x-axis) increases, the error estimates increase as well, with an approximately linear trend. **Figure 1(b)** shows that the distributed version of our algorithm DGM behaves similarly to the centralized counterpart gradient based method (GM) in the noisy range network localization problem. The localization errors of the two gradient-based modes converge to the same value which is close to the Cramer-Rao Bound (CRB) error as the update iterations τ increase.

4.3. Location Error Behavior as Range Error

We investigate how the localization error is affected by the number of nodes in a network. We fix the number of anchors $n_{anchors} = 3$, and simulate scenarios with different number of nodes $n = 30, 50, 70, 100, 150$. We test three noise levels $\sigma_d = 0.1, 0.5, 1$. **Figure 2** shows that as the number of nodes increases, the location estimates become more accurate and have lower errors. As long as the network gets almost complete communications the improvement in the estimates accuracy is low. In all cases, the position estimation errors obtained with distributed gradient-based method are almost indistinguishable from the Cramer-Rao bound errors.

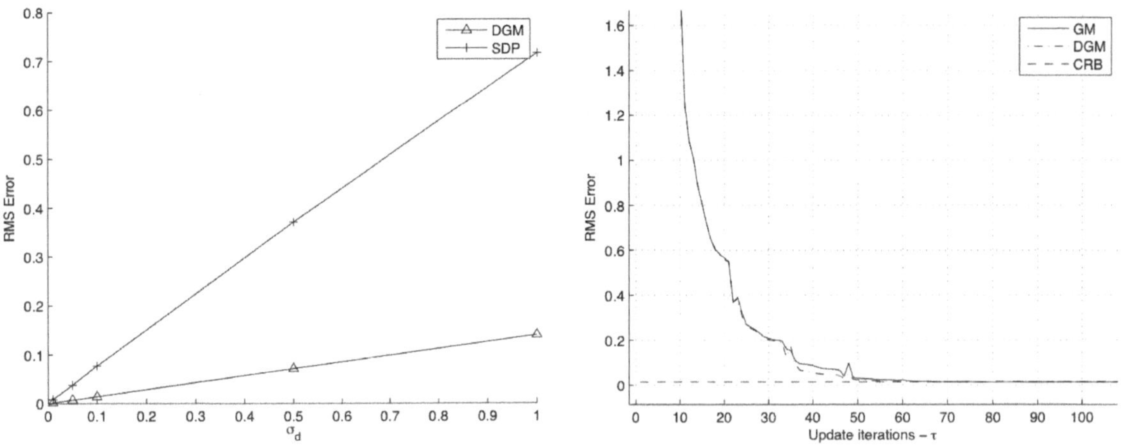

(a) (b)

Figure 1. (a) shows the location error obtained with Distributed Gradient-based Method (DGM) and the Semi-Definite Programming (SDP) relaxation-based method, for different levels of noise; (b) shows the location error of the distributed (DGM) and centralized (GM) versions, with the number of iterations, compared to the Cramer-Rao Bound (CRB), when the given distance error level is.

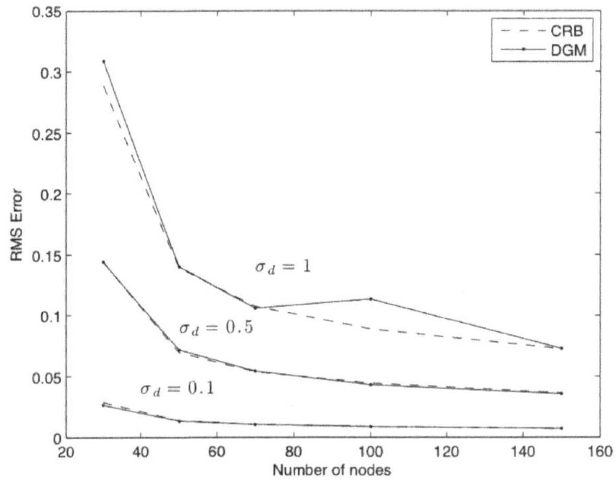

Figure 2. Location errors obtained with DGM algorithm as the number of nodes increases.

5. Conclusion

This paper focused on the analysis of distance noise effects for range-based network localization problems. A analytical tool, CRB errors, to derive the location error propagation have been discussed. A Distributed Gradient-based Method (DGM) is used to solve the noisy range localization problem. We have shown that our algorithm is robust to a wide range of noise levels in the measurements. This is not the case for several noise-free localization algorithms; particularly we have shown that Semi-Definite Programming method (SDP) is sensitive to noises.

References

[1] Patwari, N., Hero III, A.O., Perkins, M., Correal, N.S. and O'dea, R.J. (2003) Relative Location Estimation in Wireless Sensor Networks. *IEEE Transactions on Signal Processing*, **51**, 2137-2148. http://dx.doi.org/10.1109/TSP.2003.814469

[2] Savvides, A., Garber, W.L., Moses, R.L. and Srivastava, M.B. (2005) An Analysis of Error Inducing Parameters in Multihop Sensor Node Localization. *IEEE Transactions on Mobile Computing*, 567-577. http://dx.doi.org/10.1109/TMC.2005.78

[3] Anderson, B.D.O., Shames, I., Mao, G. and Fidan, B. (2010) Formal Theory of Noisy Sensor Network Localization. *SIAM Journal of Discrete Mathematic*, **24**, 684-698. http://dx.doi.org/10.1137/100792366

[4] Moore, D., Leonard, J., Rus, D. and Teller, S. (2004) Robust Distributed Network Localization with Noisy Range Measurements. *Proceedings of the 2nd International Conference on Embedded Networked Sensor Systems*, 50-61. http://dx.doi.org/10.1145/1031495.1031502

[5] Cao, M., Anderson, B. and Morse, A.S. (2006) Sensor Network Localization with Imprecise Distances. *Systems & Control Letters*, **55**, 887-893. http://dx.doi.org/10.1016/j.sysconle.2006.05.004

[6] Shames, I. and Bishop, A.N. (2011) Noisy Network Localization via Optimal Measurement Refinement Part 2: Distance-Only Network Localization. *18th World Congress of the International Federation of Automatic Control (IFAC)*.

[7] Biswas, P., Liang, T.C., Toh, K.C., Ye, Y. and Wang, T.C. (2006) Semidefinite Programming Approaches for Sensor Network Localization with Noisy Distance Measurements. *IEEE Transactions on Automation Science and Engineering*, **3**, 360-371. http://dx.doi.org/10.1109/TASE.2006.877401

[8] Calafiore, G.C., Carlone, L. and Wei, M. (2010) A Distributed Gradient Method for Localization of Formations Using Relative Range Measurements. *IEEE Multi-Conference on Systems and Control (MSC)*, 1146-1151. http://dx.doi.org/10.1109/CACSD.2010.5612764

[9] Eren, T., Goldenberg, D.K., Whiteley, W., Yang, Y.R., Morse, A.S., Anderson, B.D.O. and Belhumeur, P.N. (2004) Rigidity, Computation, and Randomization in Network Localization. *IEEE INFOCOM*, 2673-2684. http://dx.doi.org/10.1109/infcom.2004.1354686

[10] Shames, I., Fidan, B.I. and Anderson, B. (2009) Minimization of the Effect of Noisy Measurements on Localization of Multi-Agent Autonomous Formations. *Automatica*, **45**, 1058-1065. http://dx.doi.org/10.1016/j.automatica.2008.11.018

16

Enhancing the Performance of Current Online Education System—A Study of Cloud Computing and Virtualization

Basheer Riskhan, Raza Muhammed

School of Computer Science, Huazhong University of Science and Technology, Wuhan, China
Email: riskhan@yahoo.com

Abstract

Online Education (OE) system is an effective and efficient way to perform the education in all sectors of government and non-government educational organization. Low performance and minimum speed are major overhead in the current ongoing OE system due to the increase of users and some system issues. Base on the previous study and recent practical issues, a model is proposed to Enhancing the Performance of Online Education System (EPOES) to examine the bare metal virtualization, isolation and virtual machine templates. Bare metal virtualization has led the native execution, isolation isolated the running application and Virtual Machine Template has help to increase efficiency, avoiding the repetitive installation and operate the server in less time. The proposed model boosts the performance of the current OE system, and examines the benefits of the adaptation of cloud computing and virtualization which can be used to overcome the existing challenges and barriers of the current OE System.

Keywords

Bare Metal Virtualization, Cloud Computing, Isolation, Virtual Machine Template

1. Introduction

The demand for computer usage has increased significantly over the past few decades. As a result, Online Education (OE) conducted by web based technology has become a crucial part of education and it has many issues based on inefficient management, slow process, less performance and security vulnerabilities [1]. To overcome the above overhead, the proposed model introduced is called "Enhancing the Performance of Online Education

System (EPOES)". Bare metal virtualization, isolation and virtual machine templates are the cloud computing and virtualization technologies which are suggested to implement to that model. Further this model will allow the user feasibility for portable network device (PND).

In the past decades, the most flaws and overheads were improved; education providers have begun to show more and more interest in web based tools including Web CT and web 2.0 [2] for the use of OE system. But even with an upgrade software system, there still have some disabilities such as security vulnerabilities, virus attack, less bandwidth, fixed location, using only computers as the main source to access to OE, limited accessible space, limited funds, maintained by the administration, frequent software update, high cost of hardware and software. As a result IT administrators face performance and speed as a major concern as the system fails to provide optimal performance and speed with the increase in the number of OE users. Enhancing the performance of present OE system is our main objective of this paper and issues regarding performance can be overcome by proposing the model EPOES with using new technologies such as bare metal virtualization, isolation and virtual machine templates based on cloud computing.

Cloud computing technology is the most preferred method as it is one of the fastest growing technologies in IT. It is Internet (cloud) base development and use of computer technology (computing), whereby dynamically scalable and often virtualized resources are provided as a service over the Internet. Hiding and abstraction of complexity, virtualized hardware and efficient distribution of resources are some of the key benefits of cloud computing [3]. Cloud computing also reduces the cost of IT, provides resource as per demands, reduces the license constrain, increases the agility, maximizes the speed, shares and allocates hardware devices. Google, Amazon, Yahoo and Microsoft are some of the cloud computing service providers in the current market; they support the OE system by providing sufficient amount of memory and applications sharing [4]; they provide the services to access and develop the OE applications in the easy manner. Although cloud computing is still at the early stage of its development in the education sector, there are many more challenging issues waiting to be explored using this versatile system.

Implementation of new technologies based on virtualization and cloud computing will allow the user to access OE system anywhere with the aid of PND such as smart phones, PDAs, surface and iPad touch. It also has a positive impact on the high performance of OE. In the case of insufficient memory, CPU, application or storage, cloud has the technology to fulfil the overheads according to the demand. Whereas, virtualization has a lot of potential for virtualized software to reduce cost, increase fast access, unify management, improve IT responsiveness, increase IT utilization and deliver IT seamlessly and securely. The above benefits given the support upgrade the drawback of OE system.

2. Related Work

Newly proposed technology mainly focuses on virtualization and cloud computing technology. It has derived from grid computing, distributed computing, utility computing, parallel computing and other computer technologies [4]. Cloud computing shares memories and resources, as compare grid computing wants to solve the task of memories and resources. Like cloud computing and virtualization technology mostly relate with virtualization compare with Grid computing. In order to minimize the human factor in the examination process of OE, the application database technology will be a question gathered into the database before each test composed of questions from the exam papers extracted, greatly reducing the workload of teachers to effectively reflect a fair test sex [5]. This is the method using for conducting examination in manually and existing online architecture. The new model is introduced to import to the cloud computing technology with adding some new features such as resource sharing, centralized hardware and software, reduce the cost, access software's with flexibility, access anywhere, anytime and anyplace, without caring updates and security. But Security, sufficient Internet access, bandwidth and acceptance of decision makers in the institutions are some of the main factors that are considered as implementation of cloud computing in the academic institutions [6].

Most of cloud computing platforms are based on virtualized environments [3]. Virtual machine user is able to run a variety of operating systems and environments as needed by the applications. Virtualization allows user to isolate workloads, improving security and reliability [7]. According to the above statement isolation and improvement of the security from hackers are given the improvements in performance of the OE system. Due to that, the proposal to implement the isolation technology to the EPOES model by support of the virtualization. From the research, the new model EPOES is introduced as a solution to the previous overheads of the current OE system. Further this model enables to prevent the unauthorized access, improve the personal identity techniques and challenge the existing architecture of OE.

3. Analysis of the Current Performance of Online Education

OE improves the efficiency and accuracy of the overall education process. In recent years most of the higher institutions are showing interest to migrate from traditional pen and paper based education pattern to OE system. However, the present OE system architecture has lots of drawbacks such as performance, processing speed, security and dishonesty of students, such as coping answers of others, searching the internet for answer, storing the data's on the local computer and using for the future exam purposes, changing the answer, discussion of answers through the email and mobile phone [8].

Figure 1 shows the ongoing architecture system which follows most of the OE system such as Learning Management System, Moodle. This architecture processed fully based on web technology with some disabilities. The main drawback of this current model was low performance. By implementing new technologies suggested in this paper can improve the advance performance of OE system in the future.

4. Challenges of Current OE System

As technology has improved, the need for OE has also increased in recent years due to various reasons. The drawbacks in the current OE have caused slight hesitation among providers to choose it as an educational tool. The following will discuss the drawbacks of the OE system.

4.1. Band Width and Internet Speed

Bandwidth and Internet Speed are the backbone for any good communication between client and server which are both technically different. Bandwidth refers to the maximum amount of data that can be supported or transferred through the channel within a fixed amount of time through the network [9]. Therefore, less Bandwidth would cause the overhead to move the data from two geographically separated networks. Speed performance relies on a large bandwidth pipe and routing implemented in a network. In the present OE system, the clients connect with the LAN Server and request to move through that Server. Bandwidth also performs an important role when the data is moved from the LAN network. The problem arises when performing bandwidth and data intensive tasks such as moving large data sets, performing replications for high availability and backup purposes, or data mining or ETL activities [10]. If the availability of bandwidth is high, then the compatibility of data movement becomes faster. Generally all the computer labs operate in a smaller bandwidth range, which results in a less performance of OE. Internet speed is the time taken for the data to reach the destination from the source. The data transfer speed varies depending on the implemented technology. Therefore, the Bandwidth and Internet speed dominate the performance of the OE Process.

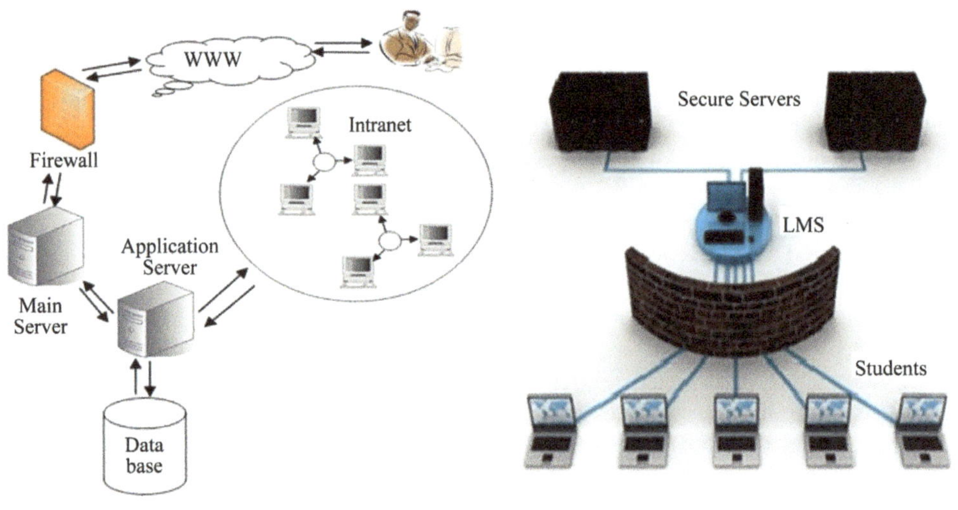

Figure 1. Technical architecture of current OE system.

4.2. Server Workload and Efficiency

Workload is the amount of processing that the computer has been set to perform at a given time. The workload consists of a number of application programs running in the computer, with a number of users connected to and interacting with the computer's applications [11]. As the number of online education user increase, the amount of processing and the amount of application programming running at the present server also increases, causing an increase in the server workload. The protocol used to send the data, network conjunction and server efficiency are factors, dependent on the network speed [9]. For example, in the scenario of attempting to transfer data from one location to another, the data transfer is considered advanced when done through a faster server than to a slower server. Given an optimal internet speed, a slow server would slow the data transfer in comparison to the fast server where the rate of transfer is high. Therefore, the performance of the OE system is dependent upon the server workload and server efficiency.

4.3. One Terminal Access

One terminal access describes the flow of signals to one single port. When the LAN users login to start the OE process, all of the user requests are directed straight to one LAN server. This process leads to a queue in the server while the requests made by the users begin to process through using FIFO or LIFO algorithm. Same scenario is applied throughout the other servers involving OE operation. Process time of these servers depend on the amount of workload submitted by the number of users through individual terminals. Once all of the tasks are received from LAN server, the data gets moved to a single terminal. This again creates another set of queue causing the process time of OE slower and the performance going to less.

4.4. More Servers inside the OE Architecture

Figure 1 clearly explore, that the Current OE systems were implemented through different servers like web server, database server, admin server, ISP Server and LAN server to uplift the standard of OE service between users and destination. Because all of the servers are depended on each other, failure of one server will cause the failure of OE system. Furthermore, the servers engaged in doing their own separate tasks require a larger bandwidth range to let the data travel through all other servers, as a result it performance down the OE process.

4.5. Traditional Security System and IP Security

Each server has its own security system for receiving and sending the request. Current OE system uses the traditional security systems such as data encryption, firewalls and data signature. High security and power hungry software's can slow down the performance of education system. Network Sniffers, IP Spoofing, Connection Hijacking, Data Spoofing are the main problems in the IP Security and an attack on a network is created in order to gain unauthorized access.

4.6. Fixed Device Access and Physical Disaster

Fix device access is another drawback of this system. At present remote access is a rising concept in IT. All the devices can be connected with the help of remote access, providing great convenience to the user. Introduction of portable network devices such as smart phones, PDAs, Surface and iPad in the current market has piqued the interest and usage of public. Although these various devices those have been introduce into the market, the operation of OE is still conducted only via the use of computers. This is also hidden technique to slow performance process of an OE. System failure of OE can also be caused by natural disasters such as fire and flood. Damages caused by these events are equally significant as the damages caused by software failure and human Errors due to above reasons the speed of process of online education might be going slow. If any early sign regarding above failures it's helpful to minimize the impact of a possible failures [12].

4.7. Malware Attack

Malware is software that infects and damages a computer system without the owner's knowledge or permission. Malware includes computer viruses, computer worms, Trojan horses, spyware, and crime ware [13]. A virus infection or security breach on the server can lead to corrupting of data, shutting down of documents, Restart and

Overload the server. Servers have run out the memory and stop the operating services in the servers, reduction in server efficiency and server breakdown. Thus, large amount of funding is required to maintain and run the servers securely from the virus attack. Furthermore virus intent to processing more unwanted task inside the server, reduce the server efficiency. One of the major concerns in the OE operation is the malicious programming attacks on the networks by hackers. Once the worms are left in a server, they can replicate and spread themselves in a short period of time. This causes the user to be automatically affected by the entire wired or wireless device excluding the proper production device which is connected to the LAN or network. Slow internet, slow network connection and slow data transfer are signs of worm infection [13]. Due to that above causes, the OE system shows less performance.

5. Methodology of the Model of Enhancing the Performance of Online Education System (EPOES)

In this section, the EPOES model is introduced to minimize the above obstacles. This model contains physical server, physical resources and virtual machine which will be briefly discussed in sub sections, cloud computing and virtualization technologies were implemented to that model. The users access to the EPOES model remotely by using any PND.

5.1. Implement Resources

In this section recommendations to implement some resources to the EPOES from the previous study and observation like Physical Server, Physical Resources and Virtual Machine based on cloud computing and virtualization, as they both have an inter-relationship. Virtualization is the key technique for Cloud Computing. Virtualization is a software that manipulates the hardware and cloud computing is the service that results from the manipulation. Virtualization is a foundational element that helps cloud computing deliver on its value. Cloud computing is currently on demand as it provides share of computing resources, software and data as a service [14]. Through virtualization and other resource sharing mechanisms, cloud computing can dramatically reduce the examination cost, increase fast access and create virtual machines. When the request is received, the physical server has the capability to create and execute the Virtual machine automatically and import the physical resources according to the user demand, based on Cloud computing.

5.1.1. Physical Server (The Host)

This physical server forms the public cloud environment, provides the virtual machine and controls its physical resources [15]. The public cloud environment provides the resources as Software and hardware according to the demand of the user, and executes the OE application interface.

5.1.2. Physical Resources (Hardware)

Figure 2 shows, the physical resources consist of CPU, memory, network, OS and etc. These resources will be allocated to the virtual machine.

5.1.3. Virtual Machine

The virtual machine is the virtual computing platform on a Physical server, by virtualized software and it is directly connected to the physical hardware shown on **Figure 2**.

5.2. Implement New Techniques

Consolidation, Isolation and Migration are the main factors in the system level virtualization and it will be used for the number of reasons [16]. The main objective of this paper is to demonstrate, how the process of performance of the OE System can be improved by the use of cloud base virtualization system. So Bare metal virtualization, Isolation and Virtual machine templates are the techniques of cloud base system level virtualization and they are the most appropriate techniques to increase the performance of present OE system.

5.2.1. Isolation

Isolation is the property of virtualization. It is an approach to improve the performance of the process and

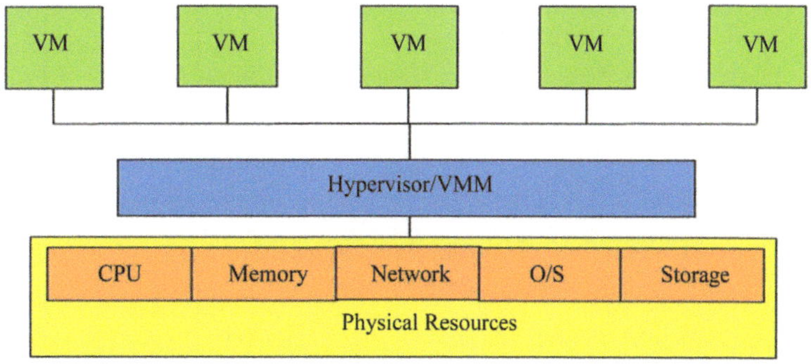

Figure 2. Virtual computing platform environment.

reliability of the system by isolating the execution environment for resources, application and OE related database [16] shown on **Figure 3**. This isolation enables multiple virtual environments to run securely while sharing the resources and applications [17]. Generally the virtual machines are the container that consists of resources, application and OE related database and they are designed to be isolated from one another.

If any execution failure occurs in any virtual machine, the other virtual machine on the same host will continue to run due to those stated reasons of isolation. However, resources or application failure has no effect on the performance of the virtual machine, the ability to access the other virtual machine or the operation to access the required resources [18]. Each virtual machine is isolated through the virtualization technology while running on the same hardware. Virtual machines share physical resources such as CPU, memory, I/O device and application, in order to utilize virtual machine isolation to concurrently execute several virtual machines in a single server. To protect the virtual machine from attracting virus or worm, safeguards such as antivirus software, firewalls can be applied to the physical machine [18]. Virtual machines requiring hypervisor/VMM and batch system can effectively map this view of the system [7].

5.2.2. Bare Metal Virtualization

Hypervisor is a program that is included in the host operation system. It can be activated through several virtual machines using a single hardware. Hypervisor is also known as a Virtual Machine monitor. This controls the host resources and allocates the required resources to each virtual machine. **Figure 4**, explores the categories of hypervisors in two types as bare metal virtualization and hosted virtualization [16]. In bare metal virtualization, VMM and VM are direct mapping with the hardware resources and the process is comparable to the native execution. Hypervisor is used only for bypass [19] as the guest OS running within the virtual machine, manages and maps the resources in order to enable the execution of an application [7].

Hosted virtualization Hypervisor only appears between a host OS and VM. The Hypervisor mapping shares resources and application according to the requirement of VM. In a hosted virtualization between host OS and VM, VMM is the bottle-neck of this proposed platform. This approach is the most suitable for the development purpose [18]. Direct connection with Physical Resources help to sustain the time conception, processing speed and high performance that are the key factors of OE system.Thus, the recommended technology for the OE is system is bare metal virtualization.

5.2.3. Virtual Machine Templates

Virtual machine template functionality is developed in VM ware Infrastructure 3. The key benefits of this is that it avoids repetitive installation, uses less processing time and is efficiency [20].

Assume 2 students are to sit for an exam. In the Manual Installation of Virtual Machines, it creates two virtual machines as in **Figure 5** and these process takes minimum of 5 steps for a single user. For 2 users this will be minimum of 10 steps, as each task is repeated by every user, it is time consuming. Whereas, Template Based Deployment of Virtual Machines creates only one virtual machine and it can reduce 10 steps to 7 steps for the entire process as in **Figure 5**. When the users increase, large differences occur within the process steps between Manual Installation of Virtual Machines and Template Based Deployment of Virtual Machines. Therefore, less time and faster performance occur through avoiding the repeated installation.

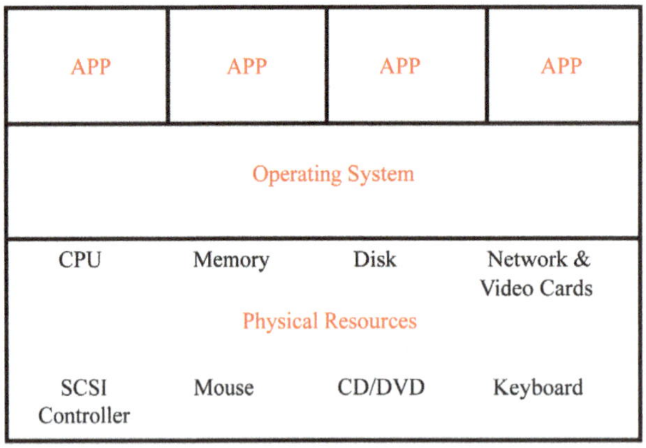

Figure 3. Virtual machine isolation.

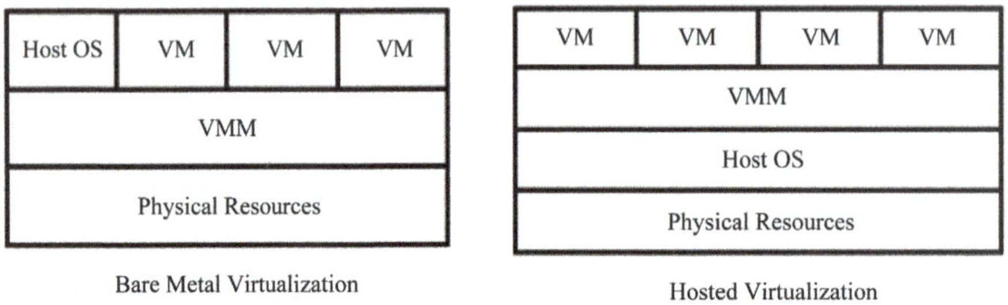

Figure 4. Types of virtualization.

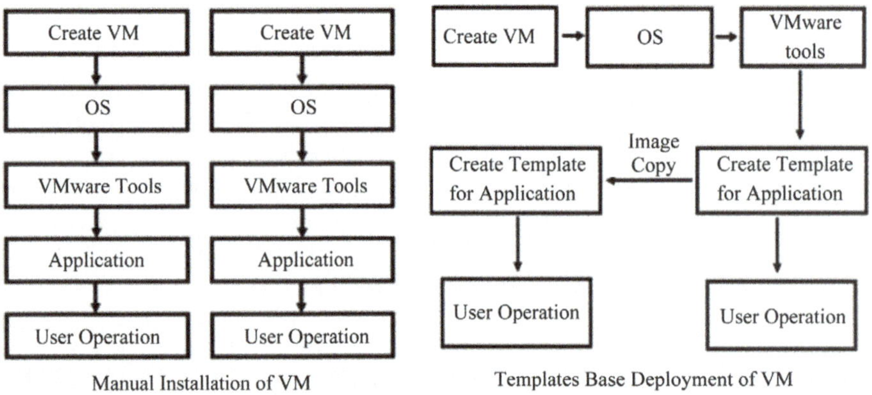

Figure 5. Manual and templates base VM.

6. Impact of Cloud Computing and Virtualization for the Current Performance of Online Education

Main key challenges with the current OE system are bandwidth, server workload, one terminal access, malware attack, more servers, IP security and fixed devices. Bare metal virtualization, Isolation and Virtual Machine Template are the technologies that have been proposed to implement the current system in order to avoid slow performance. The following section will discuss how the proposed technologies can be used to resolve the stated challenges.

For a data to be successfully transferred from one location to another (client to ISP, ISP to ISP and ISP to Server), more bandwidth and good network speed are crucial. Cloud providers usually have high bandwidth and high network speed due to good technologies being used for communication media and for other networks [21]. Recent "Computerworld Report" said that the cloud support to the lack bandwidth. Server workload is reduced through isolation, where the share resources and application in a single VM for a single user is separated. Consider most students accessing the OE system at the same time by directly connecting with the server. In this scenario, a lot of bottle-neck is created to both the server and student. This leads to slow down the performance, increase the chances of hacking questions, time limitation and there is a risk of striking the server. In order to overcome the above drawbacks, the present OE System uses the Isolation technique based on cloud based virtual environment which is one of the good solution.

One terminal access and fixed device issues can be solved by using cloud base virtualization technology. In the current technology world the portable network devices can provide convenient ways to access the network at anywhere and at anytime. Technology for these portable network devices has been extended continuously by 2G, 3G, 4G and etc. So by letting portable network devices access OE system, the issues with the fixed devices can be minimized. Also by using the portable devices, the users avoid the need to be at a fixed location such as computer labs and etc, help to reduce the one terminal access. Consider 10,000 users accessing the OE at the same time. During this occasion, the server's workload will increase along with power consumption, building space and amount of available resources will increase. Thus by using virtualization technique, it enables the users to access the OE from any location and at any time, while keeping the number of computers, building space and maintenance cost down. If any disabilities were to occur in the portable network devices due to power consumption, memory and hardware resources then the cloud computing will provide the necessary assistance with the needed on demand.

Bare metal virtualization is one of the techniques used by virtualization that has a direct connection between the Virtual Machine and Physical resource. This will assist in getting the native execution in the virtual machine resources. Therefore the number of users is not a concern, according to the number of users and the server machines create the virtual machines and share the device with the native execution. In addition, cloud computing solves the IP Security issue as the end user has no knowledge of the underlying technology [22]. The end user is one who uses the product after it has been fully developed and marketed which includes Hackers. As no knowledge of the designation point (IP) in the cloud computing in available, it causes the hackers to face major difficulties in creating interruption to the IP Security.

Virtual machine templates provide the opportunity to get high benefits with less VMs via the use of image templates. Time conception is another key concept in the OE system, as in a usual system 5 VM is required for 5 OE users to access the OE applications. However, by using Virtual machine templates technique, 5 users can access the OE in a single VM through image templates. The process is very fast, repetitions are avoided, reduced issues of Malware Attack and fewer servers are use among the benefits. If the above technique is implemented to the proposed model EPOES using cloud based system level virtualization, the barriers mentioned can be overcome.

7. Conclusion

A number of difficulties and constrains such as low performance and less speed have been raised in the current OE system due to some barriers and increased the number of users. The paper focused on low performance of OE system and this issue could be resolved by the proposed model EPOES with the implementing some new technology with the support of system level virtualization and cloud computing technology. Bare metal virtualization, isolation and virtual machine templates are the important techniques required to implement the proposed model. It provides the opportunity to enhancing the performance of OE system by sharing physical resources with the native execution, reducing the IP security issues, mitigating virus attacks and letting portable network devices to access the OE system in anytime, phase.

References

[1] Sarrayrih, M.A. and Ilyas, M. (2013) Challenges of Online Exam, Performances and Problems for Online University Exam. *IJCSI International Journal of Computer Science Issues*, **10**.

[2] Rogers, C.F. (2006) Faculty Perceptions about e-Cheating during Online Testing. *Journal of Computing Sciences in Colleges*, **22**, 206-212.

[3] Tian, W.H., Sun, X.S., Jiang, Y.Q. and Wang, H.Y. (2013) CRESS: A Platform of Infracture Resource Sharing for Educational Cloud Computing. *Communication*, *China*, **10**, 43-52. http://dx.doi.org/10.1109/CC.2013.6623502

[4] Leloglu, E., Ayav, T. and Aslan, B.G. (2013) A Review of Cloud Deployment Models for E-Learning Systems. Dependable Systems and Networks (DSN). *3rd Annual IEEE/IFIP International Conference.*

[5] Hang, B. (2011) The Design and Implementation of On-Line Examination System. *ISCCS* 11 *Proceedings of the* 2011 *International Symposium on Computer Science and Society*, Kota Kinabalu, 16-17 July 2011, 227-230. http://dx.doi.org/10.1109/ISCCS.2011.68

[6] Mokhtar, S.A., Ali, S.H.S., Al-Sharafi, A. and Aborujilah, A. (2013) Cloud Computing in Academic Institutions. *ICUIMC* 13 *Proceedings of the 7th International Conference on Ubiquitous Information Management and Communication*, **2**. http://dx.doi.org/10.1145/2448556.2448558

[7] Vallee, G., Naughton, T. and Ong, H. (2008) System-Level Virtualization for High Performance Computing. *16th Euromicro Conference on Parallel, Distributed and Network-Based Processing*, Toulouse, 13-15 February 2008, 636-643. http://dx.doi.org/10.1109/PDP.2008.85

[8] Naresh, C., Deepthi, C.H. and Shekhar, T.P. (2011) A Novel Approach to Enhance Security for Online Exams. *International Journal of Computer Science and Technology*, **2**, 85-90.

[9] Ferro, G. (2015) Home. The Difference between Bandwidth and Speed. http://etherealmind.com/basics-difference-bandwidth-speed/

[10] Treadway, J. (2010) Home. Optimizing the Cloud: Solving the Bandwidth Issue. http://www.cloudtp.com/2010/08/19/optimizing-the-cloud-solving-the-bandwidth-issue/

[11] Center, Search Data (2006) Workload Definition. [Online] http://searchdatacenter.techtarget.com/definition/workload

[12] Microsoft (2015) Library, Causes of Failure. Microsoft System Center. [Online] http://technet.microsoft.com/en-us/library/cc180043.aspx

[13] Logan, D. (2013) Tech. What Is Malware? [Online] http://www.ehow.com/facts_4857164_what-is-malware.html

[14] Angeles, S. (2014) Technology. Virtualization vs. Cloud Computing: What's the Difference? [Online] http://www.ebates.com/index.htm?src=marketing&acct=tracsion&camp=paid&utm_source=tracsion&utm_medium=banner&utm_campaign=paid&site=104

[15] Navarro, T. (2013) Home. What's the Difference between Virtualization and Cloud Computing. [Online] https://www.computenext.com/blog/the-difference-between-cloud-computing-and-virtualization/

[16] Scott, S.L., Vallee, G., Naughton, T., Tikotekar, A., Engelmann, C. and Ong, H. (2008) System-Level Virtualization Research at Oak Ridge National Laboratory. Oak Ridge National Laboratory, Oak Ridge.

[17] Sun, D.W., Chang, G.R., Guo, Q., Wang, C. and Wang, X.W. (2010) A Dependability Model to Enhance Security of Cloud Environment Using System-Level Virtualization Techniques. Pervasive Computing Signal Processing and Applications (PCSPA), 305-310.

[18] VMware (2015) VMware vSphere Introduction. VMware vSphere 4-ESX and vCenter Server. [Online] https://pubs.vmware.com/vsphere-4-esx-vcenter/index.jsp

[19] Liu, J.X., Huang, W., Abali, B. and Panda, D.K. (2006) High Performance VMM-Bypass I/O in Virtual Machines. Computer Based Learning Unit, University of Leeds, 1-7.

[20] Dodge, J. (2006) Introduction. Virtual Center 2. [Online] http://www.vmware.com/pdf/vc_2_templates_usage_best_practices_wp.pdf

[21] Momjian, B. (2015) Home. Increasing Website Bandwidth Using Cloud Service. [Online] https://momjian.us/main/writings/pgsql/cloud_bandwidth.pdf

[22] Hurwitz, J., Bloor, R., Kaufman, M. and Halper, F. (2012) Cloud Computing for Dummies. John Wiley & Sons, Inc., New York.

17

Design and Implement of Low Power Consumption SRAM Based on Single Port Sense Amplifier in 65 nm

Shunrui Li, Jianjun Chen, Zuocheng Xing, Jinjin Shao, Xi Peng

Collage of Computer Science of National University of Defense Technology, Changsha, China
Email: lisr824467100@163.com, cjj192000@163.com

Abstract

With the rapid development of integrated circuits [1], low power consumption has become a constant pursuiting goal of the designer in chip design. As the memory almost takes up the area of the chip, reducing memory power consumption will significantly reduce the overall power consumption of the chip; according to ISSCC's 2014 report about technology trends discussions, there two points of the super-low power SRAM design: 1) design a more effective static and dynamic power control circuit for each key module of SRAM; 2) ensure that in the case of the very low VDD min, SRAM can operating reliably and stably. This paper makes full use reliable of 8T cell, and the single-port sense amplifier has solved problems in the traditional 8T cell structure, making the new structure of the memory at a greater depth still maintain good performance and lower power consumption. Compared with the designed SRAM the SRAM generated by commercial compiler, as the performance loss at SS corner does not exceed 10%, the whole power consumption could be reduced by 54.2%, which can achieve a very good effect of low-power design.

Keywords

Single Port Sense Amplifier, SRAM Design, Low Power Design, 8T SRAM

1. Introduction

For dynamic power consumption, the most effective way to reduce power is the use of very low power supply voltage, for the relationship between the power consumption and the supply voltage is: the power consumption is proportional to the square of the supply voltage; for static power consumption, the low-voltage design techniques also is the most direct and effective way, because leakage power will be low enough when SRAM power supply voltage is low enough.

In this thesis, based on a 65 nm process single-port sense amplifier, designed an 8T_SRAM circuit, and compared this SRAM with the generated SRAM by memory compiler in performance, on the premise of speed quite, the former has much more advantage in power and stability.

2. Entire Circuit Structure

Full custom SRAM 512 × 32 memory structure is composed of 8T_SRAM. The circuit structure is shown in **Figure 1**. The whole circuit is divided to five parts, such as clock, latch, decoder, storage arrays and IO. The use of single port IO sense amplifier circuit in this paper, making this new SRAM acquire lower power consumption, higher performance and higher reliability compared with the conventional circuit configuration.

The key of SRAM circuit design is the reading circuit, because the performance of the read circuit determines the reading speed of the SRAM, and the read circuit is the critical path in the SRAM, so reducing the delay on the critical path can effectively improve the timing performance of memory.

2.1. Bit Cell

Due to the characteristics of 6-T SRAM [2] memory's structure, the data output are all using double-port sense amplifier to detect bit line voltage difference, and amplified voltage difference to output. But with the process size reduction and lower power supply voltage, the stability of the 6-T cell is getting worse, more and more weak of noise immunity, noise margin getting smaller and smaller, so it is difficult for sense amplifier circuit to control the time of turning on and off, and the worst is there often rise a noise voltage turning up on the bit line, and amplified by sense amplifier to output which results in erroneous operation. The design of the SRAM structure proposed in this paper is using 8-T cell, because of its better stability than6-T, 8T is a big trend of future development. Although compared with 6-T cell, 8-T cell has another two transfer transistors, the transistor size is not limited to read without breaking the rules and can follow the minimum rules to design, and 8T cell area below 65 nm process can be done fairly compared with 6T cell area, so the 8T_cell area does not increase. 8T_cell structure is as follows **Figure 2**.

2.2. Sense Amplifier IO Block

Traditional 8T memory cell read operation use single-port sense amplifier, and the conventional domino logic single-port sense amplifier circuit just uses this sense amplifier. Domino circuit structure is shown as **Figure 3**.

Domino sense amplifier logic circuit is the most typical one of those single-port sense amplifier circuits, and is also the most common single- port sense amplifier logic. Though this single-port sense amplifier logic in the small SRAM has a good performance, with the increase of cell number on the bit line RBL, the performance and power consumption will be a sharp deterioration. That is because as the mount too many cells, the RBL load

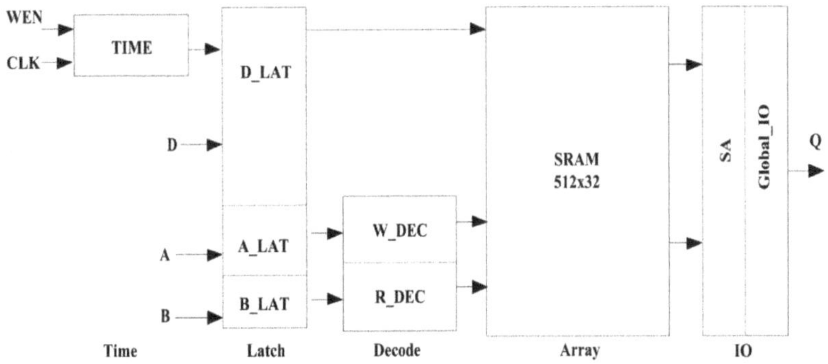

Figure 1. SRAM circuit configuration.

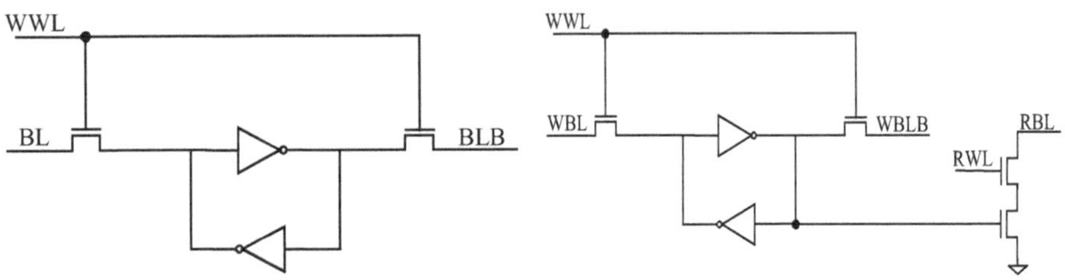

Figure 2. 6-T cell and 8-T cell circuit.

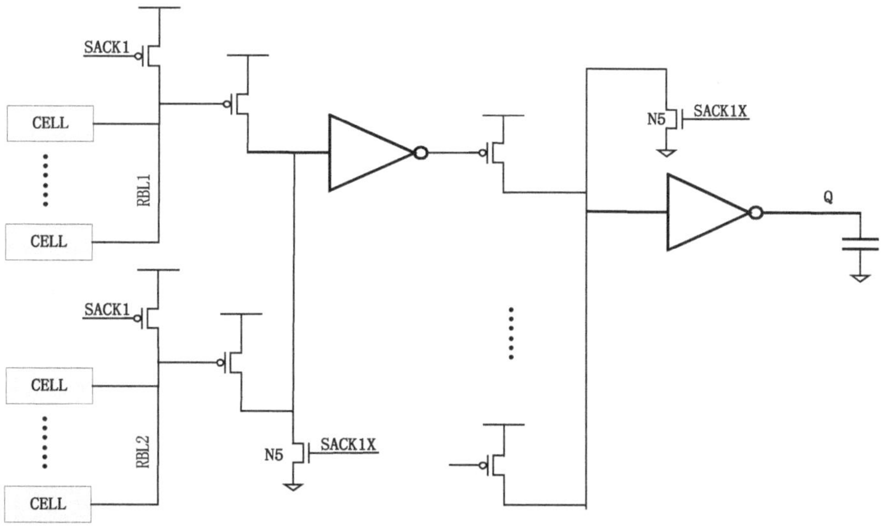

Figure 3. Domino logic circuit diagram.

capacitance is too big, the pre-charge time is longer, and the discharge time is also longer. Additionally, the single-port sense amplifier logic bit line voltage is full swing between 0-VDD, each discharge will have a great dynamic power. Therefore, power consumption also seriously deteriorated.

Based on those drawbacks of the single-port sense amplifier, TBP (Trip point Bit-line) solves this kind of problems better, make the sense amplifier remains high performance and can keep very low power consumption in condition of the high capacity and a long bit line load. **Figure 4** is TBP-sensitive amplifier circuit [3] structure.

Firstly introduce about the structure of the sensitive amplifier circuit. This sensitive amplifier circuit is a single port sense amplifier, which in the sense amplifier module is made up of a static complementary inverter INV, a balance PMOS transfer P3, a transmission gate T, the NMOS transistor N3 and a precharge transistor N1. Among those components, the inverter, in which the pull-down network of it is controlled by the enable signal REN. This can effectively reduce the static leakage current of the sensitive amplifier. So the read enable signal of the sensitive amplifier is effective. At the rest of the time, REN = 0 the sense amplifier doesn't work, N3 and N2 can form the DIBL effect which can reduce the static leakage current.

The TBP sensing scheme is to directly precharge RBL to the trip point of the inverter INV. During the precharge phase, the SACK1 turns on the precharge NMOS transistor N1and N5, for charging RBL and Z respectively. At the same time, transmission gate T is turned on, and the INV input, LBL, is connected to the output Y. This precharges LBL near the trip-point of the sense amplifier by the precharge NMOS transistor N1 and the diode-connected PMOS transistor P1, and thus, the offset of the sense amplifier is compensated. The working process of the sensitive amplifier is also divided into two stages, the pre-charge stage (SACK1 = 1) and the evaluation phase (SACK1 = 0).

When reading 0, supposed that the value 0 will appeared on the RBL1, then the bit line RBL1 will discharging, due to the bit line voltage value is at the trip-point of the inverter, so the voltage gain is very big at this point. When the bit line voltage drops, then P3 MOSFET opens, the output voltage of Y point is quickly pulled high, Z point is pulled high, then output GRBL = 0, and P2 is turned on at the same time, making the voltage of Z pulled up faster to the steady high level state. P2 formed a positive feedback circuit [4], which can make the sense amplifier speed up. When reading 1, it will be auto-zero [5], the voltage of bit line will not drop, the voltage of the X point will not up or down, so the voltage of the Y point will not drop, P3 will not open, the voltage of Z point will remain low, outputs GRBL = 1, the feedback circuit is closed. This is calling regenerative sensing logical [6]. The operational waveforms of the TBP sensing scheme are shown in **Figure 5**.

3. Results Contrast

The data on the **Table 1** and **Table 2** shows the sensitive single-port amplifier circuit performance and memory compiler SRAM and tradition in the case of a considerable delay, power consumption was greatly reduced,

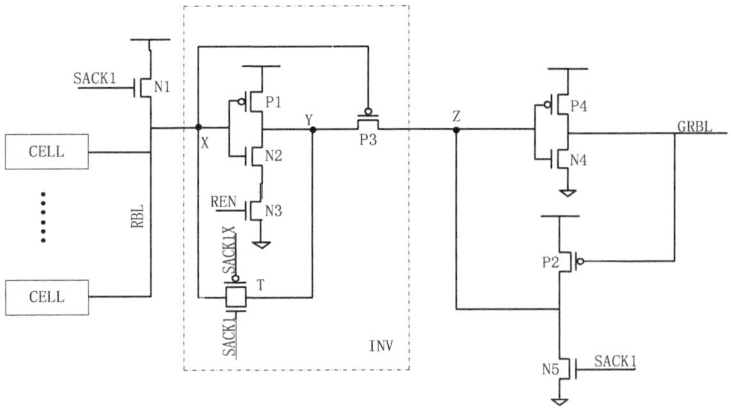

Figure 4. TBP-sensitive amplifier circuit structure.

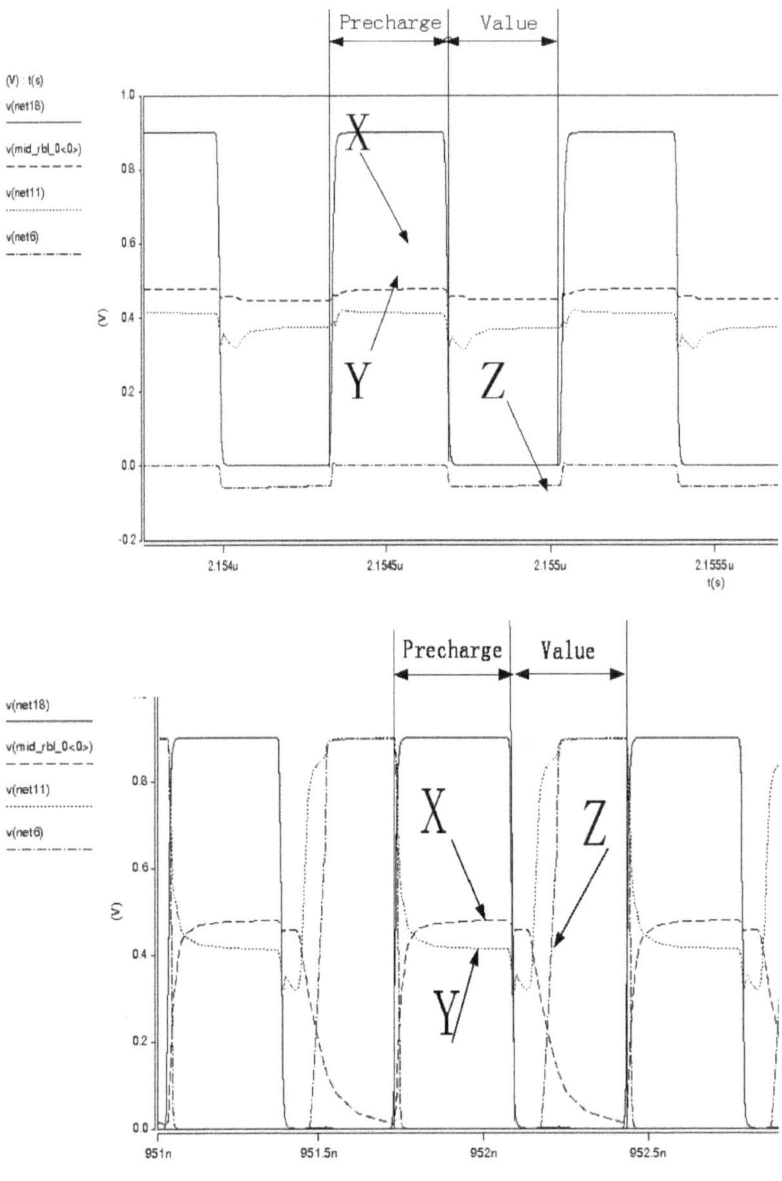

Figure 5. The waveform of sensitive amplifier read 0 and read 1.

Table 1. Dynamic power.

Power consumption	Compiler SRAM (W)	TBP-Sense-amplifier SRAM (W)	Percentage
SS corner	8.1e−3	3.85e−3	−52.4%
TT corner	1.3e−2	7.03e−3	−45.9%
FF corner	1.55e−2	1.16e−2	−25.2%

Table 2. CLK-Q_delay.

CLK-Q	Compiler SRAM (ns)	TBP-Sense-amplifier SRAM (ns)	Percentage
SS corner	0.694	0.768	+9.01%
TT corner	0.477	0.505	+5.82%
FF corner	0.291	0.316	+8.6%

which reduces power consumption at FF corner (25.2%), the TT corner under power reduction (45.9%), at SS corner reduce power consumption (52.4%). Because sensitive single-port amplifier circuit is reduced by 50% of the bit line voltage swing, additional use of decoding module, reducing the length of the bit line, and can be decoded by the address block, reducing unnecessary decoding circuit higher over so decoding module efficiency, dramatically reducing the overall power consumption of SRAM.

4. Summary

This paper finished a 512 × 32 SRAM circuit design; the aim was highly reliable, high speed, and low power consumption design, and separately with the commercial compiler SRAM performance comparison by circuit design and verification. At an equal speed, power consumption was lower than memory compiler SRAM. It shows that this design has achieved satisfactory results; the full-custom design of SRAM is very successful. Some good design methods were also proposed; it was worth to be used in other reference full-custom design.

Acknowledgements

This work was supported by the National Science Foundation of China (Grant No. 61504169) and the Preliminary Research Program of National University of Defense Technology of China (Grant No. 0100066314001).

References

[1] Rabaey, J.M. and Chandrakasan, A. (2004) Digital Integrated Circuits: A Design Perspective. 2nd Edition.

[2] Nalam, S., Chandra, V., Pietrzyk, C., Aitken, R.C. and Calhoun, B.H. (2010) Asymmetric 6T SRAM with Two-Phase Write and Split Bitline Differential Sensing for Low Voltage Operation. *Proc. 11th Int. Symp. Qual. Electron. Des. (ISQED)*, March 2010, 139-146. http://dx.doi.org/10.1109/isqed.2010.5450400

[3] Jeong, H., Kim, T., Song, T., Kim, G. and Jung, S.-O. Trip-Point Bit-Line Precharge Sensing Scheme for Single-Port SRAM. *IEEE Transactions on VLSI Systems*.

[4] Javanifard, J. (2008) A 45 nm Self-Aligned-Contact Process 1 Gb NOR Flash with 5 MB/s Program Speed. *IEEE Int. Solid-State Circuits Conf. Dig. Tech. Papers*, February 2008, 424-624.

[5] Giridhar, B., Pinckney, N., Sylvester, D. and Blaauw, D. (2014) 13.7 A Reconfigurable Sense Amplifier with Auto-Zero Calibration and Pre-Amplification in 28 nm CMOS. *IEEE Int. Solid-State Circuits Conf. Dig. Tech. Papers*, February 2014, 242-243.

[6] Verma, N. and Chandrakasan, A.P. (2008) A High-Density 45 nm SRAM Using Small-Signal Non-Strobed Regenerative Sensing. *IEEE Int. Solid-State Circuits Conf. Dig. Tech. Papers*, February 2008, 380-621.

18

Recovery of Image through Alamouti Channel with Incorporation of RSA Algorithm

Aninda Majumder, Mohammad Raihan Ruhin, Tahsina Hashem, Md. Imdadul Islam

Department of Computer Science and Engineering, Jahangirnagar University, Dhaka, Bangladesh
Email: ani_1860@yahoo.com, raihanruhin@gmail.com, tahsina.hashem@juniv.edu, imdad@juniv.edu

Abstract

In many applications, it is necessary to transmit images at a remote station, where wired Internet service is not available. In this case, wireless local loop (WLL) can help in making wireless link between one end node of the internet and remote service center. In such link, the communication is heavily affected by large and small scale fading; hence the received signal experiences huge distortion in case of forward error correction. Otherwise, huge service delay arises due to frequent negative acknowledgements. To combat the situation, we can choose Alamouti channel of full rate and fully orthogonal space-time block code (OSTBC). Our aim is to transmit images through Alamouti channel and to observe the quality of the recovered image, in context of bit error rate (BER). We have also observed the impact of fading and additive white Gaussian noise (AWGN) on the image without application of error correction or detection technique of channel coding. To ensure security, we apply the RSA algorithm on each pixel prior transmitting and decrypt them at the receiving end, where we found no impairment from the algorithm. Finally, we observe that the relative performance of the system changes digital modulation schemes.

Keywords

Quotient and Remainder Matrix, Alamouti Simulator, Multiple-Input Single-Output (MISO), Image Encryption and Decryption, BER and Discrete Wavelet Transform (DWT)

1. Introduction

Two most important limitations of wireless communication are: the range or length of link and data rate (related to channel capacity). To overcome these limitations, researchers have given their effort on space diversity and multicarrier parallel transmission called orthogonal frequency division multiplexing (OFDMA) of [1] [2]. A number of algorithms are proposed on transmitting side in [2]-[7] for attaining space diversity hence getting high signal-to-noise ratio (SNR) at receiving end. A communication model, where space-time block code (STBC) at the physical layer, with channel coding (linear block code) at the data link layer is shown in [8]-[10] and performances were measured in context of BER. A diversity coding can be used in multiple-input multiple-output (MIMO) to provide high data rate without requiring additional bandwidth and transmit power. It is noted that Alamouti space-time block coding has showed good performance than other techniques comparatively. This motivates us to use Alamouti channel for transmitting different types of images like "medical images", "secured watermark", "biometric identification" at a remote station.

Several digital modulation schemes *i.e.* binary phase-shift keying (BPSK), quadrature phase-shift keying (QPSK), quadrature amplitude modulation (QAM) are used to measure the performance of the system in [11] [12]. It is observed that a well-known cryptographic algorithm RSA [13] is used by researchers [7] for ensuring security of the transmitted message. Research [14]-[16] on MIMO with maximal ratio combining (MRC), selection combining (SC) and equal gain combining (EGC) diversity reception over several correlated and uncorrelated fading channels is also going on.

In this research, Alamouti space-time block coding is applied on transmission channel with incorporation of RSA, for successful recovery of an image. We summarize our key contributions as follows:

- We proposed an algorithm to transmit image through Alamouti channel because of its full spatial diversity gain and observe the quality of recovered image by measuring the bit error rate.
- Furthermore, to ensure security we apply the RSA algorithm on each pixel of the image.
- We perform a number of experiments over Rayleigh fading channels by varying SNR and by applying 16-QAM modulation schemes in order to measure the performance of the system.

The remainder of this paper is organized as follows. Section 2 presents our proposed algorithm of recovering image through Alamouti scheme with incorporation of RSA. In Section 3, we present our experimental results in details with respect to a variety of parameters. Finally, Section 4 concludes the paper with future research directions.

2. System Model

The space-time block code (STBC) is used to achieve orthogonality among transmitted symbols of individual antenna of the MIMO system [17] [18]. Alamouti scheme is a full rate OSTBC where two transmits and one receive antenna are used provided the receiver has the knowledge of channel gain. When more than two antennas are used, then orthogonality is possible but transmission rate become 1/2 or 3/4, for example, sporadic code.

2.1. Full Rate STBC

Full rate STBC is also possible for the case of transmitted antennas, $n_T = 2^k,\ k = 2,3,4,\cdots$; but full orthogonality is not possible and the scheme is called QSTBC explained in [19]-[21]. We can apply zero forcing (ZF) on QSTBC to convert its full orthogonal [21] at the expense of noise. In OSTBC the multiplication channel matrix and its conjugate transpose produces a diagonal matrix *i.e.* the off-diagonal elements are zero. In case of quasi-orthogonal, the result produces a few non-zero off-diagonal elements like below:

$$\begin{bmatrix} h_1 & h_2 & h_3 & h_4 \\ h_2^* & -h_1^* & h_4^* & -h_3^* \\ h_3 & h_4 & h_1 & h_2 \\ h_4^* & -h_3^* & h_2^* & -h_1^* \end{bmatrix} \begin{bmatrix} h_1 & h_2 & h_3 & h_4 \\ h_2^* & -h_1^* & h_4^* & -h_3^* \\ h_3 & h_4 & h_1 & h_2 \\ h_4^* & -h_3^* & h_2^* & -h_1^* \end{bmatrix}^H = \begin{bmatrix} \alpha & 0 & \beta & 0 \\ 0 & \alpha & 0 & \beta \\ \beta & 0 & \alpha & 0 \\ 0 & \beta & 0 & \alpha \end{bmatrix}$$

where, each element of the matrix h_i indicates the channel gain, $\alpha = |h_1|^2 + |h_2|^2 + |h_3|^2 + |h_4|^2$ and $\beta = h_1 h_3^* + h_3 h_1^* + h_2 h_4^* + h_4 h_2^*$

In zero forcing decoding technique, the quasi-orthogonal reception is made fully orthogonal. Here we can eliminate interferences at the expense of noise. From MIMO scheme, the received signal vector is

$$\boldsymbol{r} = [r_1 r_2 r_3 r_4] = \boldsymbol{Hx} + \boldsymbol{n}$$

The estimated received signal vector under ZF is

$$\hat{Y} = \left(\boldsymbol{H}\boldsymbol{H}^H\right)^{-1}\boldsymbol{H}^H\boldsymbol{r} = \left(\boldsymbol{H}\boldsymbol{H}^H\right)^{-1}\boldsymbol{H}^H\left(\boldsymbol{H}\boldsymbol{x}+\boldsymbol{n}\right)$$
$$= \left(\boldsymbol{H}\boldsymbol{H}^H\right)^{-1}\left({}^1\boldsymbol{H}^H\boldsymbol{H}\right)\boldsymbol{x} + \left(\boldsymbol{H}\boldsymbol{H}^{\mathrm{H}}\right)^{-1}\boldsymbol{H}^{\mathrm{H}}\boldsymbol{n}$$
$$= \boldsymbol{I}_4\boldsymbol{x} + \left(\boldsymbol{H}\boldsymbol{H}^H\right)^{-1}\boldsymbol{H}^H\boldsymbol{n}$$
$$= \boldsymbol{x} + \boldsymbol{n}'; \text{ where } \boldsymbol{n}' = \left(\boldsymbol{H}\boldsymbol{H}^{\mathrm{H}}\right)^{-1}\boldsymbol{H}^{\mathrm{H}}\boldsymbol{n}$$
$$= \left[x_1x_2x_3x_4\right]^{\mathrm{T}} + \left[n_1'n_2'n_3'n_4'\right]^{\mathrm{T}}$$

Now we have to go for better STBC technique, which provides orthogonality among the received symbols at the same time noise is not enhanced like ZF algorithm of above. In this case, we can choose the simplest possible orthogonal STBC of two antenna systems, called Alamouti scheme discussed in next subsection.

2.2. Alamouti Scheme

Alamouti scheme is a space-time block code under MISO, where the transmitting end uses two antennas and the receiver side has only one antenna, hence the system has no scope of using combining scheme. It is the only full rate orthogonal STBC discussed in [22]. In Alamouti scheme two symbols $\tilde{S}_1$ and $\tilde{S}_2$ are transmitted simultaneously by the antennas at instant t and $t + T$ according to **Table 1**; where T is the duration of a symbol as discussed in [23].

The gain of two wireless links or multipath fading factors is complex expressed as: $h_1 = \alpha_1 \mathrm{e}^{j\theta_1}$ and $h_2 = \alpha_2 \mathrm{e}^{j\theta_2}$. The entire scenario of Alamouti scheme is visualized from **Figure 1**.

The complex received signal at time $t' > t$ is expressed as

$$\tilde{x}_1 = \alpha_1 \mathrm{e}^{j\theta_1}\tilde{S}_1 + \alpha_2 \mathrm{e}^{j\theta_2}\tilde{S}_2 + W_1 \quad (1)$$

The corresponding received signal at time $t' + T$ is

$$\tilde{x}_2 = -\alpha_1 \mathrm{e}^{j\theta_1}\tilde{S}_2^* + \alpha_2 \mathrm{e}^{j\theta_2}\tilde{S}_1^* + W_2 \quad (2)$$

where W_1 and W_2 are the AWGN noise.

From (1) and (2),

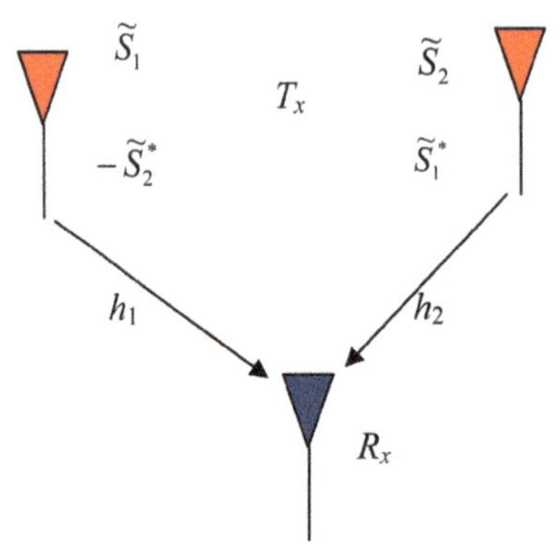

Figure 1. Alamuoti scheme in simplest form.

Table 1. Symbol sequence of two antenna elements.

Time	T_{x1}	T_{x2}
t	$\tilde{S}_1$	$\tilde{S}_2$
$t+T$	$-\tilde{S}_2^*$	$\tilde{S}_1^*$

$$\begin{bmatrix} \tilde{x}_1 \\ \tilde{x}_2^* \end{bmatrix} = \begin{bmatrix} \alpha_1 e^{j\theta_1} & \alpha_2 e^{j\theta_2} \\ \alpha_2 e^{-j\theta_2} & -\alpha_1 e^{-j\theta_1} \end{bmatrix} \begin{bmatrix} \tilde{S}_1 \\ \tilde{S}_2 \end{bmatrix} + \begin{bmatrix} \tilde{W}_1 \\ \tilde{W}_2^* \end{bmatrix} \quad (3)$$

Now, the output of the linear combiner using the concept of channel knowledge of [24] [25],

$$\begin{bmatrix} \tilde{y}_1 \\ \tilde{y}_2 \end{bmatrix} = \begin{bmatrix} \alpha_1 e^{-j\theta_1} & \alpha_2 e^{j\theta_2} \\ \alpha_2 e^{-j\theta_2} & -\alpha_1 e^{j\theta_1} \end{bmatrix} \begin{bmatrix} \tilde{x}_1 \\ \tilde{x}_2^* \end{bmatrix} \quad (4)$$

$$\Rightarrow \begin{bmatrix} \tilde{y}_1 \\ \tilde{y}_2 \end{bmatrix} = \begin{bmatrix} \alpha_1 e^{-j\theta_1} & \alpha_2 e^{j\theta_2} \\ \alpha_2 e^{-j\theta_2} & -\alpha_1 e^{j\theta_1} \end{bmatrix} \left(\begin{bmatrix} \alpha_1 e^{j\theta_1} & \alpha_2 e^{j\theta_2} \\ \alpha_2 e^{-j\theta_2} & -\alpha_1 e^{-j\theta_1} \end{bmatrix} \begin{bmatrix} \tilde{S}_1 \\ \tilde{S}_2 \end{bmatrix} + \begin{bmatrix} \tilde{W}_1 \\ \tilde{W}_2^* \end{bmatrix} \right)$$

$$\Rightarrow \begin{bmatrix} \tilde{y}_1 \\ \tilde{y}_2 \end{bmatrix} = \left(\alpha_1^2 + \alpha_2^2\right) \begin{bmatrix} 1 & 0 \\ 0 & 1 \end{bmatrix} \begin{bmatrix} \tilde{S}_1 \\ \tilde{S}_2 \end{bmatrix} + \begin{bmatrix} \alpha_1 e^{j\theta_1} & \alpha_2 e^{j\theta_2} \\ \alpha_2 e^{-j\theta_2} & -\alpha_1 e^{-j\theta_1} \end{bmatrix} \begin{bmatrix} \tilde{W}_1 \\ \tilde{W}_2^* \end{bmatrix}$$

$$\therefore \begin{bmatrix} \tilde{y}_1 \\ \tilde{y}_2 \end{bmatrix} = \left(\alpha_1^2 + \alpha_2^2\right) \begin{bmatrix} \tilde{S}_1 \\ \tilde{S}_2 \end{bmatrix} + \begin{bmatrix} n_1 \\ n_2 \end{bmatrix} \quad (5)$$

In $\tilde{y}_1$, there is no component of $\tilde{S}_2$ and in $\tilde{y}_2$, there is no component of $\tilde{S}_1$, hence the system is fully orthogonal. The simulation of entire Alamouti scheme under fading channel can be represented by the flowchart of **Figure 2**.

2.3. Algorithm

This subsection provides the algorithm of image processing and transmission of the image through fading channel.

1. Select an RGB image and convert it to grayscale. Resize the image as the size of $N \times N$.
2. Covert the image to a column vector $\mathbf{\Gamma}$ of size of $1 \times N^2$.
3. Divide the amplitude of each pixel of column vector by 16 and store the remainder $R(i)$ and quotient $Q(i)$ of ith pixel where $i = 1, 2, 3, \cdots, N^2$ using the concept of [26].
4. Since the amplitude of each pixel is in the range of 0 - 255 therefore both $R(i)$ and $Q(i)$ has the value in the range of 0 - 15 hence we can use the values as the constellation points of 16-QAM.
5. Assign each value of $R(i)$ and $Q(i)$ against a constellation point of 16-QAM.
6. Transmit the constellation points or symbols against corresponding values of $R(i)$ and $Q(i)$ thorough a Alamouti simulator with Rayleigh fading channel contaminated by AWGN. The flowchart of simulator is shown in **Figure 2**.
7. Recover each constellation points using orthogonal space diversity algorithm at the receiving end.
8. Convert the constellation points to $Q(i)$ and $R(i)$.
9. Evaluate the amplitude of pixel using $\Gamma(i) = 16 * Q(i) + R(i)$.
10. Convert the column vector $\mathbf{\Gamma}$ to the matrix $N \times N$ in the reverse way of step 2.
11. Smooth the image using DWT with a threshold value.

We can apply RSA algorithm of [27] [28] on $R(i)$ and $Q(i)$ at step 5 to achieve secured communication.

3. Results

Let us first select a test image (main gate of Jahangirnagar University campus) and apply the algorithm on it for the fading channel of average SNR of 12, 15, 20, 25 and 30 dB. The corresponding results are shown in **Figures 3-7**. The recovered image is found appreciable at a glance for 20 dB or above. The quality of received image improves at higher SNR is visualized from the figures. We can get better results by applying lower modulation scheme like, BPSK, QPSK, 8-PSK instead of 16-QAM, although throughput is higher at higher modulation scheme. We measure the BER of the same image (main gate of JU campus) for four different modulation schemes shown in **Figure 8**, where we get the conventional results, *i.e.* performance is better for lower modulation scheme. Actually the algorithm has very small effect on different images in context of BER. The phenomenon is shown in **Table 2**, where we select 6 different images and measure BER under 16-QAM taking SNR

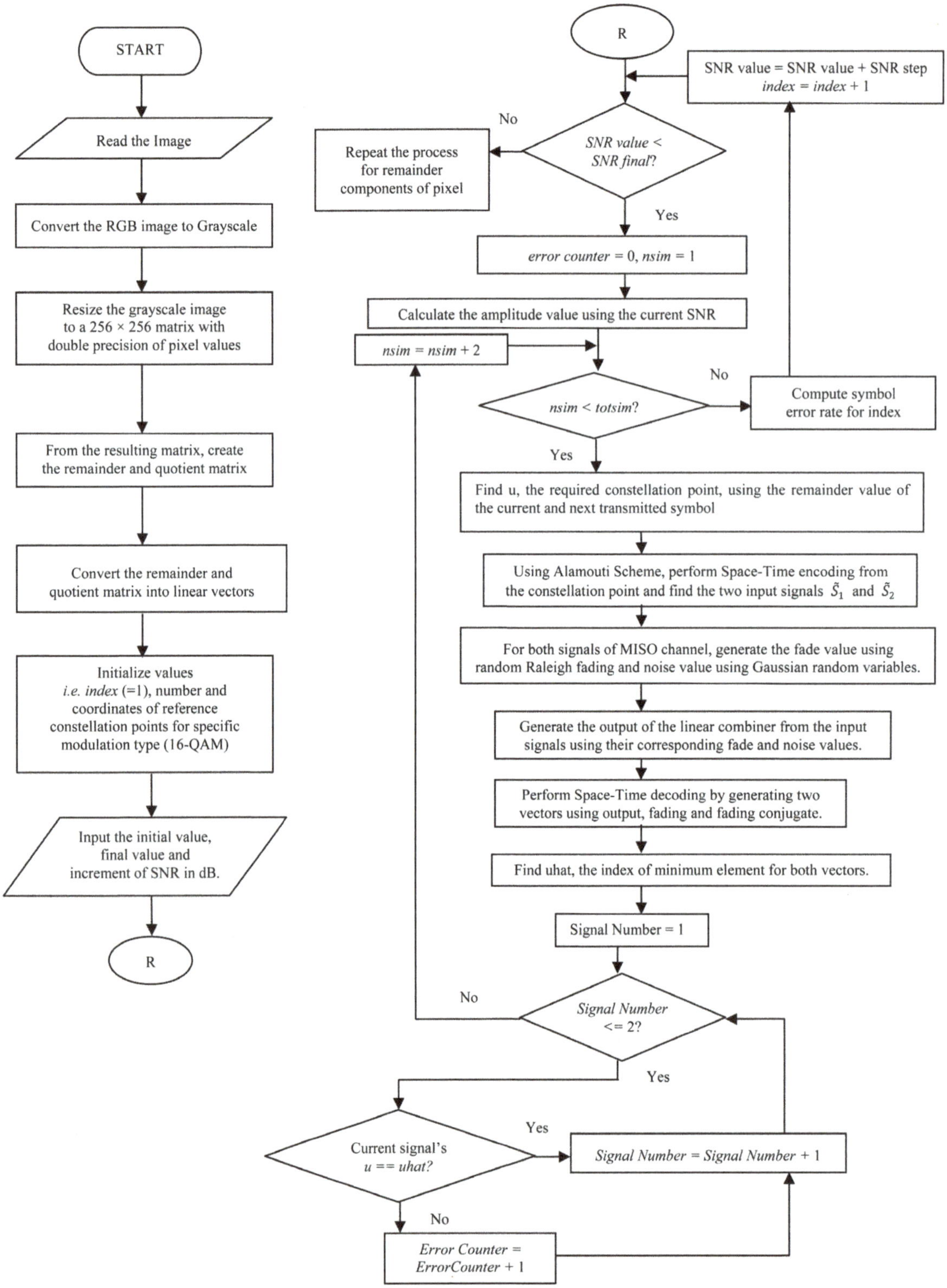

Figure 2. Flowchart of Alamouti Simulator.

from 12 to 30 dB. Finally, for secured communication we apply RSA algorithm on $Q(i)$ and $R(i)$ of four images. The corresponding results are shown in **Figure 9**, where we show the original RGB image, its grayscale version, encrypted image and recovered/decrypted image.

Figure 3. Image recovery at SNR of 12dB using 16-QAM.

Figure 4. Image recovery at SNR of 15dB using 16-QAM.

Figure 5. Image recovery at SNR of 20dB using 16-QAM.

Figure 6. Image recovery at SNR of 25dB using 16-QAM.

Figure 7. Image recovery at SNR of 30dB using 16-QAM.

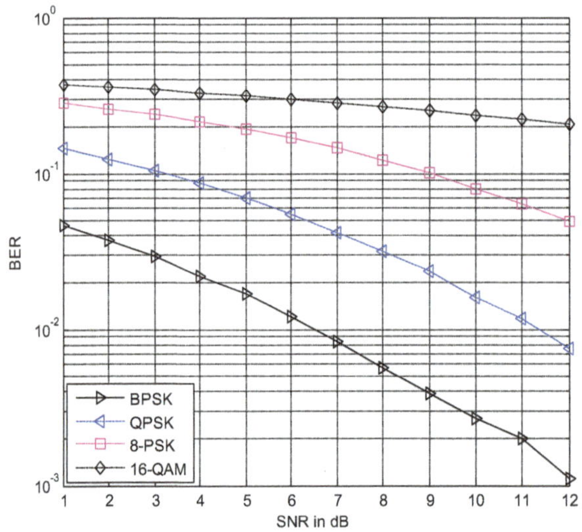

Figure 8. BER of the image of JU Gate under four different modulation schemes.

Figure 9. Encryption and decryption of image.

Table 2. BER of six test images under 16-QAM.

SNR (dB)	BER (image 1) 	BER (image 2) 	BER (image 3) 	BER (image 4) 	BER (image 5) 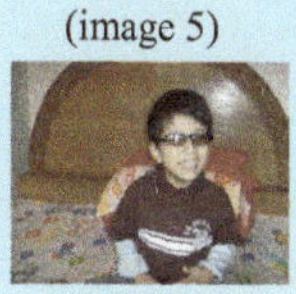	BER (image 6) 
12	0.2082	0.2074	0.2092	0.2059	0.2076	0.2107
14	0.1835	0.1813	0.1826	0.1803	0.1830	0.1862
16	0.1598	0.1587	0.1600	0.1579	0.1612	0.1629
18	0.1370	0.1366	0.1374	0.1349	0.1386	0.1399
20	0.1158	0.1154	0.1156	0.1146	0.1158	0.1164
22	0.0920	0.0911	0.0921	0.0919	0.0930	0.0932
24	0.0692	0.0686	0.0685	0.0676	0.0695	0.0704
26	0.0490	0.0477	0.0487	0.0483	0.0486	0.0492
28	0.0310	0.0305	0.0317	0.0296	0.0304	0.0309
30	0.0179	0.0177	0.0176	0.0176	0.0178	0.0181

4. Conclusion

In this paper, we are able to recover an image transmitting through Alamouti channel with incorporation of RSA algorithm; where the quality of the image depends on the SNR of the channel. Although we reveal the results for grayscale images, we can also do the similar job for RGB images as well, sending the R, G and B components separately, then combining and at receiving end. We can apply channel coding scheme like linear block code, convolutional code or CRC (Cyclic Redundancy Code) scheme to enhance the performance of the system. MIMO and combining scheme (maximal ratio combining, equal gain, selection combining) can be considered an alternate technique of improvement of the system.

References

[1] Zhang, J.K., Chen, S., Mu, X.M. and Hanzo, L. (2011) Joint Channel Estimation and Multiuser Detection for SDMA/OFD Based on Dual Repeated Weighted Boosting Search. *IEEE Transactions on Vehicular Technology*, **60**, 3265-3275. http://dx.doi.org/10.1109/TVT.2011.2161356

[2] Zhu, H.L., Xia, B. and Tan, Z.H. (2011) Performance Analysis of Alamouti Transmit Diversity with QAM in Imperfect Channel Estimation. *IEEE Journal on Selected Areas in Communications*, **29**, 1242-1248.

[3] Alamouti, S. (1998) A Simple Transmit Diversity Technique for Wireless Communications. *IEEE Journal on Selected Areas in Communications*, **16**, 1451-1458. http://dx.doi.org/10.1109/49.730453

[4] Wang, X.Y. and Wang, J. (2004) Effect of Imperfect Channel Estimation on Transmit Diversity in CDMA Systems. *IEEE Transactions on Vehicular Technology*, **53**, 1400-1412.

[5] Derryberry, R.T., Gray, S.D., Ionescu, D.M., Mandyam, G. and Raghothaman, B. (2002) Transmit Diversity in 3G CDMA Systems. *IEEE Communications Magazine*, **40**, 68-75. http://dx.doi.org/10.1109/35.995853

[6] Alam, M.M., Islam, A.Z.M.T. and Ullah, S.E. (2011) Performance Analysis of a Concatenated LDPC Coded Alamouti Based STBC-OFDM System on Text Message Transmission. *International Journal of Computer Science*, *Engineering and Applications* (*IJCSEA*), **1**, 1-9.

[7] Ahmed, T., Ali, M.M. and Ullah, S.E. (2012) BER Performance Analysis of a STBC Encoded Secured Multiuser MIMO-OFDM Wireless Communication System. *International Journal of Hybrid Information Technology*, **5**, 19-30.

[8] Anoh, K.O.O., Ochonogor, O., Abd-Alhameed, R.A.A., Jones, S.M.R. and Mapuka, T.T. (2014) Improved Alamouti STBC Multi-Antenna System Using Hadamard Matrices. *International Journal of Communications*, *Network and System Sciences*, **7**, 83-89. http://dx.doi.org/10.4236/ijcns.2014.73010

[9] Sklar, B. (2001) Digital Communications: Fundamentals and Applications. 2nd Edition, Prentice Hall, Upper Saddle River.

[10] Ryan, W.E. and Lin, S. (2009) Channel Codes: Classical and Modern. Cambridge University Press, Cambridge, UK, 4.

[11] Xia, B. and Wang, J. (2005) Effect of Channel Estimation Error on QAM Systems with Antenna Diversity. *IEEE Transactions on Communications*, **53**, 481-488.

[12] Gu, D. and Leung, C. (2003) Performance Analysis of Transmit Diversity Scheme with Imperfect Channel Estimation. *IEEE Electronics Letters*, **39**, 402-403. http://dx.doi.org/10.1049/el:20030243

[13] Stallings, W. (2005) Cryptography and Network Security Principles and Practices. 4th Edition, Prentice-Hall, New Jersey.

[14] Saeed, A., Xu, H.J. and Quazi, T. (2014) Alamouti Space-Time Block Coded Hierarchical Modulation with Signal Space Diversity and MRC Reception in Nakagami-m Fading Channel. *IET Communications*, **8**, 516-524. http://dx.doi.org/10.1049/iet-com.2013.0519

[15] Fang, L., Bi, G. and Kot, A.C. (2000) New Method of Performance Analysis for Diversity Reception with Correlated Rayleigh-Fading Signals. *IEEE Transactions on Vehicular Technology*, **49**, 1807-1812. http://dx.doi.org/10.1109/25.892585

[16] Ahmed, T., Anower, M.S., Sarkar, M.Z.I. and Ali, M.M. (2013) Investigation of Correlated Rayleigh Fading Channel with Alamouti's STBC-MRC System. *International Journal of Advanced Science and Technology*, **58**, 65-74. http://dx.doi.org/10.14257/ijast.2013.58.06

[17] Perişoară, L.A. (2012) BER Analysis of STBC Codes for MIMO Rayleigh Flat Fading Channels. *Telfor Journal*, **4**, 78-82.

[18] Bhatnagar, M.R., Vishwanath, R. and Bhatnagar, V. (2007) Performance Analysis of Space-Time Block Codes in Flat Fading MIMO Channels with Offsets. *EURASIP Journal on Wireless Communications and Networking*, **2007**, Article ID: 030548.

[19] Badic, B., Herdin, M., Weinrichter, H. and Rupp, M. (2004) Quasi-Orthogonal Space-Time Block Codes on Measured MIMO Channels. *Joint IST Workshop on Mobile Future*, 2004 *and the Symposium on Trends in Communications*, Bratislava, 24-26 October 2004, 17-20.

[20] Badic, B., Rupp, M. and Weinrichter, H. (2004) Quasi-Orthogonal Space-Time Block Codes for Data Transmission over Four and Eight Transmit Antennas with Very Low Feedback Rate. *5th International ITG Conference on Source and Channel Coding* (*SCC*), Erlangen, 14-16 January 2004, 157-164.

[21] Park, U., Kim, Y. and Kim, S. (2010) A New Result on Turbo Coded QO-STBC Schemes. *IEEE Communications Letters*, **14**, 199-201. http://dx.doi.org/10.1109/LCOMM.2010.03.092404

[22] Chaudhary, S. and Patil, A.J. (2012) Performance Analysis of Mimo-Space Time Block Coding with Different Modulation Techniques. *ICTACT Journal on Communication Technology*, **3**, 510-514.

[23] El-Astal, M.T.O., Abu-Hudrouss, A.M., Salmon, B.P. and Olivier, J.C. (2015) An Adaptive Transmission Protocol for Exploiting Diversity and Multiplexing Gains in Wireless Relaying Networks. *EURASIP Journal on Wireless Communications and Networking*, **2015**, 1-15.

[24] Gupta, S., Dutta, R.K. and Tiwari, A.C. (2014) Performance Analysis of Alamouti and Orthogonal Space-Time Block Codes in MIMO System under Rayleigh Fading Scenario. *International Journal of Advanced Research in Electrical, Electronics and Instrumentation Engineering*, **3**, 12423-12429.

[25] Rao, V. and Malavika, T. (2014) Performance Analysis of MIMO-OFDM for Multiple Antennas. *International Journal of Advanced Research in Electrical, Electronics and Instrumentation Engineering*, **3**, 9349-9355.

[26] Jassim, F.A. and Qassim, H.E. (2012) Five Modulus Method for Image Compression. *Signal & Image Processing*: *An International Journal* (*SIPIJ*), **3**, 19-28.

[27] Ayele, A.A. and Sreenivasarao, V. (2013) Modified RSA Encryption Technique Based on Multiple Public Keys. *International Journal of Innovative Research in Computer and Communication Engineering*, **1**, 859-864.

[28] Jamgekar, R.S. and Joshi, G.S. (2013) File Encryption and Decryption Using Secure RSA. *International Journal of Emerging Science and Engineering* (*IJESE*), **1**, 11-14.

19

Security: A Core Issue in Mobile *Ad hoc* Networks

Asif Shabbir[1], Fayyaz Khalid[1], Syed Muqsit Shaheed[2], Jalil Abbas[3*], M. Zia-Ul-Haq[3]

[1]Department of Computer Science, University of Gujrat, Gujrat, Pakistan
[2]Department of Computer Science, University of Lahore, Lahore, Pakistan
[3]Department of Computer Science& IT, Government Collage University Faisalabad (Layyah Campus), Layyah, Pakistan
Email: rabtabox4u@gmail.com, fayyazrao@yahoo.com, syedmuqsit@hotmail.com, *sjshah786@gmail.com, ziaulhaq.arain@gmail.com

Abstract

Computation is spanning from PC to Mobile devices. The Mobile *Ad hoc* Networks (MANETs) are optimal choice to accommodate this growing trend but there is a problem, security is the core issue. MANETs rely on wireless links for communication. Wireless networks are considered more exposed to security attacks as compared to wired networks, especially; MANETs are the soft target due to vulnerable in nature. Lack of infrastructure, open peer to peer connectivity, shared wireless medium, dynamic topology and scalability are the key characteristics of MANETs which make them ideal for security attacks. In this paper, we shall discuss in detail, what does security mean, why MANETs are more susceptible to security attacks than wired networks, taxonomy of network attacks and layer wise analysis of network attacks. Finally, we shall propose solutions to meet the security challenges, according to our framed security criteria.

Keywords

Mobile Devices, Optimal Choice, MANET Security, Wireless Network, Taxonomy, Intrusion Detection, Network Security Threats, Layer Wise Network, Vulnerabilities of Wireless, Security Criteria

1. Introduction

Nowadays, computing is the need of everyone, everywhere. Everyone is looking for on the spot handy computa-

*Corresponding author.

tional solutions. In this need based scenario, advancement in Hardware Engineering made it possible, the invention of a number of mobile computing devices. PDA's, Pocket PC's and smart phones can be seen everywhere. In the last decade, the massive growth in mobile computing devices brought a revolutionary change in computing, the evolution of ubiquitous computing. At present, the concept of ubiquitous computing is a research hot spot in Computer Science society [1]. In ubiquitous computing environment, the individual users may retrieve information smoothly whenever and wherever they are by utilizing heterogeneous electronic platforms simultaneously [2].

MANETs are the best choice for practical implementation of the ubiquitous computing and wireless medium is a natural ally of MANETs for ubiquitous computing because wired link does not support heterogeneous devices, on demand connectivity and device mobility. There are numbers of practical applications of MANETs [3] such as:

- Personal area networks
- People sitting in airport lounge
- Kids playing in ground
- Rescue and emergency services
- Military movement on vehicles
- Healthcare services
- Stock market brokers

The word "MANET" is a self descriptive. It is a temporary wireless network of arbitrary self organized mobile nodes that is why it is viewed as Mobile *Ad hoc* Network [4].

MANET is an art of networking without formal networking which is quite easy to develop. In the infrastructure less network environment, there is no concept of node dominancy or centralized control. The nodes within the radio range of each other directly communicate with each other while the node(s) which are not in direct communication range, form a multi-hop communication environment, where intermediate nodes are used for multi-hop communication.

Multi-hop routing is the core of multi-hop communication. An efficient and power aware routing protocol with reduced traffic over head is required for this multi-hop routing [5]. Different routing protocols like Dynamic Source Routing (DSR) [6], *Ad hoc* On-Demand Distance Vector (AODV) [7], The Optimized Link State Routing (OLSR) [8] and Destination-Sequenced Distance-Vector (DSDV) [9] are available multi-hop routing.

Formation of the MANET is quite easy and interesting. Heterogeneous devices with varying capability and processing power are allowed to join the network, even at runtime. There is no specific topology and limit of nodes in a network. Nodes are free to join, leave and move with any mobility pattern, on the fly. Following are the typical key features of the MANETs [10] [11].

- Absence of Infrastructure
- Peer to peer connectivity/communication
- Dynamic topology
- Shared wireless medium
- Scalability

The above mentioned features make the MANETs highly susceptible to security attacks. In the section two of this paper we discuss in detail, what does security mean? In section three vulnerabilities of MANETs, in section four taxonomy of network attacks and in section five layer wise analyses of network attacks are discussed. We proposed some solutions to meet the security challenges in section six. At the end of the paper conclusion of our study and future research is given.

2. What Does Security Mean?

It is highly debatable that:

- What does security mean?
- Is there exists any safety measurement criteria?
- How to assess security state of mobile *ad hoc* network?

We must have to formulate some security criteria to answer the above said questions. Following are the guide lines to evaluate the security state of the MANETs.

2.1. Availability

Availability ensures the guaranteed access of all the services of the network, to all the privileged nodes, under any circumstances [10]. It is a challenging issue to provide the network services smoothly, especially, in the presence of selfish nodes and Compromised nodes and denial of service attacks.

2.2. Integrity

Integrity means a message will never altered or tampered after the transmission. At runtime a message could be truncated, replayed or even altered due to any malicious attack or even due to accidental hardware failure [12]. Integrity guarantees the protection of information against Malicious as well as Accidental altering.

2.3. Confidentiality

Confidentiality means the privacy of information. There should exist a mechanism to protect information from being exposed to unauthorized entities. Only privileged nodes should be given access to private information.

2.4. Authorization

In a network environment certain entities are assigned with specific credentials. These credentials define their privileges to access certain network services. Authorization enables the receivers to verify that message is from trusted entity. It is usually implemented via certificate or digital signature.

2.5. Authenticity

Authenticity means that both sender and receiver are genuine in nature, not the impersonate one [10]. There should be a mechanism like handshaking to verify that both the sender and receiver are legitimate to communicate.

2.6. Non Repudiation

Non repudiation means that the sender can't deny whatever it has transmitted and likewise receiver whatever it has received. They must own their mutual communication. It is necessary and helpful, especially, in tracking of compromised nodes in a network. If a node identifies erroneous messages it would alert other nodes with an evidence to be aware of expected abnormal behavior of that particular transmitter node.

2.7. Anonymity

The term anonymity is closely related to privacy. Anonymity demands that information about the ownership of current node should be kept private by the node itself as well as by the system software. It is the only way to protect a privacy of a node from being exposed to the other nodes.

3. Why MANETs Are More Exposed to Security Attacks

Wireless medium is considered more exposed to security attacks than wired links due to its vulnerabilities. Since Wireless medium is a backbone of the MANETs, therefore MANETs inherit all the features of wireless links. These inherited vulnerabilities of wireless medium make the MANETs highly vulnerable in nature and a hive for all season attacks. Let's discuss these vulnerabilities, one by one, in detail.

3.1. Transitive & Non Secure Boundary

Radio waves are transmitted in a broadcast fashion. During broadcast unguided signals spread all around and cover some specific elastic circular range. In that particular range, strength of the signal is primarily dependent upon the distance from the transmitter that is why there is no sharp boundary in wireless links [13] [14]. Since there is no sharp border line in Wireless links, therefore, they are considered more exposed to security attacks than wired links. Physical links are more secure because the malicious attack must have to pass through several lines of defense such as fire walls, gateways, routers and filters before reaching to target node [15]. MANETs are dependent upon Wireless medium for communication; therefore, it is quite easy to inject malicious code in MANETs in the absence of sharp boundary line of wireless links. There exists no safety mechanism against unauthorized network access due to mobile and *ad hoc* nature of MANETs. In the absence of the secure boundary different types of attacks may affect performance of the network *i.e.* passive eavesdropping, Active Interfering, exposure of secret information, data tempering and denial of service attacks [10].

3.2. Varying Strength of Signals

Varying strength of signals is another problem with wireless medium. Signal strength is mainly dependent on the transmitter, radio antenna and even on battery power at some extent. Signal strength is inversely proportional to the distance from the transmitter. Signals get weaken as the distance from the transmitter is increased. So on intermediate walls and building may also weaken the signals, considerably. Weak signals may leads to the unavailability of certain intermediate network nodes and ultimately the unavailability of certain network services. Varying strength of signals may cause QoS issues in network and even may generate the illusion of denial of service attacks.

3.3. Interference with External Signals

Since MANETs use wireless as a communication medium so it is highly susceptible to be attacked by external signal interference. External signal interference is, often seriously destructive and may tamper your vital information. It is also considered that external signal interference may leads to active interfering, data tampering and information leakage attacks.

3.4. Hidden Nodes and Signal Collision

In MANETs nodes flood packets in broadcast fashion. Only nodes within the specific radius of transmitter can detect the signals. This distance-dependent carrier sensing can generate a hidden/exposed node problem. The situation when a node receives the same packet from two different sources but the source nodes are hidden from each other is called a hidden node problem which leads to the signals collision [13]. As a result of collision loss of data occurs and ultimately there may generate illusion of denial of service attacks.

3.5. Lack of Infrastructure and Central Management

MANETs are infrastructure less networks hence the lack of central management is a major threat to MANET's security. There exists no network server in MANETs hence there is no centralized entity to impose network policies, trust management and authorization. In the absence of network infrastructure there is no incorporation of any security feature in MANETs. Due to the non incorporation of security features and heavy traffic flow, it is quite difficult to detect any malicious activity in highly dynamic networks like MANETs [16]. Some algorithms of MANETs are dependent upon mutual participation of nodes and require network infrastructure. In the absence of network infrastructure an adversary node may take the advantage of decentralization of MANET to break the cooperative algorithms [17].

3.6. Compromised Nodes

Wireless links make the MANETs a hot cake for security attacks. The attacker takes the advantage of vulnerable nature of MANETs and launches a compromised node inside the network. A compromised node is an adverse node inside networks that has been captured by any unfair means and executes the further malicious activities in the network. Then it is easier for the attacker to employ a compromised node inside the network to get access and control the ongoing communication [18] [19].

3.7. Adaptive Routing Protocols

Instead of designing specialized routing protocols Many of MANET routing protocols are borrowed from wireless routing protocols, which were designed and configured for static wireless environment. Mobility is the key feature of MANETs hence statically configured routing protocols are not capable to handle the security threats, caused of mobility. There are required special secure versions of the protocols to protect MANETs from security attacks. Different secure versions of protocols are recommended such as sDSR, sAODV, sOLSR and sDSDV are the secure versions of DSR, AODV, OLSR and DSDV protocols respectively.

3.8. Shared Bandwidth

Wired links always provide a better bandwidth than wireless links because they transmit data over confined physical lines. In wireless communication, radio channels are shared among many users. Wireless communication is taken place in broadcast manner hence a lot of bandwidth is shared and consumed during the broadcasting of the wireless signals. Periodic hello messages for neighbor sensing might also utilize a lot of bandwidth which ultimately leads to network overhead. Shared bandwidth is the major drawback of wireless medium that limits the data transfer rate, quality of service and successful delivery of packets. The attackers may launch the denial of service attacks by utilizing extensive bandwidth of the wireless medium.

3.9. Scalability

Unlike wired networks MANETs can grow and shrink on the fly whereas the scalability of wired networks is usually predefined at design time. Freedom to join and mobility are two key features of MANETs, therefore, it is hard to predict the scale of networks like MANET. Scalability may leads to serious routing and QoS concerns. The network services like key management and routing protocols must be compatible with varying scale of the MANETs *i.e.* OLSR is more scalable than DSR due to the use of MPR [20]. It is also founded that hierarchal and GPS assisted routing protocols are more scalable then flat routing protocols because they are more capable of affording node density than flat protocols [21]. Rapid change in scale at run times my invite security and QoS issues and destabilize the MANETs.

3.10. Limited Battery Power

Most of the devices of MANETs are mobile in nature so they strongly rely on battery power for radio transmission whereas wired networks are not battery constrained in nature because they get their power from electric power outlets.

Restricted battery power is major threat in MANETs. The adversary node may flood continuous routing requests to target any battery constrained node or it may grasp a target node to perform a kind of battery consuming activity. Such kind of tasks will exhaust the whole battery power and all the nodes beyond the target node might go out of service.

So on some nodes encountering limited power supply may behave like selfish nodes. As a result network will suffer from intentional non-cooperative behavior of selfish nodes. No doubt, selfish nodes are considered as a kind of security threat.

4. Taxonomy of Network Attacks

Network security attacks can be categorized in a number of ways on the basis of locality of attacker, nature of attack interaction and target layer of the OSI network model. Let's discuss major classifications.

4.1. External V/S Internal Network Attacks

In broader sense network attacks against MANETs can be classified into Internal and external attacks [15] [22]. External network attacks are caused due to unauthorized external entities. They may generate problems like network overhead, denial of service and destructive signal interference. External network attacks are comparatively easy to diagnose as compare to internal network attacks [23]-[25] because Internal network attacks are caused by trusted and authorized nodes which are located inside network. It is very difficult to track and diagnose the malicious behavior of internal compromised network nodes because these nodes are capable to generate valid digital signature and public keys. These nodes may broadcast invalid or modified routing information to other nodes by deceiving other nodes being as shorter path [23] [26].

4.2. Active Attacks V/S Passive Attacks

Based upon the nature of attack-interaction, the attacks against MANETs can be categorized into active and passive attacks. The word passive is self descriptive. These attacks act just like slow poising and do not perform any serious alter to packets and routine operations of the network. These attacks are difficult to detect due to mild in action. Passive attacks usually deduce some info from routed traffic instead of disturbing the routing [23] [25] [27]. On other hand active attacks are very hazardous in action and affect the network very badly. Active attacks may be either internal or external in penetration. These attacks do not hide their identity and prove their presence by altering packets or misrouting the information.

4.3. Stallings Classification of Routing Attacks

Routing is the core of communication in MANETs. Routing is the only mechanism which establishes the successful communication path between sender and receiver. The basic responsibility of a routing protocol is to establish an efficient path between sender and the receiver nodes. Since routing is the core of communication, therefore, its vital role invites the network attackers all the times to launch routing attacks against MANETs [14] [28]. Here we discuss the classification of attacks against MANETs that was suggested by Stallings. According to the Stallings, attacks against MANETs are classified as under [29]:

4.3.1. Modification Attacks

It is a special type of routing attack in which adversary node makes some potential changes in the routing message, as a result routing message may lose its integrity. Packet misrouting is the good example of such attacks in which a message is deviated from its original route [30]. Likewise node impersonation is another example where a malicious node hacks the identity of another node to receive all the messages of that particular node [31].

4.3.2. Interception Attacks

Someone may intercept the normal flow of routing messages by gaining an unauthorized access. In such situation there is a potential hazard of packet alteration before their further forwarding. Wormhole attacks, black hole attacks and routing packet analysis attacks are the major examples of interception attacks.

4.3.3. Fabrication Attacks

There is another technique to chaos the network operation that is fabrication. In fabrication attacks the attacker fabricates its own packets in network to cause the disorder in routine operations of the network. Lack of authentication in routing protocols of MANET leads to fabrication attacks which generate erroneous routing messages [32]-[34]. Sleep deprivation attacks and route salvage attacks are few examples of fabrication attacks.

4.3.4. Interruption Attacks

These attacks interrupt routine network traffic by blocking the routing messages before reaching to destination. In interruption attacks different techniques are used to disturb the normal flow of network traffic such as Packet dropping, flooding etc.

5. Layer Wise Network Attacks

The ISO (International Organization for Standardization) proposed a conceptual model of networking, the OSI model. According to the OSI model, internal functions of communication system are divided into abstraction of seven layers [35]. Since OSI model is a well organizes model of network operations, therefore, it will be better to analyze network attacks according to layer wise stack. Now we discuss the layer wise network attacks in **Table 1**.

5.1. Physical Layer Attacks

Since physical layer deals with physical medium and network devices so in this section we shall discuss network attacks that are originated from physical medium and the hardware. In the absence of secure boundaries [13] [14], it is quite easy to attack wireless medium at physical layer. Physical medium attacker does not need to know the network technology details and attacking techniques in depth. Physical layer is quite sensitive against network attacks; even the failure in communication hardware can generate a considerable problem at physical layer.

Table 1. Layer wise network attacks.

OSI Layer	Attacks
Physical Layer	Eaves dropping, Jamming, Active Interference
Data Link Layer	Selfish node behaviour, Malicious node behaviour, DoS (Denial of service), Integrity, Misrouting Traffic
Network Layer	Black Hole, Wormhole, Sinkhole, Replay, Link spoofing, Resource consumption, Sybil
Transport Layer	SYN Flooding, Session Hijacking
Application Layer	Malicious behavior, Data corruption, Virus

5.1.1. Eaves Dropping Attacks

In eaves dropping is a special technique where attacker makes it sure, the passive listening of messages by an unintended receiver [36]. **Figure 1** shows the attacker between the communication of Sender and Receiver.

There is a simple technique; wireless medium can easily be intercepted, just by proper tuning up the receiving node to specific frequency. The eaves dropping is usually aimed to steal secret information by unauthorized entity such as capturing private keys, public keys and password etc. it is usually done by tapping the wireless link. Captured information could be used for subsequent attacks.

5.1.2. Jamming Attacks

Network Jamming attacks are specific to wireless medium only as oppose to wire networks, where, there is no concept of signal jamming. Jamming is a special technique to degrade the performance of wireless medium by lowering the signal availability [37]-[40]. The attacker or adversary, who is known as jammer, continuously emits the radio signals to rush up partially or even fill up completely the entire wireless band. As a result of the abundance of adverse signals, the legitimate network traffic will be blocked [41] [42]. To launch jamming attacks, first the frequency of legitimate signals of network is determined and then the jamming attack is initiated by the jammer. Usually jammer blocks the legitimate traffic by flooding and filling up the available band. Jamming attacks can be detected and counter fetched by finding signal strength and applying location consistency checks.

5.1.3. Active Interference Attacks

Active interference is the major problem with wireless signals. Interference of Internal or external Electromagnetic signal can strongly disturb the wireless signals and ultimately leads to the denial of service. Active interference may change the order of messages and even may replay old messages. Active interference is the special clause of jamming attacks. During jamming, the transmitted messages are corrupted by challenger node due to electromagnetic interference with the operational frequencies of the targeted receivers [43]. The worst case of destructive interference occurs when the attacker propagate signal with the difference of half wave length of legitimate network signals. That scenario totally nullifies legitimate signals of the network. **Figure 2** gives a comparison of constructive and destructive interfence.

5.2. Data Link/Mac Layer Attacks

The Mac layer attacks are classified in terms of their penetration level and their effect on issues like route discovery failure, energy consumption and link breakage etc [44] [45]. At this layer certain algorithms are used which susceptible to denial of service attacks. The adversary nodes may behave as a just selfish or may exhibit some malicious behavior.

5.2.1. Selfish Nodes Behavior

A selfish node is special type of compromise node which does not suppose to attack the other nodes of a network. It does not cooperate with other nodes and simply refuse its services to other network nodes to save its battery life, CPU cycles and available bandwidth. Networks often encounters a situation when a node is connected to the network but it refuses to forward packets, intentionally, to save its resources like battery power and bandwidth. Such behavior is known as free riding [46]. Sometimes selfish behavior of nodes may exhibit the denial of service.

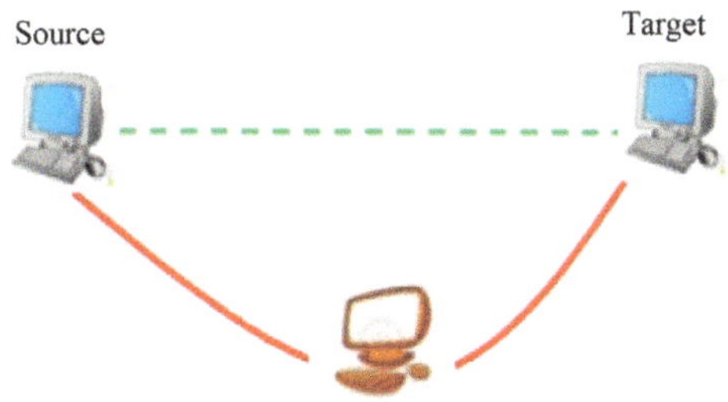

Figure 1. Attacker on communication between sender and receiver.

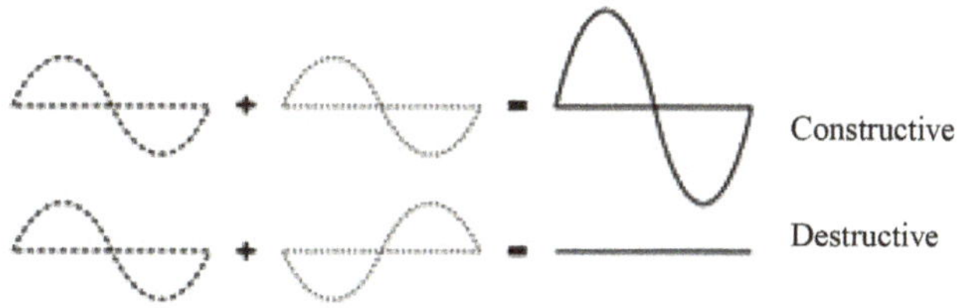

Figure 2. Constructive v/s destructive interference.

Selfish nodes can be classified into three categories according to their level of selfishness.

SN1—A node takes part in network route discovery and maintenance tasks but do not forward packets to other nodes.

SN2—A nodes does not take part in any routing and maintenance operation but just transmit its own packets.

SN3—The nodes which behave normal when the energy level is up to some threshold but if battery level lowers below the threshold value then it behave like SN2.

The level and technique of selfish behavior of a node is dependent upon a protocol in use. Such as a node using DSR protocols may exercise a number of misbehaviors to save its energy and resources like: [47] [48]

- Don't respond to route request (RREQ)
- Don't respond to route replies (RREP)
- Set hop limit/TTL to some minimum value
- Don't send acknowledgements
- Don't forward packets
- Drop data packets

5.2.2. Malicious Nodes Behavior

Another common misbehavior of nodes is identified as malicious nodes. Malicious nodes are considered more risky than selfish nodes because these are capable to launch malicious attacks to the other network nodes where as selfish nodes just refuses to take part in network operations. Malicious nodes may exhibit a number of misbehaves in addition to misbehaves exhibited by selfish nodes. Detail is as given below:

1) Denial of service (DoS) Attacks

Denial of service DoS attacks are aimed to grab the availability of the entire node or some of its services, offered to the network. In other words denial of serve means the unavailability of services to the intended receiver. Denial of service attacks are usually originated from a malicious compromised node in the network. DoS attacks usually prevent the victim to get benefit from the network services. Denial of service can be launched in a number of ways. Taking the advantage of vulnerabilities of Link Layer a attacker may utilize the binary exponential back-off scheme of IEEE 802.11 to refuse access to its local neighbors in a wireless link [49]-[51]. Most of the network services are distributed in nature. DoS attacks may target a node for intensive fake traffic routing to exhaust its processing and battery power. As a result a node may go into hibernate state and break cooperative algorithms [17]. Attackers some time use jamming techniques to launch denial of service attacks. Flooding a network with adverse signals may block the legitimate network [41] [42].

2) Attacks on network Integrity

Integrity means a message will never altered or tampered after transmission. At runtime a message can be truncated, tampered or altered due to any malicious attack. So the integrity attacks must be handled.

3) Attacking Neighbor sensing protocols

Links between neighboring nodes may be marked as broken links due to advertisement of fake error messages by a malicious node. As results of fake error messages a link between two neighboring nodes remain no more active and even neighboring nodes remain failed to sense availability of each other.

4) Traffic Analysis

In MANET the adversary node may take the advantage of traffic patterns by analyzing the traffic flow between different nodes. Confidential information about the topology and traffic flow can be extracted from traffic patterns. Traffic patterns guide the attacker about to track the active and high valued candidate target in a network. Following information can be extracted by analyzing traffic patterns:

- Location of active nodes
- Topological information
- Available source & destination nodes

5.3. Network Layer Attacks (Routing Attacks)

In MANET nodes are connected in hop-by-hop manner [44] [45] [52]. Network layer protocols mange the routing of packets, where each node takes routing decisions for packet forwarding. Routing attacks are very common in MANET. The idea of routing attacks is so simple and straight forward. Routing attacks can be launched by injecting a malicious node in between source and destination to either divert or absorb the network traffic, even a routing loop can be established to create service congestion in a network. The concept is elaborated in **Figure 3** where "M" is a malicious node which intercepts and diverts the traffic between nodes "B" and "C".

Routing is core of communication in a network. Different routing protocols such as DRS [6], AODV [7], OLSR [8], DSDV [9] are used to route the packets to their destination. Routing protocols are the major area of attack. Attackers target the routing protocols at different phases of routing. Message flooding, routing table overflow and routing loop attacks are major attacks at route discovery phase [53] [54]. Even the route maintenance and packet forwarding phases are not the exceptions. A lot of routing attacks may target packet forwarding and may exploit the packet forwarding functionality at network layer [16] [55]. Attacks on routing protocols are aimed to block the delivery of packets from source to destination. Let's discuss routing attacks in detail.

5.3.1. Black Hole Attacks

In black hole routing attacks a malicious node uses a trick and advertises itself as an optimal choice for routing from source to destination. On receiving a route request from the source node the adversary node rapidly replies with a fake shortest path [56]. On receiving fake route reply the source node forwards its packets via adversary node. Once the adversary node succeeded to get routing privilege as an intermediate node it can do anything with packets before further forwarding to next hop neighbors. To demonstrate the concept see the **Figure 4**, where node "M" is a malicious node and issues a fake reply to cheat sender node "S" as an the next hop to the shortest path to the destination "D".

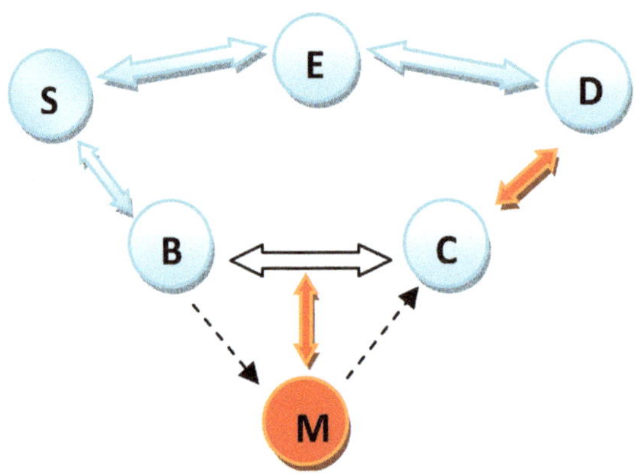

Figure 3. Routing attack by malicious attack.

Consider the AODV protocol for case scenario (**Figure 4**). When the source node "S" advertise a route request RREQ. The malicious node "M" rapidly replies with a higher value of *dest_seq_number* than existing route entry of the routing table of source node "S". In this way node "M" succeed to deceive source node "S" by claiming itself being as a fresh and optimal path towards destination "D". As a result of this black hole data packets are sent over path S->N->M->P->D instead of the path S->N->P->D.

5.3.2. Rushing Attacks

In [57] the authors introduced new routing attack called Rushing attack. Routing protocols, specifically on-demand routing protocols are more susceptible to these attacks where rout discovery is subverted by using duplicate route suppression during process of route discovery [44] [57] [58]. Flooding of the route discovery requests (RREQ) is a major issue in on-demand protocols therefore the protocols like DSR and AODV follow a mechanism to control this flooding. In order to control the flooding, nodes forwards only those request that arrive first. All later same RREQs are simply discarded. The rushing attacks take the advantage of this vulnerability. Let's explain mechanism of rushing attacks. When a route discovery request is initiated by the sender node "S", the malicious node "M" acts very quickly. It passes the same request to all neighbors of the destination "D" very quickly before the arrival of legitimate RREQ from the legitimate route. **Figure 5** elaborates the scenario where nodes "X" and "Y" receive the RREQ from compromised node "M" earlier than any other neighbor, later on when a RREQ is received from nodes "B" and "E" , it is simply discarded by both "X" and "Y" nodes. In this way source "S" always remains fail to find any safe path other than the involvement of compromised node "M" due to the rushing attack of node "M".

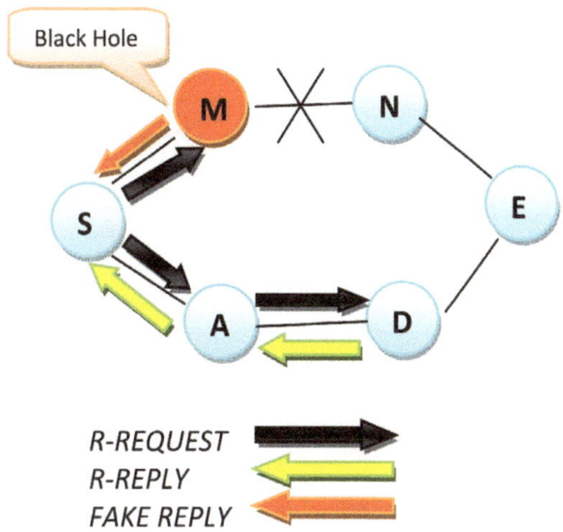

Figure 4. Black hole attack.

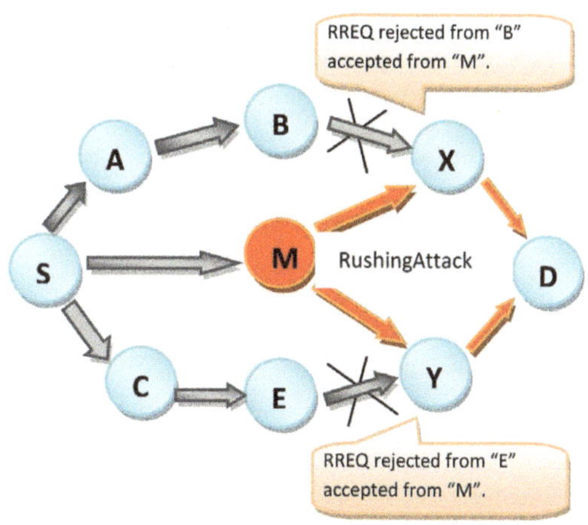

Figure 5. Rushing attack.

5.3.3. Wormhole Attacks

In wormhole attacks, there participates more than one malicious node as compare to rushing attacks, where, there was a single malicious node. In wormhole attacks when a malicious node receives packets, it sends them to another malicious node through a tunnel, that controlled tunnel between two malicious nodes is actually called a wormhole [57]. Attacker uses the wormhole to get more and more network traffic by creating an illusion of the best available path to other nodes. In this way a lot of network traffic is passed through a controlled tunnel [59] [60]. Wormhole attacks disturb the routing protocols and prevent them from finding a reliable route other than the wormhole. **Figure 6** clearly shows that there are two high speed Malicious nodes "M" and "N". There exists a wormhole between them. Route reply path will always be S->M->N->D.

5.3.4. Sinkhole Attacks

Sinkhole attack is another dangerous attack in MANET's. In sinkhole attacks a malicious node broadcasts the wrong routing information to advertize itself as a specific node. In this way the adversary node get attention the whole network traffic and alter the incoming packet before their re-forwarding or even drop the packets. Malicious node tries to capture the secure information of other nodes as shown in **Figure 7**.

Sinkhole attacks effect the performance or routing protocols such as AODV either by maximizing the Sequence_No or minimizing the hop count [52] [61]. In this way the malicious node succeeds in claiming that the best available path passes through it.

5.3.5. Replay Attacks

MANET is the network of self organizing and arbitrary moving network nodes. At run time nodes dynamically change their location. Replay attacks take the advantage of this vulnerability of MANET. In replay attacks [49] a

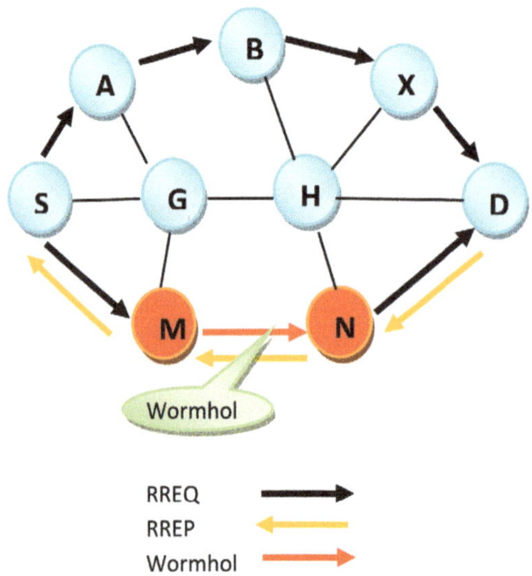

Figure 6. Wormhole attacks.

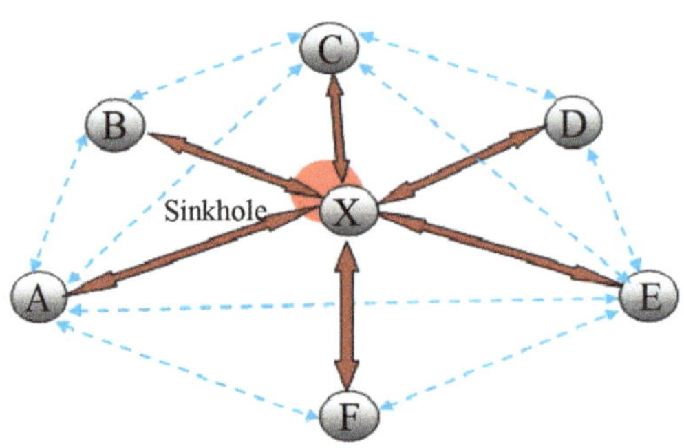

Figure 7. Sinkhole attacks.

malicious node records control packets (TC messages in case of OLSR) of other nodes, move its location and resend them on some later time. In this way other nodes are being cheated by adversary node and they record the stale routes in their routing table that even does not exist, at all. Usually packets are delayed or fraudulently repeated by the malicious node.

5.3.6. Link Withholding and Spoofing Attacks

In link spoofing attacks the malicious node withholds the link and does not broadcast routing information to the specific nodes or it advertize the fake link information to the non neighboring nodes to disturb the routing operations [59]. Kannhavong *et al.* [62] showed that in OLSR protocol the attacker broadcasts the fake link to next two hop neighbors of the target node. In this way the target is black mailed to choose the infected node as a MPR node (see **Figure 8**).

5.3.7. Resource Consumption Attacks

In resource consumption attacks, the compromised node targets a node to exhaust its resources like battery power and bandwidth by engaging the victim node in some fuzzy network operation. Usually excessive route requests are sent to the victim to consume its resource like battery power and band width [59] [63]. Flooding is the major technique to launch a resource consumption attack. S. Desilva *et al.* [64] have demonstrated their work that flooding can degrade up to 84% overall throughput of MANET. As the result of this resource consumption the victim node remain no more available for routing and cooperative network algorithms.

5.3.8. Sybil Attack

The Sybil attacks are launched in a network by generating a number of fake identities of the network node. A single malicious node exhibit itself as a many independent nodes. Actually these additional fake identities acquired by the malicious node are called Sybil nodes [65] [66]. The Sybil node may steal the identity of another legitimate node or even fabricate new identity for itself. In the presence of Sybil nodes in the network it is too much difficult to track the misbehaving nodes. The presence of these virtual Sybil nodes may generate network issues like unfair resource allocation among network nodes, absorption a lot of network traffic, denial of service by redirecting the packet of legitimate nodes to their self etc.

See the **Figure 9** where a malicious node "M" get three additional Sybil identities "X", "Y" and "Z". Actually

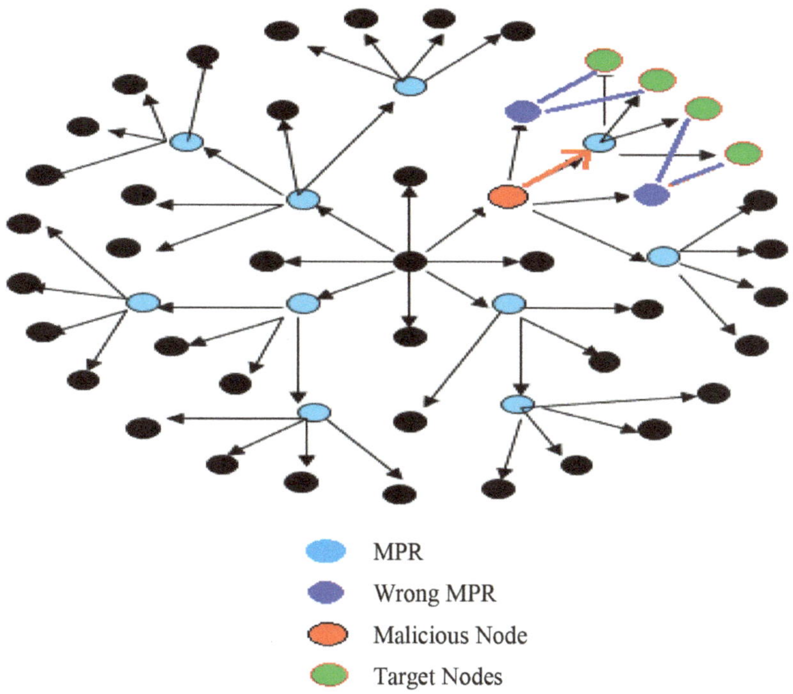

Figure 8. Wrong Selection of MPR.

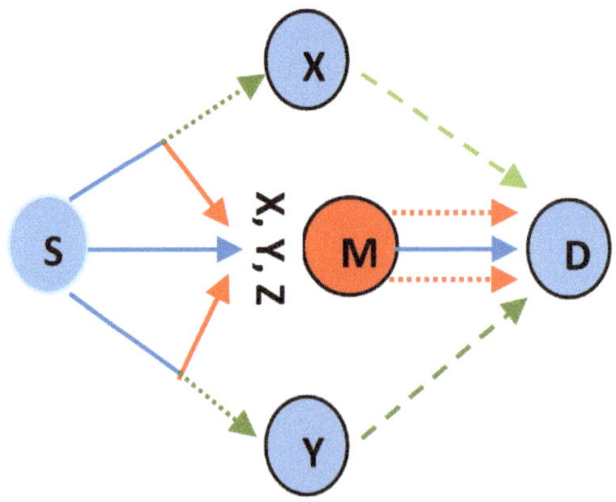

Figure 9. Malicious node M with Sybil identities X,Y,Z.

"Z" is the brand new Sybil identity while the identities "X" and "Y" are stolen from legitimate nodes to get diversion of their traffic to pass through the node "M". Now the node "M" is capable either to pass or block the traffic towards the node "D".

5.3.9. Byzantine Attacks

The Byzantine attacks may involve single or group of compromised nodes in a network. Usually byzantine attacks target the MANETs either by creating routing loops or sending packets via non-optimal path to degrade overall routing services of the network [67].

5.4. Transport Layer Attacks

Transport layer accepts messages of variable length from session layer and pack them into packets and submit them to the network layer for transmission. On reaching destination these packets are again reassembled at transport layer. So sequencing and reassembling are two key operations at transport layer. Transport layer assigns a sequence number to each packet during segmentation, prior to transmission. This sequence number is used to reassemble packets in a proper order at destination. TCP and UDP are two major protocols used at transport layer. Transport layer attacks may change the packet sequence number to disturb the reassembling.

5.4.1. Session Hijacking

In MANET there is a concept of just initial setup of the session. There exists no proper session protection mechanism during the communication. Attacker takes the advantage of this vulnerability and spoofs the IP address of the victim node and gives opportunity to the malicious node to behave as a legitimate node. After spoofing, attacker captures the intended Sequence_No from the packets and becomes in a position to launch denial of service attacks (DoS) on the basis of captured Sequence_No. usually secret information like password, logion name etc. are captured during session hijacking attacks.

TCP-ACK storm is the good example of session hijacking. See the **Figure 10** where malicious node "M" injects a TCP_ACK hijacking attack and injects a session data to node "A". The node "A" sends a acknowledgment packet (ACK) to node "B" but this packet does not contain a sequence_number that was expected by the node "B". So on receiving packet at node "B", when node "B" tries to synchronize the TCP session with node "A", it remain fails due to change in session data attack, launched by node "M". Node "B" becomes confused and sends lost ACK back to node "A" for resynchronization. In this way TCP-ACK loop is established and due to TCP- ACK storm system becomes unavailable to other nodes.

5.4.2. SYN Flooding Attacks

It is a special kind of denial of service attacks where a attacker sends a successive SYN requests to the target node to consume its resources and make it unavailable to legitimate traffic. Normally there is a three way hand shaking mechanism between sender and receiver nodes that is:

1) Node "X" requests for connection by sending SYN message to node "Y".

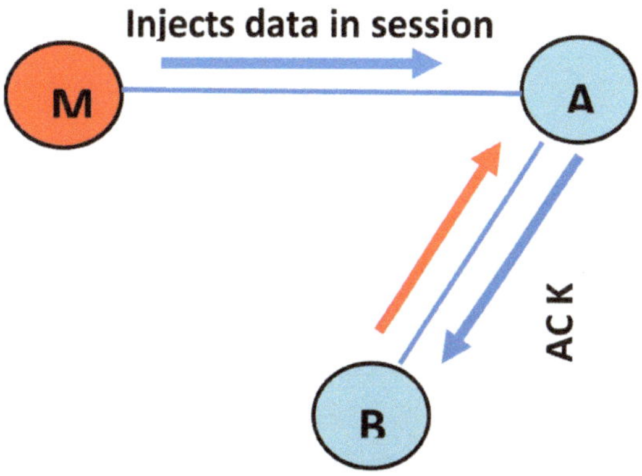

Figure 10. Session hijacking example.

2) Node "Y" acknowledges request by sending SYN-ACK message to node "X".
3) Node "X" again responds with ACK and connection is established between "X" and "Y".

But in SYN attacks the node "X" floods the SYN request at all ports of the node "Y" but does not acknowledge the SYN-ACK that was generated by node "Y" to establish the connection. In some cases it replies with the spoofed source IP. So there will form a loop of half opened connection between node "X" and "Y". The node "Y" will reply again and again and ultimately it will go to the denial of service due to non availability of free ports. This situation occurred due to flooding of SYN attacks.

5.5. Application Layer Attacks

Application layer is the actual layer where user interacts with the data. Many protocols work at this layer such as HTTP, SMTP, FTP, TELNET etc. These protocols are vulnerable in nature so they can be attacked by the expert attacker at the application layer.

5.5.1. Malicious Code Attacks

There is a general rule of data processing that is, GIGO (garbage in garbage out).This rule also applies at data communication. If we shall transmit a wrong data definitely target will also receive a wrong data. At application layer great care is needed to protect data against the malicious code, worms and virus attacks before the transmission and even after the reception at target. Viruses, spywares, trogon horses and other malicious codes can damage application, data and even operating system. Data corruption attacks are very common at application layer. These attacks may change the format of data even convert data into some unreadable format. Therefore certain measures should be taken to protect data at application layer and counter fight against application layer attacks.

5.5.2. Repudiation Attacks

Repudiation means the denial from active participation in network communication and abstain from the taking the ownership of what has been transmitted or received by the node. In the presence of repudiation attacks it becomes difficult to isolate the origin of malicious activity if no one will accept the responsibility. At the level application layer still there are chances of network attacks. Many of security measures taken at different layers are still not sufficient to protect the packets. Application layer programs such as antivirus and firewalls must be configured properly to avoid such attacks. Firewalls and logical port security is also very important. Attacker may take advantage of open ports. So, mutual cooperation of running application and operating system should guarantee the security and availability of under lying logical communication ports for smooth network communication.

MANET attacks are summarized in **Table 2**.

6. How to Meet Security Challenges?

As so for we have done a lot of deal with MANET, its vulnerabilities, different types of MANET attack. After the detailed analysis of wireless network vulnerabilities and network attacks, we are now in a position to propose

Table 2. Summary of MANET attacks.

Attacks			
External and Internal Attacks			
Active and passive attacks			
Stalling classification of routing attacks	Modification of attacks		
	Interception attacks		
	Fabrication attacks		
	Interruption attacks		
Layer Wise Network attacks	Physical Layer attacks	Eaves Dropping	
		Jamming attacks	
		Active interference Attacks	
	Data/Mac Layer attacks	Selfish Nodes behavior	
		Malicious nodes behavior	DoS Attacks
			Attacks on network integrity
			Attacking neighbor sensing protocols
			Traffic Analysis
	Network Layer Attacks	Black hole attacks	
		Rushing Attacks	
		Warm hole attacks	
		Sinkhole Attacks	
		Replay attacks	
		Link withholding and spoofing attacks	
		Resource consumption attacks	
		Sybill Attacks	
		Byzantine Attacks	
	Transport Layer Attacks	Session Hijacking	
		SYN Flooding Attacks	
	Application Layer	Malicious code Attacks	
		Repudiation Attacks	

some solutions. It is the time to meet the security challenges to make the MANET more and more secure. In broader sense there are three possible techniques of solutions:

- Cryptography/Encryption
- Intrusion detection system
- Secure routing

6.1. Cryptography

Cryptography is the one the earlier solutions that were proposed for the data security during its transmission from sender to receiver. In cryptography data is usually encrypted before transmission at source and a mutual agreed key is shared to decrypt it, reaching at destination. At reaching the destination data is again reformed into understandable format using decryption. Many techniques are used in the field of cryptography such as a public key, authentication and digital signature etc.

6.2. Intrusion Detection

Once there was a time when cryptography was considered enough for the intrusion prevention but very soon it was realized that it was not more than just a first aid to security threats. With the passage of time the idea of a proper intrusion detection system was introduced in network security. Intrusion detection is mechanism to monitor the activities of the network at run time. An intrusion detection system (IDS) performs this duty. It collects the data about network activities and evaluates data to trace any security violation in a network. If it finds any security hazard, it alerts the whole network and launches defense line to combat against the threat. Intrusion detection systems are developed on the basis of two basic assumptions that [68]:

- It is possible to monitor the activities of the user and the program.
- A clear line of differentiation can be drawn between normal and intrusive network activities.

Historically early intrusion detection systems were developed for wired networks, where traffic must go through several network devices like routers and gateways, hence an IDS can easily be implemented and hard coded into these devices [69] [70]. These IDS's were not compatible with MANET due to mismatching characteristics of wired and wireless networks. Therefore, a modified version of IDS was required for the MANET. On the basis of auditing data intrusions detection system can be classified into two types:

- Host based IDS
- Network based IDS

A host based intrusion detection system depends upon operating system and application logs for intrusion detection while on other hand network based intrusion detection systems rely on packets captured from network traffic. IDS's are classified into three categories on the basis of the techniques of intrusion detection [71].

1) Anomaly detection system: it keeps the record of normal network behavior and periodically compares the captured data with the recorded normal behaviors of the network. If there found any deviation from the baseline recorded data it is treated as intrusion.
2) Misuse detection system: it keeps the definitions (patterns) of the known attacks and compares these patterns with captured data to find any intrusion.
3) Specification based detection: it is the definition based intrusion detection where set of constraints define a normal behavior of network. The system continuously monitors the network operations according to the defined parameters.

Now the issue is that which intrusion system is best for the MANETs? Different architectures of IDS's are available but the optimal IDS architecture for a MANETs may depend on the network configuration itself [72]. Following is the list of major architectures of MANET IDS's.

- Stand alone IDS
- Distributed and cooperative IDS
- Hierarchal IDS
- Mobile agent

6.2.1. Stand Alone Intrusion Detection System

In stand-alone architecture, the IDS run on each node independently. The IDS just monitors the behavior of that particular node only, on which it is installed. There is no exchange of data and cooperation among different nodes of the network even in the same network nodes remain unaware of what is happening with neighboring nodes. Although this architecture is not so much effective due to its limitations but it may be implemented where each node is capable of running IDS. It is not recommended for MANET.

6.2.2. Distributed and Cooperative Intrusion Detection System

Keeping in mind the vulnerable nature of MANET, the distributed and cooperative model of IDS was first proposed by Zhang and Lee [73]. They stated that the intrusion detection system may be, both, distributed as well as cooperative. Every node participates actively by having an intrusion detection agent running on it. An intrusion detection agent is responsible monitoring local event, data collection, identifying the possible intrusion, as well as launching response independently. Although each intrusion detection agent works well locally, however, neighbouring nodes also share their investigations with each other for mutual cooperation. See **Figure 11**

Especially Inter node cooperation occurs when a node detects an anomaly but don't have enough evidence to figure out the kind of intrusion that was occurred. In that scenario, a node shares its data with other nodes within the communication range to check their security logs to trace the possible intruder. Let's us now discuss the internal structure of the IDS agent. Here is the conceptual model comprising of four major functional modules as in **Figure 12**.

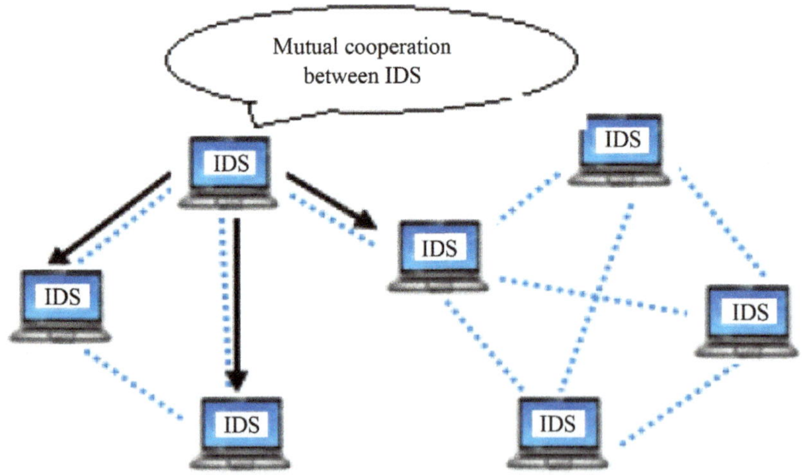

Figure 11. Mutual cooperative IDS architecture.

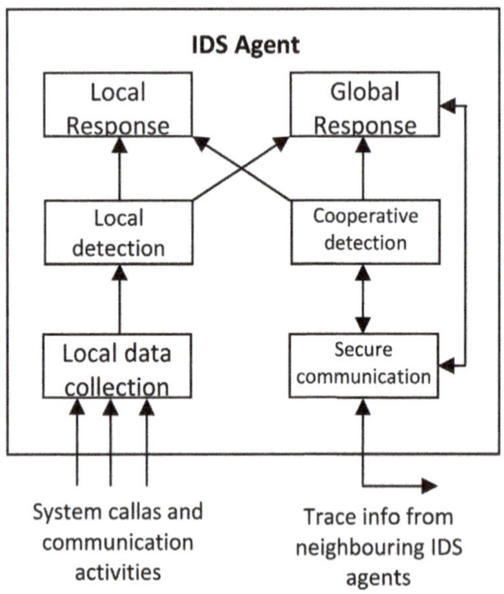

Figure 12. Conceptual model of an IDS agent.

1) Local data collection module: it is responsible to handle the collection of real time auditing data coming from different sources.
2) Local detection engine: it examines and evaluates the collected data by local data collection module. It inspects the data for any possible anomaly. The IDS should not rely on misuse detection technique to find intrusion because it is based upon the comparison of collected data with the known patterns of the attacks. Successful IDS should relay on statistical Anomaly Detection Technique rather than the use of misuse Detection Technique, which is not capable detect novel attacks.
3) Cooperative detection engine: this module is responsible for inter node sharing of suspicious anomalies. If there found any ambiguous anomaly at any particular node, the cooperative detection engine initiates a cooperated intrusion detection process by advertizing its state information to all nearby neighbors. As a result of this initiation all other nodes also check their current states and reach to consciences about intrusion and its origin.
4) Intrusion response module: this module is responsible for launching anti response against the possible intrusion. The anti intrusion response is dependent upon the nature and level of the intrusion found. Some time it simply reassigns the keys or rearranges the network nodes. The communication with compromised nodes can be blocked or even these nodes may be removed from the network community.

6.2.3. Cluster-Based Intrusion Detection System

As we discussed in the previous section about distributed and cooperative IDS, where all the nodes equally participate in the process of cooperative intrusion detection even then it was not necessary, the participation of each node. It may leads to a lot of power consumption in network. As a result of such consumption of energy a node may behave as selfish node to save its energy and regret to participate in cooperation network operations. To solve this problem Huang *et al.* presented the idea of cluster-based intrusion detection system [74]. In cluster-based IDS the network is divided into cluster so that the cluster head, usually, have to perform more functionality than member nodes. The cluster head functions as control point like router or gateway of wired networks. The IDS agents of the member nodes are responsible for local detection and response but the cluster head is responsible for local detection as well as inter cluster intrusion detection. In this way only cluster heads participate in network Intrusion while member nodes only monitor their local intrusion. All the nodes in the network are organized in such a way that every node becomes a member of at least one cluster. Usually a cluster is composed of the one hop neighbors within the direct communication range of each other. There will be a single node in a cluster that will play a role of cluster head for some period of time. Cluster head is chosen on fairness and efficiency basis. By fairness we mean that every node have equal chances for the selection being as a cluster head and each node will act as cluster head for some period of time. There should exist a mechanism to choose a cluster head periodically, a node with higher efficiency. Here is the finite state machine of the cluster formation protocol in the **Figure 13**.

There are four states in cluster formation protocol *i.e.* initial, clique, done and lost. At first all the network nodes will be in initial state, which means they will monitor their behavior for intrusion detection locally. There is a pre requisite prior to the selection of cluster head that is the clique computation. A clique is defined as the group of nodes where each pair has a direct link with each other. After the clique computation every member in a clique is aware of its clique fellows. Then a node with higher efficiency will be selected as cluster head, at random. There are further two protocols that assist cluster for doing validation and recovery tasks.

1) Cluster valid assertion protocol: A node periodically uses this protocol to maintain its link with cluster-head. If the link is broken with cluster-head, the node looks for another cluster-head for link establishment. In case of failure it goes into a LOST state and generates a route recovery request. There should re-election timeout for cluster-head to ensure fairness in cluster-head selection mechanism. At the timeout expiry all the nodes go into INITIAL state from the DONE state.
2) Cluster Recovery protocol: this protocol is used when a cluster member loses its connection with cluster-head and goes into LOST state. It helps the node in discovering a new cluster-head. It is proven by Huang *et al.* [74] that cluster-based IDS performs very well with low CPU utilization and power consumption as compare to distributed and cooperative IDS's.

6.2.4. Mobile Agents for IDS's

There is another idea of mobile agents for the intrusion detection systems. Due to its ability to move through in a

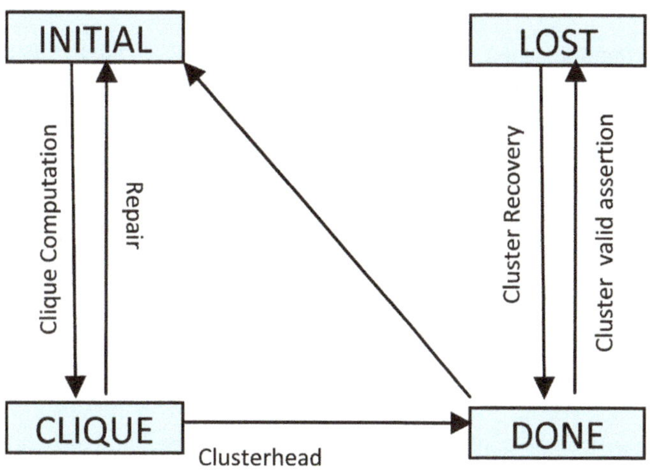

Figure 13. Finite state machine of cluster.

large network, each mobile agent is assigned a single particular task. Different mobile agents are distributed over a large network to distribute the intrusion detection task among all nodes. There are several advantages of mobile agents. Since each mobile agent is assigned a single task so the intrusion detection overhead is distributed among the whole network and a lot of energy is saved from consumption. IDS also become a fault tolerant. In case of the failure of any node the mobile agent of the other node handles the situation.

6.2.5. Intrusion Detection through Cross Layer Analysis

Multi layered IDS may produce better results than single layered intrusion detection systems because different attackers use different network layers to attack. So multi layered IDS is the better approach because on the spot detection and remedy is more effective than later diagnosis and therapy but multi layered IDS has its own pores and coins. For example it may produce more processing overhead on the nodes than the single layered IDS's. The idea of cross layer analysis and intrusion detected was proposed by *Parker and his colleagues* [75]. In their paper they elaborated that how smartly an attacker takes the advantage of multiple vulnerabilities at multiple layers to stay below the detection threshold of the single layered IDS. In multi layered IDS data from different layer stacks is analyzed comprehensively. There is no debate about the efficiency of cross layered IDS, definitely a cross layered IDS is better in performance against security threats as compare to single layered IDS but we have to do something to balance the processing overhead.

6.2.6. Watch Dog Path Rater

Watch dog rating is a special technique to improve the performance of the network even in the presence of nasty nodes. Let's have a look at working principal of the watch dog as it explained with detail in [76] [77]. Watch dog detects the misbehavior of the adjacent nodes by keeping a copy of sent packets in a buffer of the node. It continuously spies the neighboring nodes to determine whether they have forwarded the packet as it is or forwarded a modified version of packets. If the packets are forwarded without modification than they are removed from the buffer of observing node otherwise after some timeout they are considered as dropped are modified. In that case the neighboring node that was responsible for packet forwarding is assumed being as malicious node. Each successive violation by the specific node increases its chances to be declared as malicious one. If the number of violations crosses the some predefined threshold then the node is declared as malicious one and its information is passed to the path rater component. At each node the path rater component rates all the known neighbor nodes with respect to their reliability regarding violations. Initially nodes start with a neutral rating but ratings are updated at run time according to the perspective of any particular node. The nodes which are found misbehaving by the watchdog are immediately rated as −100. A packet modification by the node is considered as misbehavior while link breakage is considered as unreliability. Nodes are selected for the routing based upon their rating and reliability so watch dog is a good way to detect and boycott the malicious nodes in the network to improve the network security.

6.3. Secure Routing Techniques

Routing is the core of communication in MANETs so routing protocols are the hot target for security attacks. In the sub-section of section 5.3 we have discussed in detail different security attacks against routing protocols. Now we discuss the proposed solutions to these attacks.

6.3.1. Defense against Wormhole Attacks

Wormhole attack is the one of the major attacks against the routing protocols [78]. For detailed information about wormhole attacks please refer to section 5.3.4 of this paper. In [79] Y. Hua, A. Perrig and D. Johnson presented a solution of wormhole attacks that is a packet leashing. It is a mechanism to detect and launch a defense against wormhole attacks [78] [79]. Leash is the additional information that is added to packet to restrict it maximum life and distance covered by the packet. There are two types of leashes (Temporl leash and Geographical leash). Both of these leashes protect the network by restricting the packets, not to go away from the certain distance limit.

1) Temporal leash: it restricts the upper bound on life time of the packet. Every node calculates the expiry time "t_e" and padded it into its packets so that it may not travel far farther than certain distance "L". At the arrival of packets each node evaluates the "t_e" contained in packets, compare it with the current time and decide

about, whether the RREQ was possibly tunneled through a malicious high speed link? To implement temporal leashes the TIK protocol is used. TIK stands for TESLA with instant key disclosure and it is the extension of TESL protocol [80]. TIK protocol provides defense against wormhole attacks by preventing the packet to travel longer distance than the nominal radio range.

2) Geographical leash: it ensures that the receiver is within the certain distance range from the sender. Every node must be provided with respective geographical position and the transmission time of the packets. This can be used along with signature to catch attackers that are supposed to be residing on multiple locations.

Based upon geographical location, a similar solution was also proposed by D. Dhillon *et al.* [81]. However, it requires the integration of the public-key infrastructure with the time synchronization between all nodes. According to them in MANETs using OLSR as a routing protocol, each node should insert its current location and time stamp in HELLO messages. At the arrival of messages this embedded information is used compute the travelled distance. If the distance exceeds some maximum limit, the messages are considered highly suspicious. Some authors have proposed different detection techniques of wormhole attacks. The L. Hu and D. Evans in [82] have shown the antenna based detection techniques of wormhole attacks that do not require clock synchronizations among nodes. So on Qian *et al.* [83] introduced a new technique of detection that is statistical analysis over multiple path routing but this technique is not valid for AODV because it does not maintain multiple paths.

6.3.2. Defense against Rushing Attacks

Rushing attack is kind of denial of service attacks that acts against routing protocols On-demand protocols are more susceptible to these attacks where rout discovery is subverted by using duplicate route suppression during process of route discovery [44] [57] [58]. For detailed discussion about rushing attacks, refer to section 5.3.2 of this paper. When a route discovery request is initiated by the sender, the attacker acts very quickly. Attacker passes the same request to all neighbors of the destination very quickly before the arrival of legitimate request from other legitimate route. There is a problem with on demand routing protocols that is they restrict an intermediate node to choose at most single route request for further forwarding to the destination. As a result of this efficiency of the attacker and the limitation of protocol all the legitimate request are discarded by the neighbors of the destination because they have already forwarded such request sent by the attacker node. So the denial of service occurs. To undermine the rushing attacks a frame work of techniques was introduced by Y. Hu, A. Perrig and D. Johnson [57]. This frame work is comprised of secure neighbor detection, secure route delegation and randomized route request forwarding.

1) Secure neighbor detection: it allows the neighbors to verify each other that they are within the maximum transmission range of each other. Now if a node "X" determines in advance that node "Y" is neighbor (within allowable communication range), it sign a *Route delegation message* allowing the "Y" for further forwarding of the request. Now when "Y" determines that "X" is within a allowable range it accepts the delegation message of node "X". in this way genuine neighborhood of the nodes can be guaranteed.
2) Randomize Selection of R_REQ: it ensures the randomize selection of R_REQ for further forwarding which replaces the duplicate suppression in on-demand protocols. So the path with low latency is more likely to be selected but not guaranteed. Relationship between different security mechanisms is elaborated in **Figure 14**.

6.3.3. Defense against Black Hole Attacks

Different efforts were made to combat against black hole attacks. Designing of security-aware *ad hoc* routing protocol SAR [84] is the one of those which based upon on-demand protocols. SAR uses two techniques to countermeasure the black hole attacks. First it inserts a security metric in RREQ packets and secondly it uses alternate route discovery mechanism. Security metric is inserted for achieving an acceptable security level. At

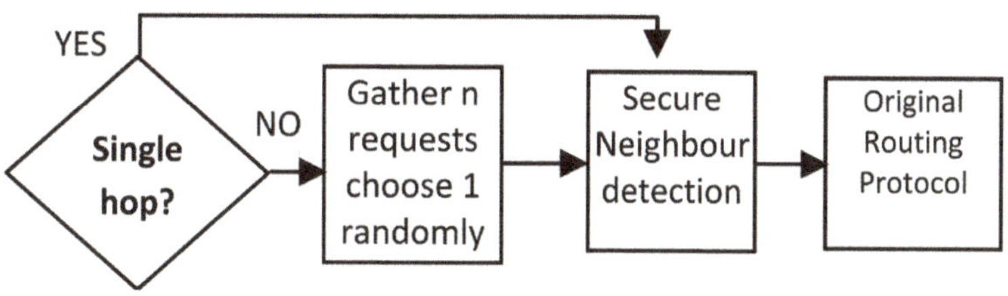

Figure 14. Combined mechanism to MANET against rushing attacks.

each trust level packets are encrypted with a shared symmetric cryptographic key. Nodes with different trust levels remain unable to read encrypted packets. Whenever a node receives a RREQ from a particular source, it verifies the trust level associated with it from the security metric information attached with packet. On satisfaction, the packet is forwarded otherwise it is dropped by intermediate nodes. Likewise when the packets reach at destination, they are evaluated. If destination is satisfied of complete end-to-end path then it generates a RREP otherwise source is notified to adjust its security level to any other alternative route. Another approach to combat against black hole attacks was introduced by S. Lee *et al.* [36]. They introduced the idea of sending confirmation CRREQ and CRREP prior to sending routine RREQ and RREP requests to verify the validity of the path from source to destination. However this approach is fail against pair of nodes working in collusion. Al-Shurman *et al.* [85] proposed a solution of this problem that the source node should wait for arrival of RREP from multiple nodes and then decide the accuracy of path information.

6.3.4. Defense against Byzantine Attacks

In [86] Robust source routing (RSR) was introduced by Cr' epaue *et al.* as the solution of byzantine attacks. RSR uses a fore runner (FR) packet to notify the intermediate nodes with expected route and data flow time frame. If the intermediate node does not receive any data flow within the notified time period it informs the source node about this event. In this way the links with selfish and malicious behavior could be isolated.

6.3.5. Defense against Resource Consumption Attacks

In [63] Yi *et al.* proposed an effective security mechanism to control resource consumption attacks. This is particularly effective for MANETs, using AODV as a routing protocol. According to the proposed mechanism each node computes and records the RREQs arriving from neighboring nodes. If the RREQ rate from the particular nodes exceeds from some particular threshold rate, the node is blacklisted and its further requests are simply discarded. The technique seems a good but there is a problem with this technique. It may even block some legitimate because it is prone to false positive.

Similarly in [64] S. Desilva and R. V. Boppana introduced an anomaly-based detection mechanism, where, threshold rate of RREQs is calculated on the fly based upon statistical analysis of RREQs.

6.3.6. Defense against Link Spoofing Attacks

B. Kannhavong *et al.* [62] showed that a link withholding attack against (TC) messages of OLSR protocol can isolate and prevent a particular node from communication. They also proposed a solution to detect link withholding and spoofing attacks. Their solution is based upon a hypothesis that if a node receives "hello" messages form its MPR but does not receive "TC" messages then the node is supposed to be a suspicious one. In that situation the node is allowed to switch to another node. However this technique works very well but it is fail to detect attacks launched by two malicious partners, where first partner pretend to advertize "TC" messages while the second one discard them. D. Raffo *et al.* In [87] introduced a new detection scheme of link spoofing attacks which relies on the information obtained from GPS and encrypted time stamps. In this scheme every node is bound to advertise its GPS coordinates along with time-stamp. From these two pieces of information, It is possible to detect link spoofing by calculating inter nodal distance of particular nodes. Likewise it could also be verified whether the nodes are in between or beyond the maximum transmission range. The major limitation of this solution is that each node must be equipped with GPS all the times.

6.3.7. Defense against Replay Attacks

In order to protect MANETs against replay attacks C. Adjih *et al.* [55] proposed a solution that is based upon time-stamps and asymmetric encryption. Every receiver node compares its current time with the embedded time-stamp of the control messages. If the time difference is much more than some threshold duration the packet is considered as replayed and discarded by the receiver. In this way routing protocol avoids to update its routing table with stale routes.

7. Conclusion

Throughout the paper we addressed each and every aspect of the MANET that is concerned with the security in any way at any extent. We started from the scratch; we discussed the architecture of MANET, vulnerabilities of the MANET, different security threats and even different types of security attacks in this paper. Non-stable architecture of MANETs and wireless vulnerabilities helped us to understand, why the MANETs are easy to attack. Layer wise network attacks and their proposed solutions enlightened us to understand the way of action and execution plan of different network attacks. From the whole scene one thing is crystal clear that MANETs will go on the way in the same fashion without any major change. Wireless is their natural ally as a communication medium; there is no substitution or alternative of wireless medium. These things will persist at least in near future unless the emergence of any new technology. Even though there is the invention on any substitution, it will take long to switch to that particular technology. At present we have to accept the MANETs and wireless vulnerabilities as a good evil. We can make effort to improve the things by keeping in mind these vulnerabilities. For a moment if we think positively, in fact we are blessed with a great room, from research point of view due to these vulnerabilities of the MANET and wireless medium. A lot of research work has been done by the researchers but still there is a lot to do. Network security is a dynamic issue. New and new attacks are getting introduced. So the constant efforts are required to make the MANET more and more secure. There is a lot of research scope in the field of secure routing. Intrusion detection and its healing is another research hot spot for the network security researchers. Most of the intrusion detection systems and techniques look very beautiful and convincing on papers but still applied research work is waiting for the researchers in certain areas of network security. There is need to implement, evaluate and improve these intrusion detection systems practically.

8. Open Issues and Future Research

MANETs are the networks of the day and wireless is their natural ally as a communication medium. We discussed different vulnerabilities of wireless medium in detail. For a moment if we think positively, in fact we are blessed with a great opportunities, from research point of view due to these vulnerabilities of the MANETs and wireless medium. A lot of research work has been done by the researchers but still there is a lot to do. There is a lot of research scope in the field of MANETs security. Routing is the core of communication in networks therefore routing protocols are high valued target for the attackers. Most of the known routing attacks against MANETs take the advantage of the different vulnerabilities of the routing protocols therefore *development of new secure versions of the routing protocols is most demanding area for the researchers*, now a days.

A lot of work has been done by the researchers in finding and combating against different attacks against routing protocols but still there is *wide research scope in detection of newly born security attacks and their solutions*. Up till now, most of the known attack detection techniques are attack specific. Even their proposed solutions are either specific to particular attack or protocol. Likewise most of the proposed solutions are not efficient in the presence of multiple cooperative malicious nodes. *From research point of view it is the open issue to develop a generic attack detection and prevention mechanism to handle a variety of network attacks.* System based intrusion detection system (IDS) is another attractive research hot spot in MANETs. Most of the intrusion detection techniques and combating systems are designed to handle security issues at particular network layer of the OSI model. *There are great research opportunities to develop intrusion detection systems based upon cross layer analysis of security attacks.*

We have discussed in detail different architectures of the intrusion detection systems. Intrusion detection systems utilize different strategies to find out possible intrusion. *Development of a specification based smart intrusion detection system is another demanding area for the researchers.* Researchers may conduct a research to find new constraints to improve the defined set of normal behaviors of the network. Known anomaly detection and healing techniques look very beautiful and convincing on papers but still *applied research work is waiting for the researchers in certain areas of network security.* A lot of work has been done in the field of intrusion detection systems but *running cost of detection systems is still an open research issue for the researchers.* Researchers may conduct their research to minimize the running cost of the intrusion detection system in the presence of battery constrained nodes.

Acknowledgements

I, personally feel extremely beholden to Prof(R) Ghulam Qasim Shah who with self-abnegation graciously spared his time and reviewed this paper.

References

[1] Conti, M. (2003) Body, Personal, and Local *Ad Hoc* Wireless Networks. In: Ilyas, M., Ed., *The Handbook of Ad Hoc Wireless Networks*, CRC Press, Inc., Boca Raton, 3-24.

[2] Weiser, M. (1991) The Computer for the Twenty-First Century. *Scientific American*, **265**, 94-104.

[3] Perkins, C.E. (2001) *Ad Hoc* Networking. Addison-Wesley, Reading.

[4] Corson, M.S., Macker, J.P. and Cirincione, G.H. (1999) Internet-Based Mobile *Ad Hoc* Networking. *IEEE Internet Computing*, **3**, 63-70. http://dx.doi.org/10.1109/4236.780962

[5] Macker, J. (1999) Mobile *Ad Hoc* Networking (MANET): Routing Protocol Performance Issues and Evaluation Considerations. RFC 2501, January1999.

[6] Johnson, D.B. and Maltz, D.A. (1996) Dynamic Source Routing in *Ad Hoc* Wireless Networks. In: Imielinski, T. and Korth, H.F., Eds., *Mobile Computing*, Springer, New York, 153-181. http://dx.doi.org/10.1007/978-0-585-29603-6_5

[7] Perkins, C., Belding-Royer, E. and Das, S. (2003) *Ad Hoc* On-Demand Distance Vector (AODV) Routing. RFC 3561, July 2003.

[8] Gellens, R. (1999) Wireless Device Configuration (OTASP/OTAPA) via ACAP.

[9] Perkins, C.E. and Bhagwat, P. (1994) Highly Dynamic Destination-Sequenced Distance-Vector Routing (DSDV) for Mobile Computers. *ACM SIGCOMM Computer Communication Review*, **24**, 234-244. http://dx.doi.org/10.1145/190809.190336

[10] Mishra, A. and Nadkarni, K.M. (2003) Security in Wireless *Ad Hoc* Networks. In: Ilyas, M., Ed., *The Handbook of Ad Hoc Wireless Networks*, CRC Press, Inc., Boca Raton, 499-549.

[11] Yang, H., Luo, H., Ye, F., Lu, S. and Zhang, L. (2004) Security in Mobile *Ad Hoc* Networks: Challenges and Solutions. *IEEE Wireless Communications*, **11**, 38-47. http://dx.doi.org/10.1109/MWC.2004.1269716

[12] Raffo, D., Adjih, C., Clausen, T. and Mühlethaler, P. (2005) Securing OLSR Using Node Locations. *Proceedings of the* 11*th European Wireless Conference* 2005—*Next Generation Wireless and Mobile Communications and Services* (*European Wireless*), Nicosia, 10-13 April 2005, 1-7.

[13] Myers, A.D. and Basagni, S. (2002) Wireless Media Access Control. In: Stojmenovic, I., Ed., *Handbook of Wireless Networks and Mobile Computing*, John Wiley & Sons, Inc., New York, 119. http://dx.doi.org/10.1002/0471224561.ch6

[14] Zhang, Y., Lee, W. and Huang, Y.A. (2003) Intrusion Detection Techniques for Mobile Wireless Networks. *Wireless Networks*, **9**, 545-556. http://dx.doi.org/10.1023/A:1024600519144

[15] Perkins, C.E. (2001) *Ad Hoc* Networking: An Introduction. *Ad Hoc Networking*, **40**, 20-22.

[16] Papadimitratos, P. and Haas, Z.J. (2002) Securing Mobile *Ad Hoc* Networks. In: Ilyas, M., Ed., *The Handbook of Ad Hoc Wireless Networks*, CRC Press, Inc., Boca Raton, 665-671. http://dx.doi.org/10.1201/9781420040401.ch31

[17] Zhou, L. and Haas, Z.J. (1999) Securing *Ad Hoc* Networks. *IEEE Network*, **13**, 24-30. http://dx.doi.org/10.1109/65.806983

[18] Vinayakray-Jani, P. (2002) Security within *Ad Hoc* Networks. *Proceeding of the PAMPAS Workshop*, London, 16-17 September 2002.

[19] Biswas, K. and Ali, M.L. (2007) Security Threats in Mobile Ad-Hoc Network. Master's Thesis, Department of Interaction and System Design School of Engineering, Blekinge, 9-26.

[20] Woo, S.C.M. and Singh, S. (2001) Scalable Routing Protocol for *Ad Hoc* Networks. *Wireless Networks*, **7**, 513-529. http://dx.doi.org/10.1023/A:1016726711167

[21] Parker, J. (2006) Discussion Record for the 1st MANET Reading Group Meeting. February.

[22] Viennot, L., Jacquet, P. and Clausen, T.H. (2004) Analyzing Control Traffic Overhead versus Mobility and Data Traffic Activity in Mobile Ad-Hoc Network Protocols. *Wireless Networks*, **10**, 447-455. http://dx.doi.org/10.1023/B:WINE.0000028548.44719.fe

[23] Amitabh, M. (2008) Security and Quality of Service in *Ad Hoc* Wireless Networks. Cambridge University Press, Cambridge.

[24] Gokhale, V., Ghosh, S. and Gupta, A. (2010) Classification of Attacks on Wireless Mobile *Ad Hoc* Networks and Vehicular *Ad Hoc* Networks: A Survey. In: Pathan, A.S.K., Ed., *Security of Self-Organizing Networks*, *MANET*, *WSN*, *WMN*, *VANET*, Auerbach Publications, Boston, 195-225. http://dx.doi.org/10.1201/EBK1439819197-12

[25] Papadimitratos, P. and Haas, Z.J. (2003) Secure Link State Routing for Mobile *Ad Hoc* Networks. *IEEE Symposium on Applications and the Internet*, 27-31 January 2003, 379-383.

[26] Ilyas, M., Ed. (2002) The Handbook of *Ad Hoc* Wireless Networks. CRC Press, Boca Raton.

[27] Burg, A. (2003) *Ad Hoc* Network Specific Attacks. In: *Seminar Ad Hoc Networking*: *Concepts*, *Applications*, *and Security*, Technische Universitat Munchen, Munich.

[28] Sanzgiri, K., Dahill, B., Levine, B.N., Shields, C. and Royer, E.M.B. (2002) A Secure Routing Protocol for *Ad Hoc*

Networks. *Proceedings of the 10th IEEE International Conference on Network Protocols*, Paris, 12-15 November 2002, 78-87. http://dx.doi.org/10.1109/icnp.2002.1181388

[29] Li, H., Chen, Z., Qin, X., Li, C. and Tan, H. (2002) Secure Routing in Wired Networks and Wireless *Ad Hoc* Networks. Technical Report, Department of Computer Science, University of Kentucky, Lexington.

[30] William, S. and Stallings, W. (2006) Cryptography and Network Security: For VTU, 4/e. Pearson Education, India.

[31] Rajavaram, S., Shah, H., Shanbhag, V., Undercoffer, J. and Joshi, A. (2002) Neighborhood Watch: An Intrusion Detection and Response Protocol for Mobile *Ad Hoc* Networks. In: *Proceedings of the Student Research Conference*, University of Maryland at Baltimore County (UMBC), Baltimore County.

[32] Douligeris, C. and Mitrokotsa, A. (2004) DDoS Attacks and Defense Mechanisms: Classification and State-of-the-Art. *Computer Networks*, **44**, 643-666. http://dx.doi.org/10.1016/j.comnet.2003.10.003

[33] Perkins, C. (1998) *Ad Hoc* On-Demand Distance Vector (AODV) Routing. *IEEE Internet Computing*, **2**, 58-69.

[34] DeCleene, B., Dondeti, L., Griffin, S., Hardjono, T., Kiwior, D., Kurose, J., *et al.* (2001) Secure Group Communications for Wireless Networks. *Proceedings of the IEEE Military Communications Conference, Communications for Network-Centric Operations: Creating the Information Force*, **1**, 113-117.

[35] Sun, B., Wu, K. and Pooch, U.W. (2004) Towards Adaptive Intrusion Detection in Mobile *Ad Hoc* Networks. *Proceedings of the IEEE Global Telecommunications Conference*, **6**, 3551-3555.

[36] Florian, D. (2008) Security Concepts for Robust and Highly Mobile Ad-hoc Networks. April.

[37] Lee, S., Han, B. and Shin, M. (2002) Robust Routing in Wireless *Ad Hoc* Networks. *Proceedings of the International Conference on Parallel Processing Workshops*, Vancouver, 20-23 August 2002, 73-78.

[38] Lazos, L., Liu, S. and Krunz, M. (2009) Mitigating Control-Channel Jamming Attacks in Multi-Channel *Ad Hoc* Networks. *Proceedings of the Second ACM Conference on Wireless Network Security*, Zurich, 16-19 March 2009, 169-180. http://dx.doi.org/10.1145/1514274.1514299

[39] Noubir, G. and Lin, G.L. (2003) Low-Power DoS Attacks in Data Wireless LANs and Countermeasures. *ACM SIGMOBILE Mobile Computing and Communications Review*, **7**, 29-30. http://dx.doi.org/10.1145/961268.961277

[40] Xu, W., Trappe, W., Zhang, Y. and Wood, T. (2005) The Feasibility of Launching and Detecting Jamming Attacks in Wireless Networks. In: *Proceedings of the 6th ACM International Symposium on Mobile Ad Hoc Networking and Computing*, ACM Press, New York, 46-57. http://dx.doi.org/10.1145/1062689.1062697

[41] Xu, W., Wood, T., Trappe, W. and Zhang, Y. (2004) Channel Surfing and Spatial Retreats: Defenses against Wireless Denial of Service. In: *Proceedings of the 3rd ACM Workshop on Wireless Security*, ACM Press, New York, 80-89. http://dx.doi.org/10.1145/1023646.1023661

[42] Wood, A.D., Stankovic, J. and Son, S.H. (2003) JAM: A Jammed-Area Mapping Service for Sensor Networks. *Proceedings of the 24th IEEE Real-Time Systems Symposium*, Cancun, 3-5 December 2003, 286-297. http://dx.doi.org/10.1109/REAL.2003.1253275

[43] Pullen, J. and Wood, D. (1995) Networking Technology and DIS. *Proceedings of the IEEE*, **83**, 1156-1167.

[44] Omura, J.K. (1994) Spread Spectrum Communications Handbook. Volume 2, McGraw-Hill, New York.

[45] Fadlullah, Z.M., Taleb, T. and Schöller, M. (2010) Combating against Security Attacks against Mobile *Ad Hoc* Networks (MANETs). In: Pathan, A.S.K., Ed., *Security of Self-Organizing Networks, MANET, WSN, WMN, VANET*, Auerbach Publications, Boston, 173. http://dx.doi.org/10.1201/ebk1439819197-11

[46] Gokhale, V., Ghosh, S.K. and Gupta, A. (2010) Classification of Attacks on Wireless Mobile *Ad Hoc* Networks and Vehicular *Ad Hoc* Networks. In: Pathan, A.S.K., Ed., *Security of Self-Organizing Networks, MANET, WSN, WMN, VANET*, Auerbach Publications, Boston, 195-225. http://dx.doi.org/10.1201/EBK1439819197-12

[47] Feldman, M., Papadimitriou, C., Chuang, J. and Stoica, I. (2006) Free-Riding and Whitewashing in Peer-to-Peer Systems. *IEEE Journal on Selected Areas in Communications*, **24**, 1010-1019. http://dx.doi.org/10.1109/JSAC.2006.872882

[48] Johnson, D.B. (2003) The Dynamic Source Routing Protocol for Mobile *Ad Hoc* Networks.

[49] Zhang, Y. and Lee, W. (2005) Security in Mobile Ad-Hoc Networks. In: Mohapatra, P. and Krishnamurthy, S.V., Eds., *Ad Hoc Networks*, Springer, New York, 249-268. http://dx.doi.org/10.1007/0-387-22690-7_9

[50] Gupta, V., Krishnamurthy, S. and Faloutsos, M. (2002) Denial of Service Attacks at the MAC Layer in Wireless *Ad Hoc* Networks. *Proceedings of the IEEE MILCOM*, **2**, 1118-1123. http://dx.doi.org/10.1109/milcom.2002.1179634

[51] Kyasanur, P. and Vaidya, N.H. (2003) Detection and Handling of MAC Layer Misbehavior in Wireless Networks. *Proceedings of the 2003 International Conference on Dependable Systems and Networks*, San Francisco, 22-25 June 2003, 173-182. http://dx.doi.org/10.1109/dsn.2003.1209928

[52] Radosavac, S., Benammar, N. and Baras, J.S. (2004) Cross-Layer Attacks in Wireless *Ad Hoc* Networks. *Proceedings of the* 2004 *Conference on Information Sciences and Systems*, Princeton, 17-19 March 2004, 1266-1271.

[53] Zimmermann, H. (1980) OSI Reference Model—The ISO Model of Architecture for Open Systems Interconnection. *IEEE Transactions on Communications*, **28**, 425-432. http://dx.doi.org/10.1109/TCOM.1980.1094702

[54] Lou, W. and Fang, Y. (2004) A Survey of Wireless Security in Mobile *Ad Hoc* Networks: Challenges and Available Solutions. In: Cheng, X.Z., Huang, X. and Du, D.-Z., Eds., *Ad Hoc Wireless Networking*, Springer, New York, 319-364. http://dx.doi.org/10.1007/978-1-4613-0223-0_9

[55] Hu, Y.C. and Perrig, A. (2004) A Survey of Secure Wireless *Ad Hoc* Routing. *IEEE Security & Privacy*, **2**, 28-39. http://dx.doi.org/10.1109/MSP.2004.1

[56] Gideon, N. and Edwin, B.K. (2014) Clustering Effects on Wireless Mobile *Ad-Hoc* Networks Performances. *International Journal of Computer Science & Information Technology* (*IJCSIT*), **6**, 1-19.

[57] Yi, S., Naldurg, P. and Kravets, R. (2001) Security-Aware *Ad Hoc* Routing for Wireless Networks. In: *Proceedings of the 2nd ACM International Symposium on Mobile Ad Hoc Networking & Computing*, ACM Press, New York, 299-302. http://dx.doi.org/10.1145/501416.501464

[58] Saini, A. and Kumar, H. (2010) Effect of Black Hole Attack on AODV Routing Protocol in MANET. *International Journal of Information Technology, Modeling and Computing* (*IJITMC*), **2**, 9-17.

[59] Li, W. and Joshi, A. (2008) Security Issues in Mobile *Ad Hoc* Networks—A Survey. Department of Computer Science and Electrical Engineering, University of Maryland, Baltimore County, 1-23.

[60] Jawandhiya, P.M., Ghonge, M.M., Ali, M.S. and Deshpande, J.S. (2010) A Survey of Mobile *Ad Hoc* Network Attacks. *International Journal of Engineering Science and Technology*, **2**, 4063-4071.

[61] Chirala, A.P. (2010) Analysis and Diminution of Security Attacks on Mobile *Ad Hoc* Network. IJCA Special Issue on "Mobile Ad-Hoc Networks", MANETs, 105-110.

[62] Adjih, C., Raffo, D. and Mühlethaler, P. (2005) Attacks against OLSR: Distributed Key Management for Security. *Proceedings of the 2nd OLSR Interop/Workshop*, Palaiseau, 28-29 July 2005.

[63] Kannhavong, B., Nakayama, H., Kato, N., Nemoto, Y. and Jamalipour, A. (2006) Analysis of the Node Isolation Attack against OLSR-Based Mobile *Ad Hoc* Networks. *Proceedings of the* 2006 *International Symposium on Computer Networks*, Istanbul, 16-18 June 2006, 30-35. http://dx.doi.org/10.1109/iscn.2006.1662504

[64] Yi, P., Dai, Z., Zhang, S. and Zhong, Y. (2005) A New Routing Attack in Mobile *Ad Hoc* Networks. *International Journal of Information Technology*, **11**, 83-94.

[65] Kim, K. and Kim, S. (2007) A Sinkhole Detection Method Based on Incremental Learning in Wireless *Ad Hoc* Networks.

[66] Hu, Y.C., Perrig, A. and Johnson, D.B. (2003) Rushing Attacks and Defense in Wireless *Ad Hoc* Network Routing Protocols. In: *Proceedings of the 2nd ACM Workshop on Wireless Security*, ACM Press, New York, 30-40. http://dx.doi.org/10.1145/941311.941317

[67] Desilva, S. and Boppana, R.V. (2005) Mitigating Malicious Control Packet Floods in *Ad Hoc* Networks. *Proceedings of the IEEE Wireless Communications and Networking Conference*, **4**, 2112-2117. http://dx.doi.org/10.1109/wcnc.2005.1424844

[68] Mishra, A., Nadkarni, K. and Patcha, A. (2004) Intrusion Detection in Wireless *Ad Hoc* Networks. *IEEE Wireless Communications*, **11**, 48-60. http://dx.doi.org/10.1109/MWC.2004.1269717

[69] Hu, Y.C., Johnson, D.B. and Perrig, A. (2003) SEAD: Secure Efficient Distance Vector Routing for Mobile Wireless *Ad Hoc* Networks. *Ad Hoc Networks*, **1**, 175-192. http://dx.doi.org/10.1016/S1570-8705(03)00019-2

[70] Jou, Y.F., Gong, F., Sargor, C., Wu, X., Wu, S.F., Chang, H.C. and Wang, F.Y. (2000) Design and Implementation of a Scalable Intrusion Detection System for the Protection of Network Infrastructure. *Proceedings of the DARPA Information Survivability Conference and Exposition*, Hilton Head, 25-27 January 2000, 69-83.

[71] Douceur, J.R. (2002) The Sybil Attack. In: *Peer-to-Peer Systems*, Springer, Berlin Heidelberg, 251-260. http://dx.doi.org/10.1007/3-540-45748-8_24

[72] Chan, E.Y., Chan, H.W., Chan, K.M., Chan, V.P., Chanson, S.T., Cheung, M.M., *et al.* (2004) IDR: An Intrusion Detection Router for Defending against Distributed Denial-of-Service (DDoS) Attacks. *Proceedings of the 7th International Symposium on Parallel Architectures, Algorithms and Networks*, Hong Kong, 10-12 May 2004, 581-586. http://dx.doi.org/10.1109/ISPAN.2004.1300541

[73] Brutch, P. and Ko, C. (2003) Challenges in Intrusion Detection for Wireless Ad-Hoc Networks. *Proceedings of the* 2003 *Symposium on Applications and the Internet Workshops*, Orlando, 27-31 January 2003, 368-373. http://dx.doi.org/10.1109/saintw.2003.1210188

[74] Parker, J., Patwardhan, A. and Joshi, A. (2006) Cross-Layer Analysis for Detecting Wireless Misbehavior. *Proceedings of the IEEE Consumer Communications and Networking Conference*, Las Vegas, 8-10 January 2006, 6-9. http://dx.doi.org/10.1109/ccnc.2006.1592977

[75] Zhang, Y., Lee, W. and Huang, Y.A. (2003) Intrusion Detection Techniques for Mobile Wireless Networks. *Wireless Networks*, **9**, 545-556. http://dx.doi.org/10.1023/A:1024600519144

[76] Perrig, A., Canetti, R., Tygar, J.D. and Song, D. (2000) Efficient Authentication and Signing of Multicast Streams over Lossy Channels. *Proceedings of the* 2000 *IEEE Symposium on Security and Privacy*, Berkeley, 14-17 May 2000, 56-73. http://dx.doi.org/10.1109/SECPRI.2000.848446

[77] Marti, S., Giuli, T.J., Lai, K. and Baker, M. (2000) Mitigating Routing Misbehavior in Mobile *Ad Hoc* Networks. In: *Proceedings of the* 6*th Annual International Conference on Mobile Computing and Networking*, ACM Press, New York, 255-265. http://dx.doi.org/10.1145/345910.345955

[78] Hu, Y.C., Perrig, A. and Johnson, D.B. (2003) Packet Leashes: A Defense against Wormhole Attacks in Wireless Networks. *Proceedings of the Twenty-Second Annual Joint Conference of the IEEE Computer and Communications*, **3**, 1976-1986. http://dx.doi.org/10.1109/infcom.2003.1209219

[79] Huang, Y.A. and Lee, W. (2003) A Cooperative Intrusion Detection System for *Ad Hoc* Networks. In: *Proceedings of the* 1*st ACM Workshop on Security of Ad Hoc and Sensor Networks*, ACM Press, New York, 135-147. http://dx.doi.org/10.1145/986858.986877

[80] Hu, Y.C., Perrig, A. and Johnson, D.B. (2006) Wormhole Attacks in Wireless Networks. *IEEE Journal on Selected Areas in Communications*, **24**, 370-380. http://dx.doi.org/10.1109/JSAC.2005.861394

[81] Awerbuch, B., Holmer, D., Nita-Rotaru, C. and Rubens, H. (2002) An On-Demand Secure Routing Protocol Resilient to Byzantine Failures. In: *Proceedings of the* 1*st ACM Workshop on Wireless Security*, ACM Press, New York, 21-30. http://dx.doi.org/10.1145/570681.570684

[82] Dhillon, D., Zhu, J., Richards, J. and Randhawa, T. (2006) Implementation & Evaluation of an IDS to Safeguard OLSR Integrity in MANETs. In: *Proceedings of the* 2006 *International Conference on Wireless Communications and Mobile Computing*, ACM Press, New York, 45-50. http://dx.doi.org/10.1145/1143549.1143560

[83] Hu, L. and Evans, D. (2004) Using Directional Antennas to Prevent Wormhole Attacks. *Proceedings of the Network and Distributed System Security Symposium*, San Diego, 5-6 February 2004, 131-141.

[84] Schreiber, F.R. (1973) Sybil. Regnery, Chicago.

[85] Qian, L., Song, N. and Li, X. (2005) Detecting and Locating Wormhole Attacks in Wireless *Ad Hoc* Networks through Statistical Analysis of Multi-Path. *Proceedings of the IEEE Wireless Communications and Networking Conference*, **4**, 2106-2111.

[86] Al-Shurman, M., Yoo, S.M. and Park, S. (2004) Black Hole Attack in Mobile *Ad Hoc* Networks. In: *Proceedings of the* 42*nd Annual Southeast Regional Conference*, ACM Press, New York, 96-97. http://dx.doi.org/10.1145/986537.986560

[87] Crepeau, C., Davis, C.R. and Maheswaran, M. (2007) A Secure MANET Routing Protocol with Resilience against Byzantine Behaviours of Malicious or Selfish Nodes. *Proceedings of the* 21*st International Conference on Advanced Information Networking and Applications Workshops*, Niagara Falls, 21-23 May 2007, 19-26.

20

Animal Classification System Based on Image Processing & Support Vector Machine

A. W. D. Udaya Shalika, Lasantha Seneviratne

Sri Lanka Institute of Information Technology, Malabe, Sri Lanka
Email: udayashalika@gmail.com, Lasantha.s@sliit.lk

Abstract

This project is mainly focused to develop system for animal researchers & wild life photographers to overcome so many challenges in their day life today. When they engage in such situation, they need to be patiently waiting for long hours, maybe several days in whatever location and under severe weather conditions until capturing what they are interested in. Also there is a big demand for rare wild life photo graphs. The proposed method makes the task automatically use microcontroller controlled camera, image processing and machine learning techniques. First with the aid of microcontroller and four passive IR sensors system will automatically detect the presence of animal and rotate the camera toward that direction. Then the motion detection algorithm will get the animal into middle of the frame and capture by high end auto focus web cam. Then the captured images send to the PC and are compared with photograph database to check whether the animal is exactly the same as the photographer choice. If that captured animal is the exactly one who need to capture then it will automatically capture more. Though there are several technologies available none of these are capable of recognizing what it captures. There is no detection of animal presence in different angles. Most of available equipment uses a set of PIR sensors and whatever it disturbs the IR field will automatically be captured and stored. Night time images are black and white and have less details and clarity due to infrared flash quality. If the infrared flash is designed for best image quality, range will be sacrificed. The photographer might be interested in a specific animal but there is no facility to recognize automatically whether captured animal is the photographer's choice or not.

Keywords

Image Processing, Support Vector Machine (LIBSVM), Machine Learning, Computer Vision, Object Classification

1. Introduction

Object detection and recognition based on image processing is vastly concentrating field in research. The motivation for this project is to build a system for automatically detecting and recognizing wild animals for the animal researchers & wild life photographers. The animal detection and recognition is an important area which has not been discussed rapidly. However this project is targeted to build a system which would help animal researchers and wild life photographers who devote their precious time for studying animal behavior. Technology used in this research can be modified further to use in applications such as security, monitoring purposes, etc.

Wild life photography is regarded as one of the most challenging forms of photography. It needs sound technical skills such as being able to capture correctly. Wild life photographers generally need good field of crafting skills and a lot of patience. For example, some animals are difficult to approach and thus knowledge of the animals' behavior is needed in order to be able to predict their actions. Sometimes photographers need to stay calm and quiet for many hours until the exact time. Photographing some species may require stalking skills or the use of hide/blind for concealment. A great wild life photograph also is the result of being in the right place at right time. Because of the highly consuming nature of the time it costs too much and same time life can be in danger involving in the business. Usually animal researchers often travel to remote locations all over the world. Hostile conditions are often the norm in the life of photojournalist. Some will sit for hours and hours before snapping a shot which is worthy to be sold. The photographers should be brave enough to stay in hostile environment without comfort and with great patience till animals appear. To have a perfect scene or outcome, it is always coming by fewer disturbances to the natural behavior of animal. Due to their high sensitivity animal can easily identify the presence of human being. So photographers need to be ready for facing any critical moment in the jungle because we cannot predict what will happen in next moment. The DSLR cameras which use in the industry are also quite expensive and there are limitations in shutter on-off cycle. So proper recognition must exist in such equipment. The higher the quality of the images, the more space in memory it takes. So it also encourages to having proper recognition in wild photography.

In this paper, we selected to be conservative and limited our endeavor to only three kinds of animals. This is not a random selection because it is always a challenging part to collect a set of data in animals to create proper data base. Our motivation is to achieve the recognition part, and thus we used the ACE media mpeg project for feature extraction, Support vector machine (LIBSVM) to get the probabilistic outcome. For the feature extraction we used several descriptors like Color layout descriptor (CLD), Color structure descriptor (CSD), Edge histogram descriptor (EHD), Homogeneous texture descriptor (HTD), Region shape, Contour shape. Using these descriptors, research was carried out about how the individual descriptors perform and the performances of combined descriptors.

1.1. Literature Review

The low level audio visual feature extraction was used for retrieval & classification. In this paper we present an overview of a software platform that has been developed within the ace Media project, termed the ace Toolbox that provides global and local low level feature extraction from audio-visual content. The toolbox is based on the MPEG-7 experimental Model (XM), with extensions to provide descriptor extraction from arbitrarily shaped image segments, thereby supporting local descriptors reflecting real image content. This paper will describe the architecture of the toolbox as well as provide an overview of the descriptors supported to date. It will also briefly describe the segmentation algorithm provided. Then will demonstrate the usefulness of the toolbox in the context of two different content processing scenarios: similarity-based retrieval in large collections and scene-level classification of still images [1].

This paper will concern on observing animal behavior in wild life using face detection and tracking. It will present an algorithm for detection and tracking of animal faces in wildlife videos. As an example the algorithm is applied to lion faces. The detection algorithm is based on a human face detection method, utilizing Haar-like features and Ada Boost classifiers. The face tracking is implemented using the Kanade-Lucas-Tomasi tracker and by applying a specific interest model to the detected face. By combining the two methods in a specific tracking model, a reliable and temporally coherent detection/tracking of animal faces is achieved. In addition to the detection of particular animal species, the information generated by the tracker can be used to boost the priors in the probabilistic semantic classification of wildlife videos [2].

Rapidly detecting animals in natural scenes, evoked potential studies indicate that the corresponding neural signals can emerge in the brain within 150 msec of stimulus onset (S. Thorpe, D. Fize, & C. Marlot, 1996) and eye movements toward animal targets can be initiated in roughly the same timeframe (H. Kirchner & S. J. Thorpe, 2006). Given the speed of this discrimination, it has been suggested that the underlying visual mechanisms must be relatively simple and feed forward, but in fact little is known about these mechanisms. A key step is to understand the visual cues upon which these mechanisms rely. Here we investigate the role and dynamics of four potential cues: two-dimensional boundary shape, texture, luminance, and color. Results suggest that the fastest mechanisms underlying animal detection in natural scenes use shape as a principal discriminative cue, while somewhat slower mechanisms integrate these rapidly computed shape cues with image texture cues. Consistent with prior studies, little role for luminance and color cues throughout the time course of visual processing, even though information relevant to the task is available in these signals [3].

Face identification method can be used for non-native animals to be used for intelligent trap. They developed a face identification method to distinguish target non-native alien animals from other native animals using camera captured images. When a camera recognizes a targeted animal walked in to the cage, it traps the animal in the cage. Here, we set raccoon as target non-native animal, and detect its face region by using HOG features. However, the raccoon face detector often confused by raccoon dog, which is a native animal to be preserved. So, after detecting raccoon face candidates, we distinguish them by several features and SVM. Some experimental results show that we can completely distinguish raccoon and raccoon dog from camera captured images [4].

Object recognition system is capable of accurately detecting, localizing, and recovering the kinematic configuration of textured animals in real images. Deformation model of shape automatically from videos of animals was built and an appearance model of texture from a labeled collection of animal images, and combine the two models automatically. We develop a simple texture descriptor that outperforms the state of the art. We test our animal models on two datasets; images taken by professional photographers from the Corel collection, and assorted images from the web returned by Google. It demonstrates a quite good performance on both datasets. Comparing the results with simple baselines, it was evident that for the Google set, can recognize objects from a collection demonstrably hard for object recognition [5].

Detecting the heads of cat-like animals, adopting cat as a test case, we show that the performance depends crucially on how to effectively utilize the shape and texture features jointly. Specifically, study has proposed a two-step approach for the cat head detection. In the first step, two individual detectors were trained on two training sets. One training set is normalized to emphasize the shape features and the other is normalized to underscore the texture features. In the second step, we train a joint shape and texture fusion classifier to make the final decision. A significant improvement can be obtained by the two step approach. In addition, study also propose a set of novel features based on oriented gradients, which outperforms existing leading features, e.g., Haar, HoG, and EoH. Study evaluates its approach using a well labeled cat head data set with 10,000 images and PASCAL 2007 cat data [6].

Detection and identification of animals is a task that has an interest in many biological research fields and in the development of electronic security systems. We present a system based on stereo vision to achieve this task. A number of criteria were being used to identify the animals, but the emphasis is on reconstructing the shape of the animal in 3D and comparing it with a knowledge base. Using infrared cameras to detect animals is also investigated. The presented system is work in progress [7].

1.2. Existing Systems & Draw Backs

The wild life camera DVR utilizes infrared technology which will capture great footage at any time of the day or nights as well as being supplied in a sturdy weatherproof and camouflaged box. Record full color video footage or 8 MP photographs produced by the wild life camera is transferred to a SD card and then review on a PC (using USB cables). The built in rechargeable battery can last (depending on the activity) up to 2 weeks. Existing technology has following features

-Motion triggered and adjustable infrared (PIR) sensitivity
-Rechargeable battery life 2 weeks
-2.5" LCD Screen
-Auto switch color images in day/B&W night images
-SD card slot (2 GB)
-Multi shot of 1 - 3 pictures
-Programmable video length

-Programmable 10 sec to 990 sec delay between triggers
-No flash uses 54 IR LEDs to illuminate the coverage area
-Water proof housing
-Dimensions 160 × 120 × 50 mm [8]

The major drawback of this is the no detection of animal in different angles. Whatever passes the camera will be automatically captured. Night time images are black and white and have less details and clarity due to infrared flash quality. If the infrared flash is designed for best image quality, range will be sacrificed. The photographer might be interest about a specific animal and there is no facility to recognize automatically whether captured animal is the photographer's choice or not.

2. Methodology

As the project contains both hardware & Software, need to focus on both parts separately. But this research paper mainly focuses on classification and recognition. Considering completion of the study hardware side and motion tracking has been focused. For the detection, camera rotation and communication the initial circuit had been tested on a bread board. The 16f877a microcontroller is the heart of hardware part and it has connected via max232IC over serial to USB cable to carry data to PC. When communication setting baud rate is 9600 bits/sec PIR sensors had been tested separately by applying proper voltage. The servos which controlling the camera movement is needs to program on microcontroller to provide suitable pulse widths in appropriate time (**Figure 1**).

Above pulse (**Figure 2**) period is 20 ms and to get the 20 ms repeat rate Timer 0 generates an interrupt at regular intervals. Timer 0 is driven (in this case) from internal oscillator. This is further divided, inside the pic, by 4 (Fosc/4). Prescaler value is useful since using the maximum setting 1:256 can get longer timer overflow *i.e.* when the timer passes through the values 255 to 0.

The 256 denominator is there because Timer 0 only overflows after 256 counts because it is an 8 bit timer.

Figure 3 shows the final pcb with components. Supply voltage was send through the 7805 regulator to maintain supply voltage of 5 V. Suitable capacitors were connected for smoothing purposes. 20 MHz oscilator, max 232 ic and other resistors and capacitors connected as in the design.

2.1. Classification

For the feature extraction process ACE media project as used which was capable of handling several feature extraction processes. The design was based on the architecture of MPEG-7 experimentation Model (XM) the official reference software of the ISO/IEC MPEG-7 standard. In addition to a more "light weight" and modular design, a key advantage of the ACE toolbox over the XM was the ability to define the process regions in the case of image input. Such image regions can be created either via a grid layout that partitions the input image in

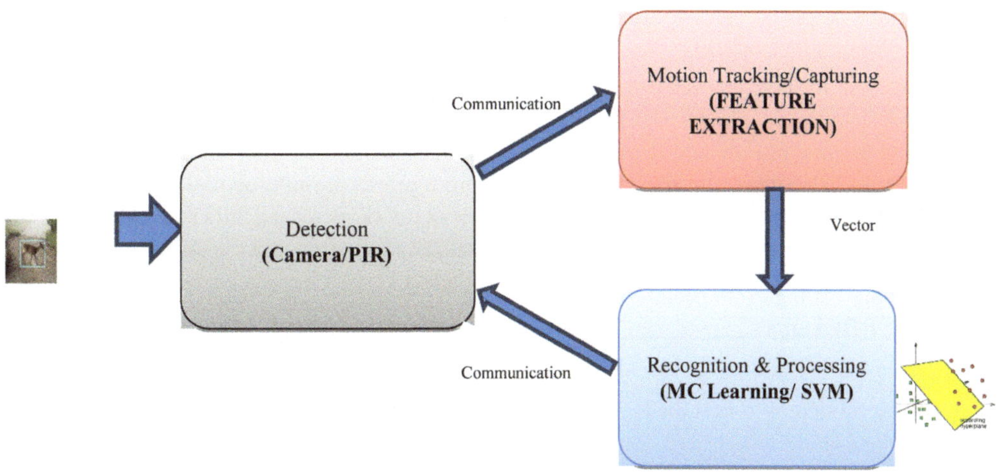

Figure 1. Whole process in nutshell. When the animal is presence detection, tracking, communication and recognition take place.

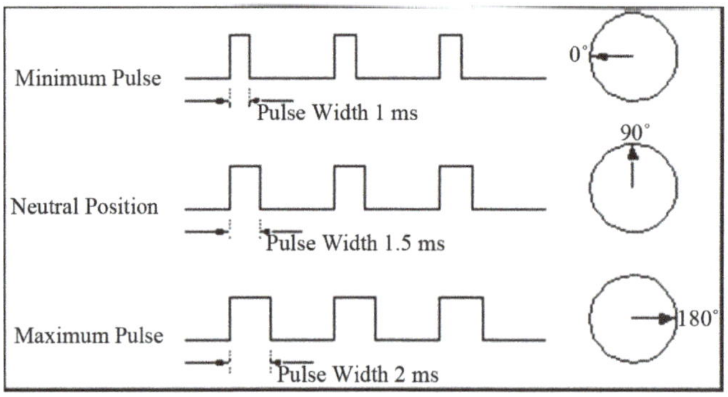

Figure 2. Standard servo controlling pulses [9].

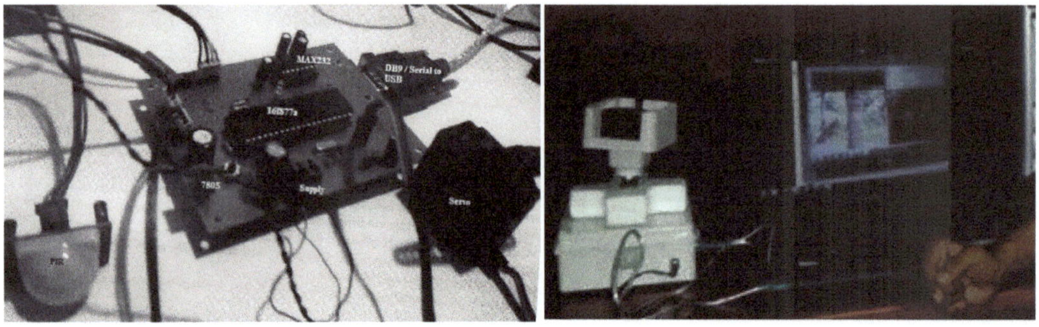

Figure 3. Final PCB with hardware.

to user defined square regions or a segmentation tool that partitions the image in to arbitrary shaped image regions, that reflect the structure of the object present in the scene.

The current version of the ace tool box supports the descriptors listed in **Figure 4**. The visual descriptors are classified in to four types: color, texture shape & motion (for video sequences). Currently there is only a single video based and audio descriptor supported, but this will be extended in the future. The output of the ace Toolbox was a XML file for each specified descriptor, which for image input relates to either the entire image or separate areas within the image. An example of typical output is shown in **Figure 5**. The tool box adopts a modular approach, whereby APIs were provided to ensure that the addition of new descriptors was relatively straightforward. The system has been successfully complied and execute on both windows-based and Linux based flatforms [1].

So initially with a aid of C programming file handling techniques and the extracted features need to be in format in **Figure 5** in to a text file. Reason is that it can be use in LIBSVM.

Formats should be as per the **Figure 6** there for use in LIBSVM. +1 in the above dataset determines that it is positive dataset thus for negative dataset then it need to be begin with −1. Since its necessity to extract features of thousands of images C code was used. It facilitated numbering the total images in sequence and accessing them in iteration by time to time and features were saving in one particular text file.

After the feature extraction concern was on developing a proper training model. So LIBSVM had been used and it was pretty straightforward. LIBSVM is a library for support vector machines (SVM). LIBSVM has gained wide popularity in machine learning and many other areas. The followed process as below

-Conduct simple scaling on the data
-Consider the RBF kernal

$$K(x,y) = e^{-\gamma\|x-y\|2}$$

-Use the cross validation to find the best parameter C and γ
-Test

The original data maybe too huge or small in range, thus it was rescaling to a proper range so that training and predicting will be faster. The main advantage of scaling was to avoid attributes in greater numeric ranges.

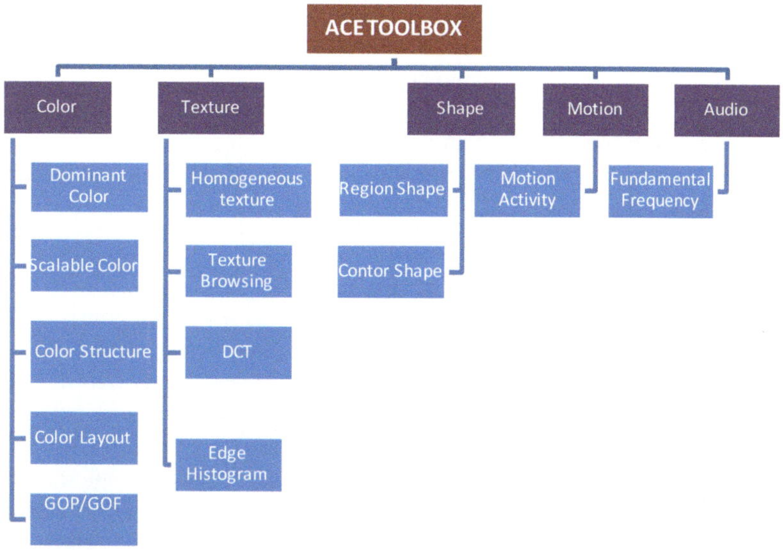

Figure 4. Supporting descriptors.

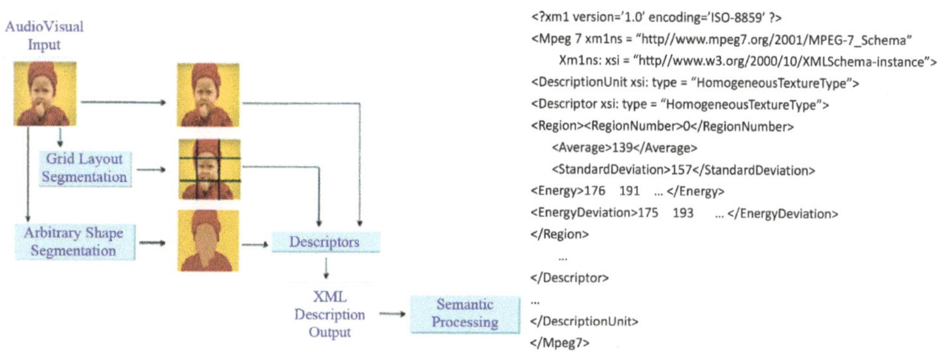

Figure 5. Overview of ace tool box & output.

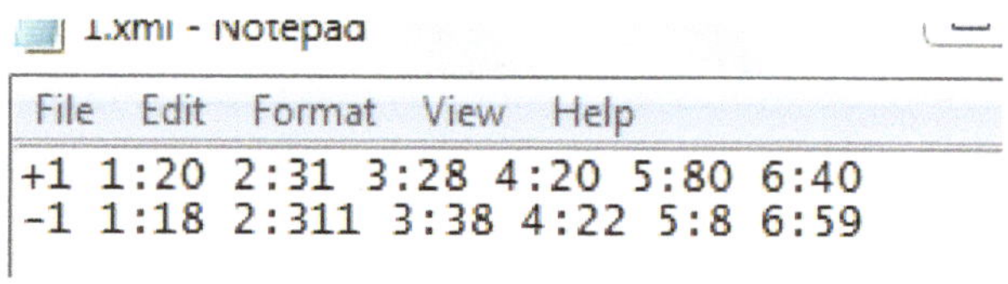

Figure 6. LIBSVM readable format.

Another advantage was to avoid numerical difficulties during the calculations. Linearly scaling for each attribute to the range [−1 to +1] or [0 to 1] was done. Below code should be run in command prompt for scaling the training set.

After scaling the dataset, kernel function was chosen for creating the model. Four basic kernels are linear, polynomial, radius basis function and sigmoid. In general RBF kernel is reasonable first choice. A recent result shows that if RBF kernel is used with model selection, then there is no need to consider linear kernel. The kernel matrix using sigmoid may not be positive definite and in general its accuracy is not better than RBF. Polynomials kernels are acceptable but if a high degree is used, numerical difficulties tend to happen. In or case also RBF kernel has been used. There are two parameters for RBF kernel: C and γ. linear kernel has a penalty parameter C. It is not known before beforehand which C and γ are best for given problem. Consequently some kind of model selection (parameter search) must be done. The goal was to identify good C and γ value, so that the classifier can accurately predict unknown data (*i.e.* testing data). For selecting the best parameter value grid.py in the libsvm-3.11 tools were used directory. Grid.py is a parameter selection tool for C-SVM classification using RBF (radial basis function) kernel. It uses cross validation (CV) technique to estimate the accuracy of each parameter combination in the specific range and helps you to decide the best parameters. To run the grid.py python interface is needed and a gun plot. In python code needed to do some modification to access the directories in LIBSVM windows and gunplot.exe.

To have higher cross validation accuracy can use 5 fold cross validation. After finding proper C and γ values need to create training model. After creating training model can see the predicted output (precision) for testing data by running SVM-predict.

2.2. Results

Based on above commands for three animals, using individual descriptors study was carry out to test how it will be the outcome. Two color descriptors, two texture descriptors, and two shape descriptors were used separately to check the ability of recognition of animals. The results gathered are summarized in **Table 1** [10].

2300 of positive images of tigers 960 of negative images like other animals, vehicle and etc. and 422 of tigers of testing were used. 2027 of positive images of dogs 752 negative images like other animals, vehicles and etc. and 1000 images of dogs of testing set was used. Cats images also taken around same amounts. As per the results on **Table 1** though it provide some good results of recognition +ve images, recognition of –ve images are very poor. When the individual descriptors were considered edge histogram descriptor performed somewhat satisfactory level compare to other descriptors.

As per the **Figure 7** it was clear that the variation between recognition of negative and positive images of individual descriptors. Most of the descriptors recognition rate was quite higher but unable to differentiate negative from images.

As per the **Figure 8** which was plotted with cross validation accuracy can identify the difference very clearly as positive and negative image efficiencies. So overall efficiency fails and it was not suitable for proper system.

Table 1. Individual descriptor performances.

Animal	Descriptor	C	γ	Cross validation accuracy	Accuracy for recognition (−) images %	Accuracy for recognition (+) images %
Tiger	Color Layout	8	0.5	85.1786	22.222	88.3581
	Color Structure	2	0.5	93.6786	44.4444	78.4615
	Edge Histogram	8	0.125	92.7143	44.44	93.0756
	Homogeneous Texture	8	0.5	97.7857	25.763	76.1538
	Region Shape	2	0.5	82.3445	0	99.74
	Contour Shape	2048	8	86.9643	0	76.6154
Dog	Color Layout	8	2	77.5285	38.913	93.3
	Color Structure	8	0.125	82.2958	27.812	91.8
	Edge Histogram	32	0.007825	82.6916	53.871	94.2
	Homogeneous Texture	8	0.5	77.7258	32.2222	92.9
	Region Shape	8	0.03125	75.2429	11.80	96.5
	Contour Shape	32,768	8	76.7182	0	97.6
Cat	Color Layout	8	0.5	80.8973	28.893	89.713
	Color Structure	8	0.5	83.6957	37.777	76.888
	Edge Histogram	8	0.125	90.9532	39.99	92.865
	Homogeneous Texture	8	0.5	80.2558	25.6771	93.25
	Region Shape	2	0.5	80.3445	11.5973	90.597
	Contour Shape	2560	8	84.9248	0	78.653

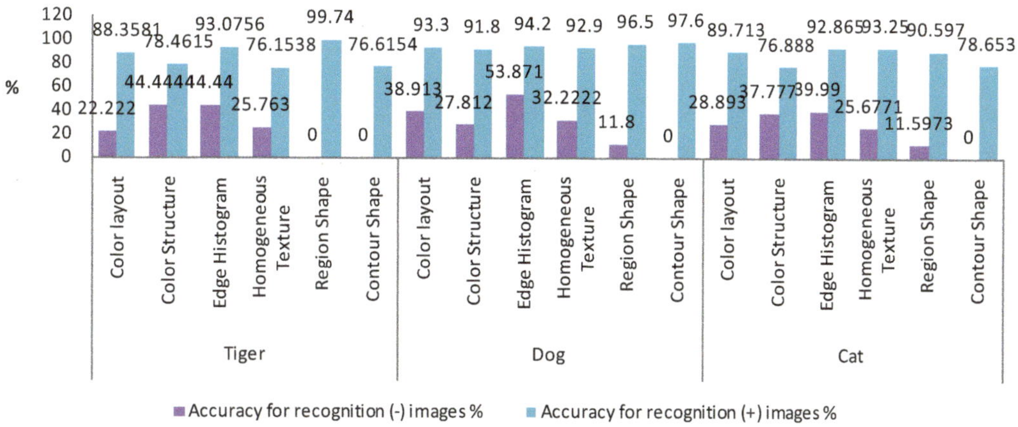

Figure 7. Performances of individual descriptors.

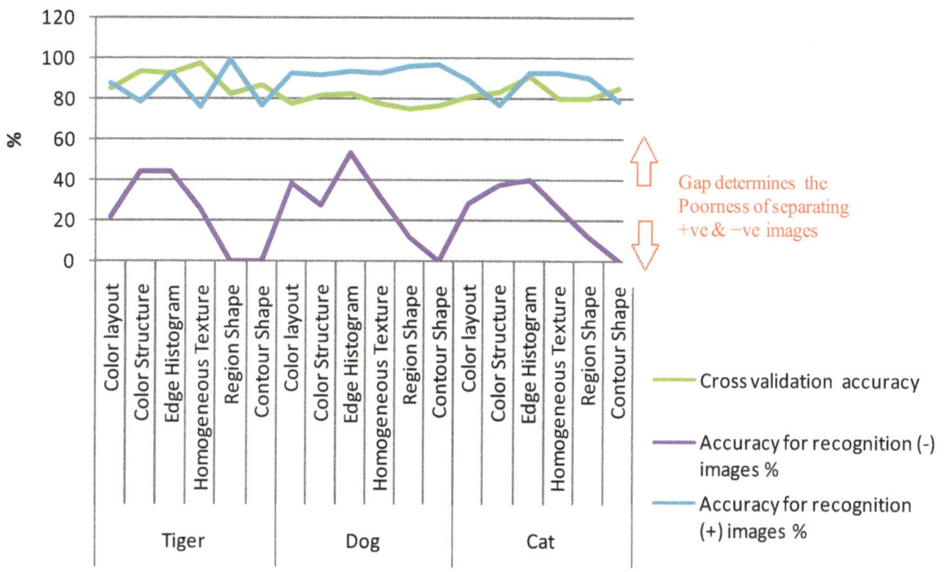

Figure 8. Performances of individual descriptors.

Then by considering above results test was carried out to identify how the output would looks like. The combination was followed by taking most efficient descriptors and all the time concerned to have single color descriptor, single texture descriptor and single shape descriptor together.

According to **Table 2** enhanced results of combined descriptors were evident and outcome had more efficiency than earlier case.

Figure 9 shows the tabulated results of combined descriptors performances.

3. Conclusion

When compared with performance of individual descriptors, recognition of positive & negative image separately was not successful. But when few descriptors combined together, very fair outcome was evident yet accuracy varied around 80%. As per the graph in **Figure 8** which represented the combined descriptor results comparably, very successful results were evident which recognize both positive & negative images fairly in equal manner. The combination results of region shape, edge histogram and color structure had very low separation between positive and negative columns. The combination results of region shape, edge histogram and color layout were given average outcome compared with the above combination. The combination of region shape, edge histogram, color layout & color structure, though it recognizes a fair amount of positive images, fails to separate negative images in a great manner. So mid configuration of both data sets, which is combination of region shape,

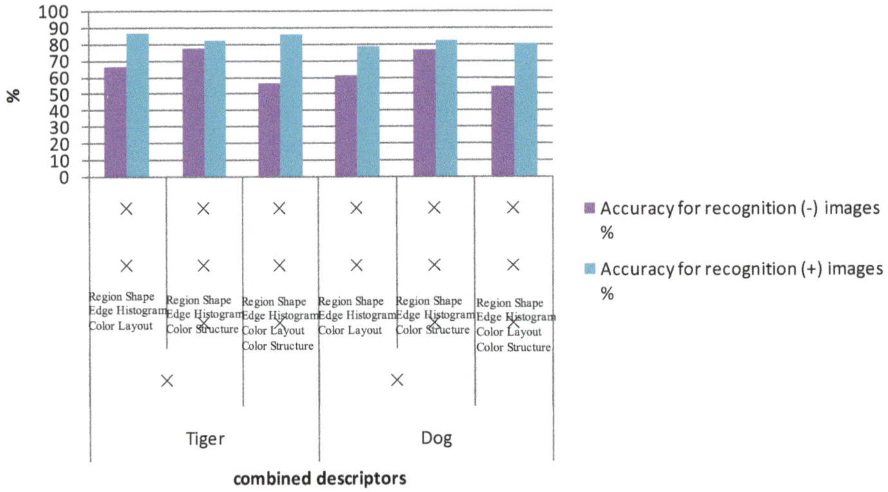

Figure 9. Performances of individual descriptors.

Table 2. Combined descriptor performances.

Animal	Color layout	Color structure	Edge Histogram	Region Shape	C	γ	Cross validation accuracy	Accuracy for recognition (−) images %	Accuracy for recognition (+) images %
Tiger	×		×	×	2	0.0312	87.6687	66.1111	86.4103
		×	×	×	2	0.25	90.8589	77.222	82.05
	×	×	×	×	2	0.125	87.2393	55.5556	85.64
Dog	×		×	×	8	2	77.5285	60.913	78.3
		×	×	×	2	0.125	82.2958	75.812	81.82
	×	×	×	×	8	2	82.6916	53.871	80.02

edge histogram & color structure, provided good results in recognition. In a summary compared with individual descriptor performances, combined descriptors provided much accurate outcome.

References

[1] O'Connor, E., *et al.* (2005) The Acetool Box: Low Level Audio Visual Feature Extraction for Retrieval and Classification. *2nd European Workshop on the Integration of Knowledge, Semantics and Digital Media Technology* (*EWIMT* 2005), 30 November-1 December 2005, 55-60. http://dx.doi.org/10.1049/ic.2005.0710

[2] Elder, J.H. and Velisavljević, L. (2009) Cue Dynamics Underlying Rapid Detection of Animals in Natural Scenes. *Journal of Vision*, **9**, 7. http://dx.doi.org/10.1167/9.7.7

[3] Kouda, M., Morimoto, M. and Fujii, K. (2011) A Face Identification Method of Non-Native Animals for Intelligent Trap. *MVA*2011 *IAPR Conference on Machine Vision Applications*, Nara, 13-15 June 2011, 426-429.

[4] Burghardt, T., Ćalić, J. and Thomas, B.T. Tracking Animals in Wildlife Videos Using Face Detection. https://www.cs.bris.ac.uk/Publications/Papers/2000186.pdf

[5] Ramanan, D., Forsyth, D.A. and Barnard, K. (2005) Detecting, Localizing and Recovering Kinematics of Textured Animals. *IEEE Computer Society Conference on Computer Vision and Pattern Recognition*, *CVPR*2005, **2**, 635-642. http://dx.doi.org/10.1109/CVPR.2005.126

[6] Zhang, W.W., Sun, J. and Tang, X.O. (2008) Cat Head Detection—How to Effectively Exploit Shape and Texture Features. *Lecture Notes in Computer Science*, **5305**, 802-816. http://dx.doi.org/10.1007/978-3-540-88693-8_59

[7] Levesque, O. and Bergevin, R. Detection and Identification of Animals Using Stereo Vision. http://homepages.inf.ed.ac.uk/rbf/VAIB10PAPERS/levesquevaib.pdf

[8] Flyonthewall. Ltd. Infrared CCD Wildlife Camera with Embedded DVR.

http://www.flyonthewall.uk.com/infrared-wildlife-camera-dvr-with-motion-sensor-8-0mp.html.

[9] https://www.servocity.com.

[10] Chang, C.-C. and Lin, C.J. (2015) LIBSVM—A Library for Support Vector Machines. https://www.csie.ntu.edu.tw/~cjlin/libsvm/

21

Discussion on Subject Knowledge Service Pattern of Agricultural University Libraries

Linlin Zhan

Heilongjiang Bayi Agricultural University Library, Heilongjiang Bayi Agricultural University, Daqing, China
Email: liblin@126.com

Abstract

Taking Heilongjiang Bayi Agricultural University as an example, the discipline service practice was explored from the aspects such as free network resources and tool use, subject navigation, subject librarians, information sharing space and information literacy education. The typical cases were provided; the optimization of utilization method of network resources was put forward; the subject theme navigation was developed; and the information sharing space of specific discipline and other working strategies were carried out. The experience and enlightenment of active service for subject construction, priority service for key subjects, subject service by lay and step and seizing of the opportunity of subject service were summarized, which developed ideas and provided reference to the establishment of deep subject services of general libraries of colleges and universities.

Keywords

Subject Service, Value-Added Service, Knowledge Service, Application Pattern

1. Introduction

Subject construction is a long-term and fundamental strategic task of institutions of higher learning, and the key of the development of colleges and universities [1]. Subject service is a new service concept launched by libraries of colleges and universities in order to adapt to the new information environment changes. It is user-centered. Relying on the national literature resource platform and public information service platform, with subject librarians as the medium, it provides highly targeted "one-to-one" service to various departments, laboratories, subject groups or individuals of colleges and universities [2]. The relevant data shows that by the end of 2010 almost all the libraries of 211 colleges and universities have subject services [3]. Subject service is the extension of traditional information consulting service. The service mode and content are more personalized, professional and educated, thus ensuring that subject teaching, subject study and individual needs are effectively met. It expands the wider survival and service space for the libraries [4].

The subject knowledge service is a proactive extension service mode, which is implemented by the university libraries under the guidance of the readers' personalized requirements. The Heilongjiang August First Land Reclamation University reorganized the department organizations in 2009 and promoted the main strategy of "subject knowledge service" for all readers in the university via a series of subject knowledge service schemes, which has achieved remarkable effect and been recognized by the readers during 5 years. This paper completely explains the subject knowledge service practice mode of the Heilongjiang August First Land Reclamation University Library in organization architecture, activity plan, service content, service pattern and implementation effect.

2. Team Cooperation Architecture of "Vertical Promotion + Organization Collaboration"

Any innovative service pattern should be based on the corresponding organization service mode. To better promote the subject knowledge service in the universities, the libraries should be organized in a vertical and horizontal manner. Different departments should provide effective service pattern for the subject knowledge service based on their features in vertical direction. Different departments should carry out the subject service team cooperation due to work requirements in the horizontal direction [5]. e.g. the library regards the reference resource development department as the main department for promoting this service. This department includes 4 subject librarians. Each subject librarian should cooperate with the collecting and editing persons to compose the subject service team, which will carry out requirement investigation, subject novelty search, reference transfer and reader training in two L2 academies of the university. The system department should cooperate with the subject librarians of reference resource publication department to complete construction of the subject knowledge service platform of the library website. e.g. "Heilongjiang reclamation farm history log database", "precise agricultural database", "network free-electronic full-text database", and "courseware navigation database" established by the library. Such vertical and horizontal organization and cooperation service pattern fully mobilizes work enthusiasm of all librarians and exerts the collective intelligence, so the librarians can further improve their existence value and improve recognition of the society. They can better participate in the subject knowledge service.

3. "Three Promotion + L2 Service" Comprehensive Activity Plan

To make the subject knowledge service achieve practical and effective role and effect and ensure that the information resources and subject services of the library can meet the requirements of the teachers and readers of the schools, the library will design promotion plan of three services under the guidance of "walk into subject·library beside you", which is described as follows:

3.1. Promotion of Foundational Service

The reference collecting and editing department effectively promotes the subject knowledge service in the foundational service of the library and carries out the book recommendation and purchase service by specifying the subject collecting and editing persons to send the subscription bibliography information to teachers in universities and carrying out "you select books and I buy order" activity, subject bibliography recommendation, and interview with the academies and departments and subject leader.

3.2. Promotion of Virtual Service

The library assists completion of subject knowledge service by establishing the subject consulting platform, characteristic subject reference resource construction, courseware navigation database construction, embedded IM online consulting, provision of subject professional website, regular sending of academy and department, and email message pushing.

3.3. Promotion of Academy and Department Service

The library perfectly transforms from the passive service to active service by walking into academies and departments, and effectively pushes the subject knowledge service to the academies and departments by cooperating with the database providers to carry out "operation method of database and reference management software" training seminar for academies and departments, subject tracking service, foreign reference pushing service, and making academic evaluation report and embedded course teaching for subject leaders and research organizations.

The library carries out different subject knowledge services for the service type and service demand of readers. e.g. for the readers with general information demands, focus on omnidirectional promotion and carry out information quality training seminar and rich cultural activities. For users with special information requirements, focus on customized services and provide deep knowledge services such as subject article pushing, academic trend tracking and conference brief report pushing.

4. Diversified Innovative Service for "Create Brand + Mine Requirement"

When the university libraries advertise the library service, generally they ignore the brand creation and are lack of the planning, advertisement and operation consciousness, so many welcome cultural activities are temporary and no characteristic brand cultural activities are formed [6]. The Agricultural University library fully considers these factors in the cultural activities with the subject knowledge service as the main line, so the cultural activities welcomed by the readers will be implemented persistently and normally at the beginning of the planning activities in order to build the characteristic cultural brand. For different requirements, with the annual reading day and reader service month as the opportunity, the agricultural university library carries out a series of subject knowledge service promotion activities to perform whole packing, series design and promotion of some independent projects, encourage readers to actively participate in and interact, and enhance the abstraction of the libraries. The library completely builds the fine cultural activities in resource, service and communication, which is described as follows:

4.1. Resources

The library holds "Book ocean treasure hunting" book retrieval skill match, I decide my resources-search beside "database Daren" match, database promotion service month, and field consulting table of agricultural university library. The resource promotion activities are held to promote understanding of extensive reads on the library, guide the readers to correctly, fully and efficiently utilize the paper resources and electronic resources of the libraries, and grasp basic methods and skills of the reference retrieval, and drive advertisement and promotion of the paper resources and use of digital resources.

4.2. Service

The library holds the graduation paper guide and training ambassador selection activities, "are you ready for postgraduate enrollment examination?" subject seminars, "walk into academies and serve subjects" activities, "love every day of library" micro-film match, good paper recommendation, and "most beautiful library-mobile library experiences". Implementation of these services will further strengthen communication and association between the library and students and teachers in different academies, so the students and teachers in the schools can know the resources and services of the libraries to improve initiative and purpose of the library information service.

4.3. Communication

The library will hold "Wang Zhen lecture hall" activity. The culture lecture hall fully shows the intension of the library to expand service fields, strengthen service function, and promote campus culture construction, and is the key platform to improve the comprehensive quality of students. The lecture hall aims to invite the well-known scholars and professors in the university to introduce the latest learning achievement and share their inspiration from the life to the readers via the lectures based on their professional knowledge and cultural background. The library held "Wang Zhen lecture hall" in Mar, 2010. Now "Wang Zhen lecture hall" has been held 100 times and the audiences reach over 6000. It has gradually formed the self-service mode and characteristics, becomes the brand culture of the library, and shows its charm in campus culture construction. The library carries out "agricultural university reads me to you·True person library" activity. This activity encourages readers in the campus to regard themselves as the true person book and share distinct experiences, specialty, skill and inspiration with others. This form not only breaks different rigorous requirements of readers for reading form, contents and range, but also encourages bi-directional exchange of students, enhances mutual understanding, and lets students believe themselves as good books. This activity not only makes the readers acquire the curious knowledge, but also provides the small show stage for the "true person book". These cultural activities are successfully held to build the library as the campus culture center, better exert the culture leading role of the library, and promote culture exchange, transfer and propagation.

5. "Stereo Association + Utmost" Linked Service Pattern

For different subject service schedule, the library cooperates with different units to carry out "3 + x" block service pattern, which is described as follows:

5.1. Cooperation with Youth League Committee of University

The digital library construction represents the development direction of the library in the information age. For university students, it is very important to grasp the retrieval method of digital resources. To increase the information acquisition channel of students and make students further recognize and know the digital resources of the library, the library started to cooperate with the Youth League Committee of the university from 2011 and has successfully held "VIP reference knowledge search skill match". In the competition activities, the Youth League Committee of the university is responsible for early advertisement and organization. The library is responsible for training and promotion. Such cooperation carries out the cultural activities of the library via the implementation strategy of "whole planning, step implementation, united guarantee and step-by-step penetration".

5.2. Cooperation with Administrative Department of University

In the university, besides the teaching and R&D service, the subject knowledge service includes the service for the university's decision department [7]. The science data is the important basis for the scientific evaluation and management. The evaluation office, science and technology division and human resource management division will require the auxiliary data as the basis for decision evaluation. For this current condition, the university library will completely launch the intelligent information service via the personalized and flexible service scheme. We provide the analysis report of "SCI paper published by Agricultural university" by using the strength in information acquisition. Depending on the Cqvip database, we establish the organization knowledge portal of the Heilongjiang August First Land Reclamation University Library to provide academic data evaluation report of the academies and departments of the university, including academic achievement output analysis, main researcher scholar and field analysis, and related research organization analysis, and offset the detailed and reliable evaluation basis for the academy and talent information collection and academic evaluation. The fact proves that that the subject knowledge service can provide reliable policy basis for subject establishment, operation strength and specialty setup, and offset favorable guarantee for long-term development of the university.

5.3. Cooperation with R&D Team

It is difficult for the library service to embed into the R&D team and provide the subject knowledge service all the time due to restrictions in specialty background and foreign language level. By combining the specialty strength and resource strength, the subject librarians will provide the subject service through the subject research of R&D team [8]. e.g. when the subject is applied, the library will assist the R&D persons to search the latest references on the subject. Many R&D persons cannot select accurate retrieval words in reference search, so they cannot reach the ideal retrieval results. At this time, the subject librarians will provide better recommendations for retrieval word selection to ensure subject novelty. In the subject research, the subject librarians will assist them to collect and acquire the latest academic trend and emergency special references by using the smart tools and deep knowledge discovery tool. After the subject ends, we can assist them to provide novelty search of subject achievements and subject achievement certificate. With the above services, the subject service of the library is trusted and welcomed by the R&D team more and more.

6. Full-Coverage Evaluation System of "Fine Test and Evaluation and Interactive Summary"

To further validate implementation effect of the subject service of the library, the library establishes the subject service evaluation system via three modes.

6.1. Reader Feedback

The library will interview the teachers and students in the academies and schools in annual reader service month to know the service requirements of teachers and students and improvement of the library in the subject service. The investigation results will provide direct reference for further development of the subject knowledge service.

6.2. Inter-Department Mutual Evaluation inside Library

In the annual summary conference of the library, the representatives from different departments and divisions will evaluate the subject service work effect in a face-to-face manner and propose the reasonable improvement comments.

6.3. Evaluation of R&D Team

For the R&D team of the subject knowledge service, when the subject is completed, the subject service persons should deeply exchange with the R&D persons of the am, and share experiences and achievements with exerts. Finally experts will propose the subjective comments and recommendations to promote the deep development of the subject service.

The Heilongjiang August First Land Reclamation University Library has successful experiences in implementation of the subject service. With repeated trial, although certain effect is achieved in a short time, it will required a long period to extensively establish equal and mutually beneficial cooperation in the reader group and successfully transform proactive pushing to proactive requirement, so the Heilongjiang August First Land Reclamation University Library should continuously innovate the subject knowledge service contents and explore the subject knowledge service fields in the subject knowledge service in future in order to provide professional, deep and personalized subject knowledge service to extensive readers.

7. Summary

In the process of subject service, general libraries of colleges and universities shall first define the positioning of subject service of their own, start from the conditions of the school and library, adjust measures to local conditions, look for the characteristic subject service methods suitable for the library, and avoid blindly copying the mode of major universities; next, they shall give full play to the subjective initiative, use the resources inside and outside the library, and open up a new prospect through focusing on construction and concentrated construction; third, they shall push from the first-level services to high-level services step by step. In addition, the subject service is not completely equal to the scientific research service. For general colleges and universities, especially, the colleges and universities without awarding power of master's and doctor's degree mainly engage in teaching; the teaching service for the undergraduate education is the top priority.

Funding

Heilongjiang Philosophy and Social Science Plan research project "Information service system building of university libraries in big data age" (project number: 14E045).

Heilongjiang Art and Science Plan co-building project "research on library service innovation development in new media age" (project number: 2014D014).

References

[1] Zhao, J.N. (2014) Information Commons-Based Subject Service in University Library. *Information Research*, **195**, 119-125.

[2] Fang, F.-Y. (2014) Research on Subject Service in University Library. *Journal of Library and Information Sciences in Agriculture*, No. 6, 145-147.

[3] Li, G.Z. (2014) Research on the Multi-Collaborative Subject Service of University Libraries. *Journal of Modern Information*, **34**, 147-151.

[4] Gao, Y.-Y. (2015) Research on the Current Situation of Discipline Service in Shaanxi's University Libraries. *Journal*

of Library and Information Sciences in Agriculture, No. 10, 155-158.

[5] Liu, H., Zeng, L. and Li, W.L. (2014) Construction and Evaluation of Subject Knowledge Service System of University Library with Nanjing University of Traditional Chinese Medicine as One Example. *University Library Service*, **34**, 24-27.

[6] Yang, Y.H. (2007) Discussion on Information Marketing and Subject Knowledge Service of University Library. *Popular Science & Technology*, No. 8, 153-154.

[7] Ma, Y.B. (2012) Discussion on Core Competition Capability of University Library in Subject Knowledge Service. *Intelligence*, **23**, 327.

[8] Tong, P. (2014) Structure and Pattern of Subject Knowledge Service System of University Library. *Journal of Anshan Normal University*, **16**, 99-102.

22

Identification of Textile Defects Based on GLCM and Neural Networks

Gamil Abdel Azim

College of Computers & Informatics Canal University, Ismailia, Egypt
Email: Gazim3@gmail.com

Abstract

In modern textile industry, Tissue online Automatic Inspection (TAI) is becoming an attractive alternative to Human Vision Inspection (HVI). HVI needs a high level of attention nevertheless leading to low performance in terms of tissue inspection. Based on the co-occurrence matrix and its statistical features, as an approach for defects textile identification in the digital image, TAI can potentially provide an objective and reliable evaluation on the fabric production quality. The goal of most TAI systems is to detect the presence of faults in textiles and accurately locate the position of the defects. The motivation behind the fabric defects identification is to enable an on-line quality control of the weaving process. In this paper, we proposed a method based on texture analysis and neural networks to identify the textile defects. A feature extractor is designed based on Gray Level Co-occurrence Matrix (GLCM). A neural network is used as a classifier to identify the textile defects. The numerical simulation showed that the error recognition rates were 100% for the training and 100%, 91% for the best and worst testing respectively.

Keywords

Image Processing, Neural Network, Gray-Level Co-Occurrence Matrices (GLCM)

1. Introduction

Texture analysis is necessary for many computer image analysis applications such as classification, detection, or segmentation of images. In the other hand, defect detection is an important problem in fabric quality control process. At present, the texture quality identification is manually performed. Therefore, Tissue online Automatic Inspection (TAI) increases the efficiency of production lines and improve the quality of the products as well. Many attempts have been made based on three different approaches: statistical, spectral, and model based [1].

In this recherche paper, we investigate the potential of the Gray Level Co-occurrence Matrix (GLCM) and neural networks that used as a classifier to identify the textile defects. GLCM is a widely used texture descriptor [2] [3]. The statistical features of GLCM are based on gray level intensities of the image. Such features of the GLCM are useful in texture recognition [4], image segmentation [5] [6], image retrieval [7], color image analysis [8], image classification [9] [10], object recognition [11] [12] and texture analysis methods [13] [14] etc. The statistical features are extracted from GLCM of the textile digital image. GLCM is used as a technique for extracting texture features. The neural networks are used as a classifier to detect the presence of defects in textiles fabric products.

The paper is organized as follows. In the next sections, we introduce a brief presentation of GLCM. In section 3, the concept of neural networks with Training Multilayer Perceptrons is described. Image Analysis (Feature Extraction & Preprocessing Data) is given in section 4. Numerical simulation and discussion are presented in section 5; in the end a conclusion is given.

2. Gray-Level Co-Occurrence Matrix (GLCM)

One of the simplest approaches for describing the texture is using a statistical moment of the histogram of the intensity of an image or region [15]. Using a statistical method such as co-occurrence matrix is important to get valuable information about the relative position of neighboring pixels of an image. Either the histogram calculation give only the measures of texture that carry only information about the intensity distribution, but not on the relative position of pixels with respect to each other in that the texture.

Given an image I, $N \times N$, the co-occurrence, and the matrix P defined as:

$$P(i,j)=\sum_{x=1}^{N}\sum_{y=1}^{N}\begin{cases}1 & \text{if } I(x,y)=i \text{ and } I(x+\Delta_x, y+\Delta_y)=j \\ 0 & \text{otherwise}\end{cases}. \tag{1}$$

For more information see [16].

In the following, we present and review some features of a digital image by using GLCM. Those are Energy, Contrast, Correlation, and Homogeneity (features vector). The energy known as uniformity of ASM (angular second moment) calculated as:

$$\text{Energy: } \sum_{i=1}^{N}\sum_{j=1}^{N}P(i,j)^2 . \tag{2}$$

Contrast measurements of texture or gross variance, of the gray level. The difference is expected to be high in a coarse texture if the gray scale contrast is significant local variation of the gray level. Mathematically, this feature is calculated as:

$$\text{Contrast: } \frac{1}{(N)^2}\sum_{i=1}^{N}\sum_{j=1}^{N}(i-j)^2 p(i,j). \tag{3}$$

Texture correlation measures the linear dependence of gray levels on those of neighboring pixels (1). This feature computed as:

$$\text{Correlation: } \frac{\sum_{m=1}^{N}\sum_{n=1}^{N} mnp(m,n)-\mu_x\mu_y}{\sigma x \sigma y} \tag{4}$$

where $\mu_x=\sum_{m=1}^{N}M\sum_{n=1}^{N}p(m,n)$, $\mu_y=\sum_{m=1}^{N}N\sum_{n=1}^{N}P(m,n)$

$$\sigma_x=\sum_{m=1}^{N}(m-\mu_x)^2\sum_{n=1}^{N}p(m,n)$$

$$\sigma_y=\sum_{n=1}^{N}(n-\mu_y)^2\sum_{m=1}^{N}p(m,n).$$

The homogeneity measures the local correlation a pair of pixels. The homogeneity should be high if the gray level of each pixel pair is similar. This calculated function as follows:

$$\text{Homogeneity: } \sum_{m=1}^{N}\sum_{n=1}^{N}\frac{P(m,n)}{(1+|m-n|)}. \tag{5}$$

3. Neural Networks Construction

Artificial neural networks (ANN) are very massively parallel connected structures that composed of some simple processing units nonlinear. Because of their massively parallel structure, they can perform high-speed calculations if implemented on dedicated hardware. In addition to their adaptive nature, they can learn the characteristics of the input signals and to adapt the data changes. The nonlinear character of the ANN can help to perform the function approximation, and signal filtering operations that are beyond optimal linear techniques see [17]-[19]. The output layer used to provide an answer for a given set of input values (**Figure 1**). In this work the Multilayer Feed-Forward Artificial Neural Networks is used as classifier, where the objective of the back-propagation training is to change the weights connection of the neurons in a direction which minimizes the error E, Which defined as the squared difference between the desired and actual results output nodes. The variation of the connection weight at the kth iteration defined by the following equation:

$$\Delta W_{ij}(k) = -\eta \frac{\partial E}{\partial w_{ij}} + \alpha \Delta W_{ij}(k-1) \quad (6)$$

where (η) is the proportionality constant termed learning rate and (α) is a momentum term for more information (see [20]-[22]).

Once the number of layers and the number of units in each layer selected the weights and network thresholds should set so as to minimize the prediction error made by the system. That is the goal of learning algorithms that used to adjust the weights and thresholds automatically to reduce this error (see [23] [24]).

4. Image Analysis (Feature Extraction & Preprocessing Data)

The feature extraction defined as the transforming of the input data into a set of characteristics with dimensionality reduction. In the other words, the input data to an algorithm will be converted into a reduced representation set of features (features vector).

In this paper, we're extracting some characteristics of a digital image by using GLCM. Those are Contrast, Correlation, Energy, and Homogeneity. The system was implemented by using MATLAB program. We put the Matlab functions that we utilized in this work between the two <<>>. The proposed system composed of mainly three modules: pre-processing, segmentation and Feature extraction. Pre-processing is done by median filtering <<medfilt2>>. Segmentation is carried out by Otsu method <<graythresh, im2bw>>. Feature extraction based on GLCM features (Contrast, Correlation, Homogeneity, and Energy) implemented by <<Graycomatrix, graycoprops>>. The extraction of the textural characteristics of the segmented image is done by using gray level co-occurrence matrices (GLCM). The textural characteristics extracted from four spatial orientations; horizontal, left diagonal vertical and right diagonal corresponding to (0o, 45o, 90o, and 135o) using <<Graycomatrix, graycoprops>> with the offsets {[0 1], [−1 1], [−1 0], [−1 −1]}, that are defined as a neighboring pixel in the four possible directions (see **Figure 2**).

In the following subsection 4.1, we present a descriptor based on GLCM computation.

4.1. GLCM Descriptorsteps

1) Preprocessing (color to gray, noise, resize (256 × 256));

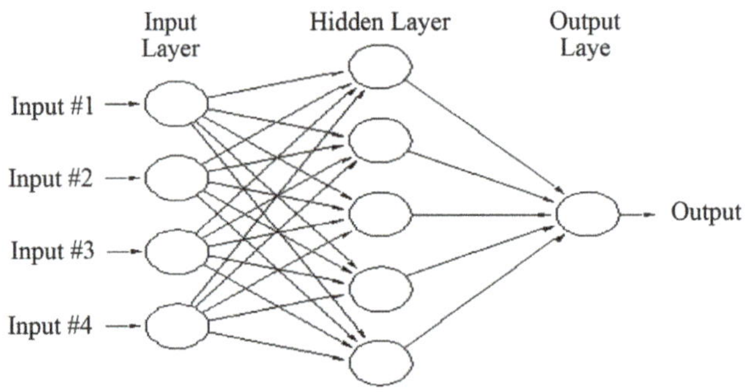

Figure 1. Feed-forward neural with three layers.

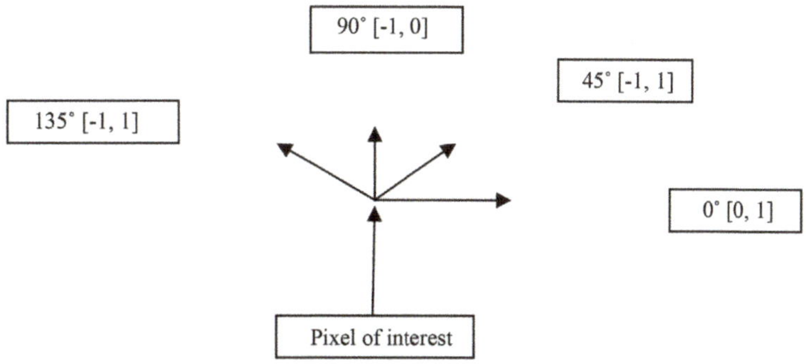

Figure 2. Adjacency of pixel in four directions (horizontal, vertical, left and right diagonals).

2) Segmentation;
3) Feature extraction based on GLCM features using <<Graycomatrix, graycoprops>> with the Offset [0 1; −1 1; −1 0; −1 −1];
4) Concatenation the Contrast, Correlation, Energy and Homogeneity.

The proposed algorithm implemented in MATLAB program developed by the author. After the application of the GLCM descriptor steps (4.1) to the image, we obtained a feature vector of dimension equal to 16 as input for neural networks classifier (**Figure 3**). The data set divided into two sets, one for training, and one for testing. Preprocessing parameters determined by using a matrix containing all the functionality that used for training or testing; the same settings were used to pre-treat the test feature vectors before transmission to the trained neural network.

A fixed number (m) of examples from each class assigned to the training set and the remaining (m - n) of the testing set. The inputs and targets are normalized, they have means equal zero and standard deviation = 1. The Forward Feed Back-propagation network trained using normalized training sets. The number of inputs of ANN equal to the number of features (m = 16). Each hidden layer contains 2 m neurons and two outputs equal to the number of classes (without defects, with defects) **Figure 4** and **Figure 5** respectively.

4.2. Simulation and Discussion

We have been developing a real database made by 1500 Images of jeans tissue. Some Images contain defects. The momentum is 0.9 and learning rate is 0.01 ((η, α in the Equation (1)). The ANN output represented by a vector belongs to $[0, 1]^n$. We describe all the features of the training set in a form matrix FM. Each column represents the pattern features. If the training set contains *m* instances for each model, which belongs to a class, the dimension of the matrix FM is equal to *m* x *n*. Where m is the number of features (feature vector size), and *n* is the number classes. The number of columns 1, 2, 3 … of the matrix FM represents the instance of the model characteristics (textile) that belongs to the class 1. The columns number $d+1, d+2, d+3, \cdots, 2d$, represents the instance features of the pattern (textile) which belongs to the class 2, etc. (see **Figure 6(a)** with $d = 6$ and **Figure 6(b)** with $d = 9$).

$$FM = {}^{l}f_{ij} \quad \text{for } l = 1,2,3,\cdots,d, i = 1,2,3,\cdots,m, \text{and } j = 1,2,3,\cdots,n \tag{7}$$

[25] and [26] exist more information about the formulation of input and outputs. The accuracy of the system is calculated by using the following equation:

$$\text{Accuracy} = \frac{n_r}{n_t} \tag{8}$$

where n_r the numbers of the correct images identification and n_t is the total number of the images testing.

Learning set contains six patterns for each category, and the testing set contains nine patterns for each type with accuracy 100%. The system learned with 7 images of each class and tested with 11, 13, 18, 21, 23, 28 images of each category. **Figure 7** shows the system accuracy, the best 100% and the worst is 91%. The best 100% results identification obtained when the sizes of each class are 11, 13, and 18.

The system learned with ten images sample of each category and tested with 1, 2… 26 images of each type.

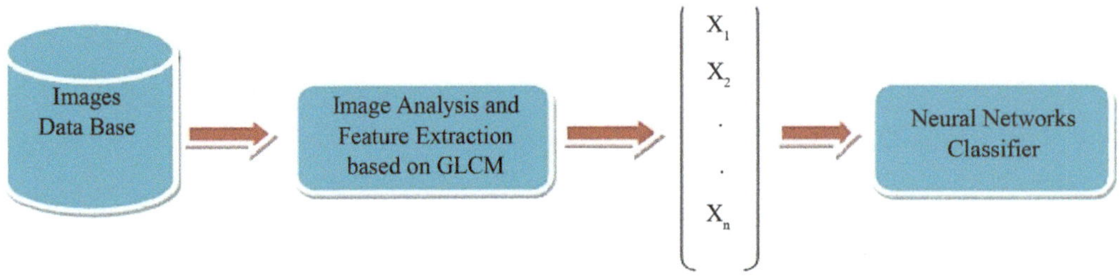

Figure 3. Flowchart of the proposed method.

Figure 4. Example of image without defect.

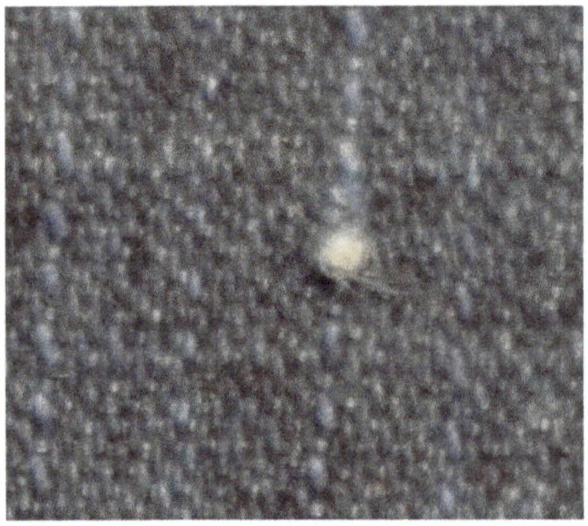

Figure 5. Example of image with defect.

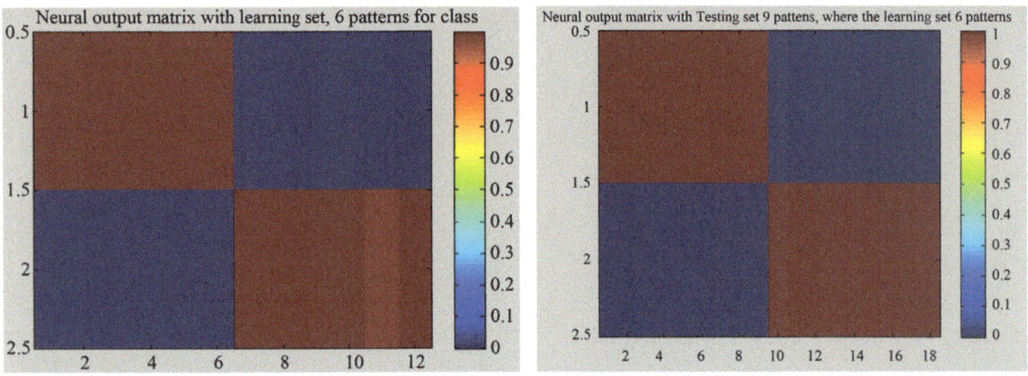

(a) (b)

Figure 6. (a) Neural output matrix with learning set, 6 patterns for class (100%); (b) Neural output matrix with Testing set, 9 patterns for class (100%).

Figure 8 shows the system accuracy, the best is 100% and the worst is 90.5%. We obtained the best results 100% identification when the sizes of class testing are 1, 2 … 15.

Figure 7 shows the correlation between the system accuracy and the size of testing data set. With a given data set of learning, the increasing numbers of patterns in the class for testing lead to decreasing the system efficiency. Increase the rate of the examples used in training to the examples used in the testing implies to increase the rate of recognition.

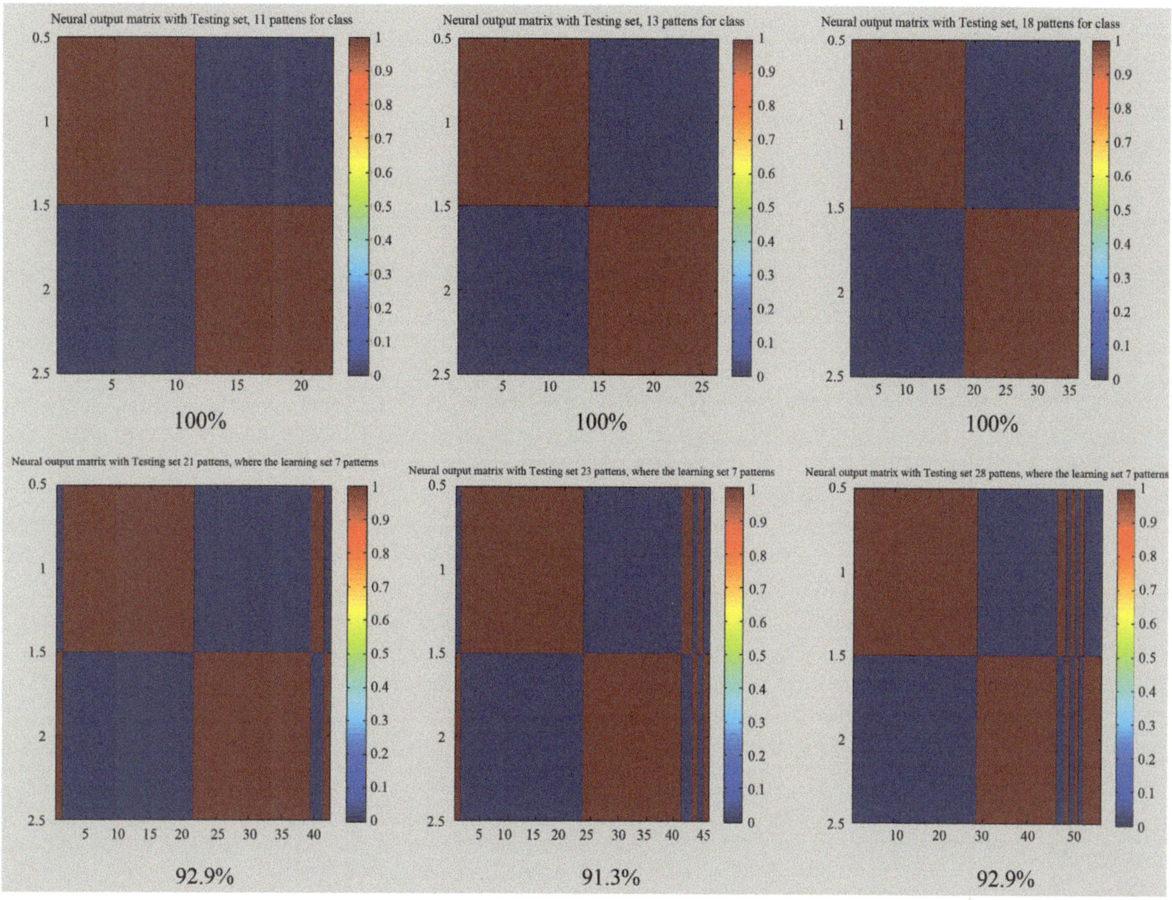

Figure 7. (a)-(f): Neural output matrix with testing data set are 11, 13, 18, 21, 23, 28 patterns for each class, where the learning data set is 7.

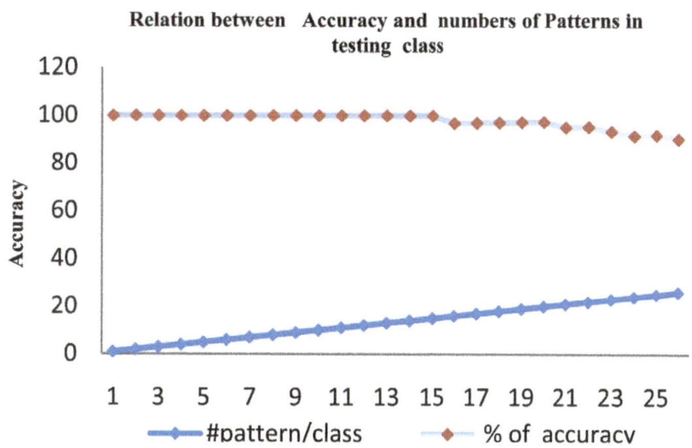

Figure 8. System efficiency with different size of testing data set (1, 2, … , 26), where size of learning data set = 10.

5. Conclusion

We studied and developed an efficient method of textile defects identification based on GLCM and Neural Networks. The descriptor of the textile image based on statistical features of GLCM is used as input to neural networks classifier for recognition and classification defects of raw textile. One hidden layer feed-forward neural networks is used. Experimental results showed that the proposed method is efficient, and the recognition rate is 100% for training and the 100%, 91% best and worst for testing respectively. This study can take part in developing a computer-aided decision (CAD) system for Tissue online Automatic Inspection (TAI). In future work, various effective features will be extracted from the textile image used with other classifiers such as support vector machine.

References

[1] Abdel Azim, G. and Nasir, S. (2013) Textile Defects Identification Based on Neural Networks and Mutual Information. *International Conference on Computer Applications Technology* (*ICCAT*), Sousse Tunisia, 20-22 January 2013, 98. http://dx.doi.org/10.1109/ICCAT.2013.6522055

[2] Davis, L.S. (1975) A Survey on Edge Detection Techniques. *Computer Graphics and Image Processing*, 4, 248-270. http://dx.doi.org/10.1016/0146-664X(75)90012-X

[3] Huang, D.-S., Wunsch, D.C., Levine, D.S. and Jo, K.-H., Eds. (2008) Advanced Intelligent Computing Theories and Applications—With Aspects of Theoretical and Methodological Issues. *Proceedings of the 4th International Conference on Intelligent Computing* (*ICIC*), 15-18 China, September 2008, 701-708.

[4] Walker, R.F., Jackway, P.T. and Longstaff, I.D. (1997) Recent Developments in the Use of Co-Occurrence Matrix for Texture Recognition. *Proceedings of the 13th International Conference on Digital Signal Processing*, Brisbane—Queensland University, 1997. http://dx.doi.org/10.1109/ICDSP.1997.627968

[5] Sahoo, M. (2011) Biomedical Image Fusion and Segmentation Using GLCM. International Journal of Computer Application Special Issue on "2nd National Conference—Computing, Communication and Sensor Network" CCSN, 34-39.

[6] Gonzalez, R.C. and Woods, R.E. (2002) Digital Image Processing. 2nd Edition, Prentice Hall, India.

[7] Kekre, H.B., Sudeep, D.T., Taneja, K.S. and Suryawanshi, S.V. (2010) Image Retrieval Using Texture Features Extracted from GLCM, LBG and KPE. *International Journal of Computer Theory and Engineering*, **2**, 1793-8201. http://dx.doi.org/10.7763/ijcte.2010.v2.227

[8] Haddon, J.F. and Boyce, J.F. (1993) Co-Occurrence Matrices for Image Analysis. *IEEE Electronics & Communication Engineering Journal*, **5**, 71-83. http://dx.doi.org/10.1049/ecej:19930013

[9] de Almeida, C.W.D., de Souza, R.M.C.R. and Candeias, A.L.B. (2010) Texture Classification Based on a Co-Occurrence Matrix and Self-Organizing Map. *IEEE International Conference on Systems Man & Cybernetics*, University of Pernambuco, Recife, 2010. http://dx.doi.org/10.1109/icsmc.2010.5641934

[10] Haralick, R.M., Shanmugam, S. and Dinstein, I. (1973) Textural Features for Image Classification. *IEEE Transactions on Systems, Man, and Cybernetics*, **3**, 610-621.

[11] Flusser, J. and Suk, T. (1993) Pattern Recognition by Affine Moment Invariants. *Pattern Recognition*, **26**, 167-174. http://dx.doi.org/10.1016/0031-3203(93)90098-H

[12] Lo, C.H. and Don, H.S. (1989) 3D Moment Forms: Their Construction and Application to Object Identification and Positioning. *IEEE Transactions on Pattern Analysis and Machine Intelligence*, **11**, 1053-1064. http://dx.doi.org/10.1109/34.42836

[13] Srinivasan, G.N. and Shobha, G. (2008) Segmentation Techniques for ATDR. *Proceedings of the World Academy of Science, Engineering, and Technology*, **36**, 2070-3740.

[14] Tuceryan, M. (1994) Moment Based Texture Segmentation. *Pattern Recognition Letters*, **15**, 659-667. http://dx.doi.org/10.1016/0167-8655(94)90069-8

[15] Gonzalez, R.C. and Woods, R.E. (2008) Digital Image Processing. 3rd Edition, Prentice Hall, India.

[16] Eleyan, A. and Demirel, H. (2011) Co-Occurrence Matrix and Its Statistical Features as a New Approach for Face Recognition. *Turkish Journal of Electrical Engineering and Computer Science*, **19**, No.1.

[17] Kröse, B. and Van Der Smagt, P. (1996) An Introduction to Neural Networks. 8th Edition. http://www.fwi.uva.nl/research/neuro

[18] Bishop, C. (1995) Neural Networks for Pattern Recognition. Clarendon Press, Oxford, UK.

[19] Freeman, J.A. and Skapura, D.M. (1991) Neural Networks: Algorithms, Applications and Programming Techniques.

Addison-Wesley, Reading.

[20] Rumelhart, D.E., Hinton, G.E. and Williams, R.J. (1986) Learning Internal Representations by Error Propagation. In: Rumelheart, D.E. and McClelland, J.L., Eds., *Parallel Distributed Processing: Explorations in the Microstructure of Cognition: Foundations*, MIT Press, Cambridge, 318-362.

[21] Rumelhart, D.E., Durbin, R. Golden, R. and Chauvin, Y. (1995) Back Propagation: The Basic Theory. In: Chauvin, Y. and Rumelheart, D.E., Eds., *Back Propagation: Theory, Architectures and Applications*, Lawrence Erlbaum, Hillsdale, 1-34.

[22] Rumelhart, D.E., Hinton, G.E. and Williams, R.J. (1986) Learning Representations by Back-Propagating Errors. *Nature*, **323**, 533-536. http://dx.doi.org/10.1038/323533a0

[23] Patterson, D. (1996) Artificial Neural Networks. Prentice Hall, Singapore.

[24] Haykin, S. (1994) Neural Networks: A Comprehensive Foundation. Macmillan Publishing, New York.

[25] Azim, G.A. and Sousow, M.K. (2008) Multi-Layer Feed Forward Neural Networks for Olive Trees Identification. *IASTED*, *Conference on Artificial Intelligence and Application*, Austria, 11-13 February 2008, 420-426.

[26] Kattmah, G. and Azim, G.A. (2013) Fig (*Ficus Carica* L.) Identification Based on Mutual Information and Neural Networks. *International Journal of Image Graphics and Signal Processing* (*IJIGSP*), **5**, No. 9.

23

Robot Global Path Planning Based on an Improved Ant Colony Algorithm

Jingang Cao

Department of Computer, North China Electric Power University, Baoding, China
Email: caojg168@126.com

Abstract

Aiming at the disadvantages of the basic ant colony algorithm, this paper proposes an improved ant colony algorithm for robot global path planning. First, adjust the pheromone evaporation rate dynamically to enhance the global search ability and convergence speed, and then modify the heuristic function to improve the state transition probabilities in order to find the optimal solution as quickly as possible; and finally change the pheromone update strategy to avoid premature by strengthening pheromone on the optimal path and limiting pheromone level. Simulation results verify the effectiveness of the improved algorithm.

Keywords

Path Planning, Ant Colony Algorithm, Grid Method

1. Introduction

The path planning is an important ability in many applications, such as UAV (Unmanned Aerial Vehicle), robotics, unmanned car and so on. Its task is to find a path from the current point (or the start point) to the target point, which is a shortest or a minimum price path without barrier in the environment which has obstacles.

The path planning is further divided into two categories [1]—global path planning and local path planning. Global path planning is required that environment should be completely known and the obstacles should be static. The robot produces a path from the starting point to the destination before it starts moving. But local path planning enables a robot to plan its path as it is moving in the environment, which means the robot will be able to generate new paths in response to the changes of environment [2].

In recent years, many researchers have studied the global path planning on various intelligent methods, such as genetic algorithms [3], particle swarm algorithms [4], neural networks [5] and ant colony algorithm [6], etc.

M. Dorigo, Italian scholar, who was inspired from ant foraging behavior and first proposed the ant colony algorithm in 1991 [6]. Ant colony algorithm is a swarm intelligent algorithm, and it has many advantages: 1) strong solved robustness; 2) easy to implement parallel processing; 3) easy to combine with other algorithms to improve [7]. However, the basic ant colony algorithm also has many disadvantages: 1) easy to fall into local optimal solution; 2) large calculation, slow convergence and so on. The paper considers the problems of ant colony algorithm and proposes an improved ant colony algorithm and uses it to the robot global path planning.

The paper is organized as follows: Section 2 and 3 describes the method of grid modeling and the basic ant colony algorithm; Section 4 describes the improved ant colony algorithm for the robot; Section 5 presents the experimental results on robot global path planning based on ant colony algorithms. Section 6 presents brief concluding remarks.

2. Grid Modeling

Grid method was proposed by the W. E. Howden in 1968, and its mainly task is to build a path grid map depending on the environment. The basic principle is to divide robot working environment into numerous tiny grid units, and the specifications of each grid is determined by the robot steps, namely one step is one grid unit. The grid is divided into two kinds: free gird and obstacle grid. Free grid is represented by white grid, and obstacle grid is represented by black grid. The grid map can be represented by a binary matrix, which 1 represents obstacle and 0 is free grid. The obstacle can occupy a grid or multiple grids, if it less than one grid is also expressed by one grid. Robot can only move in the free grid and must avoid when it encounter obstacles grid.

Depending on the position of grid, the grid can be divided into intermediate grid and boundary grid. For intermediate grid, robot may have eight directions for the next motion. Such as up, down, left, right, right-up, right-down, left-up and left-down. **Figure 1** shows the motion direction of the robot. And for boundary grid, it has to subtract inaccessible directions. Robot must avoid obstacles to select an optimization motion path moving to the target position.

The paper uses two-dimensional grid represents robot environment, and encodes the grid from left to right, from top to bottom. **Figure 2** is a 20 × 20 grid map.

In **Figure 2**, the number of grid in each row $N_x = 20$, and the number of grid in each column $N_y = 20$. The coordinate of the first grid is (0.5, 19.5), and the coordinate of the second grid is (1.5, 19.5), … Suppose $S = \{1,2,3,\cdots,N\}$ is a set of the number of grids, the coordinate of the ith grid is (x_i, y_i) and R_a is the length of each unit in axis. The relation between the coordinate and the number of the grid can be expressed as follows:

$$\begin{cases} x_i = R_a\left[\mathrm{mod}(i, N_x) - \dfrac{R_a}{2}\right] \\ y_i = R_a\left[N_y + \dfrac{R_a}{2} - ceil(i/N_y)\right] \end{cases} \tag{1}$$

where, N_x is the number of grid in each row; N_y is the number of grid in each column; i is the ith grid; *mod* is a function to get remainder, and *ceil* is a function to get integer.

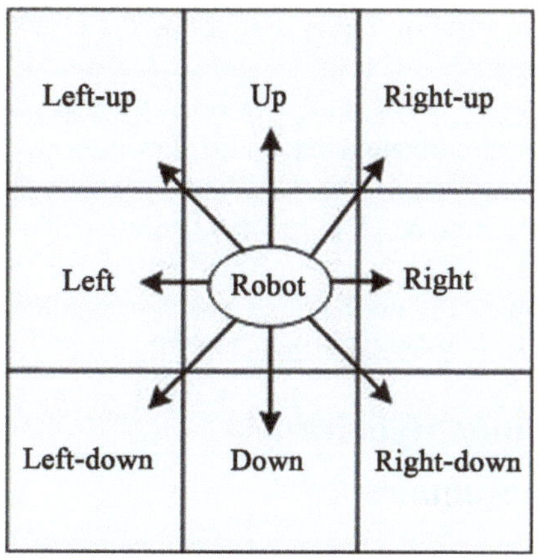

Figure 1. The motion direction of the robot.

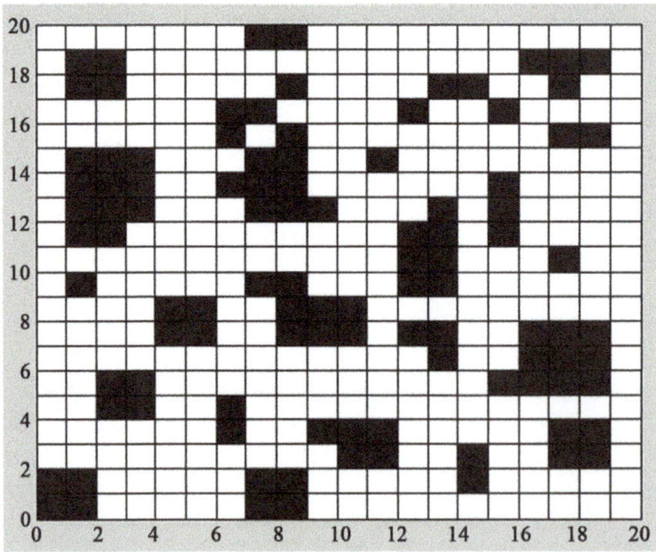

Figure 2. The grid map.

3. The Basic Ant Colony Algorithm

Ant colony algorithm (ACO) is an intelligent heuristic search algorithm by simulating the ant behavior to find the optimal path between the food source and their nest. Ants release pheromone on the path which they passed. Through these pheromones, ants can communicate with each other and find the shortest path to food finally. When the ant reaches an intersection at first time, it will randomly choose a motion direction forward. The more ants follow a given path, the more pheromone will be left on the path, and also the more attractive this path becomes to be followed by other ants. The probability of other ants selecting the path is increased. Ants can also adapt to environmental changes, when there is an obstacle on the path, they will find a new path quickly. This process can be expressed as a loop of positive feedback.

The probability of the ant *k* moving from the grid *i* to *j* is defined as follows [6]:

$$p_{ij}^{k}(t)=\begin{cases}\dfrac{\tau_{ij}^{\alpha}(t)\eta_{ij}^{\beta}(t)}{\sum\limits_{s\in \text{allowed}_k}\tau_{is}^{\alpha}(t)\eta_{is}^{\beta}(t)} & j\in \text{allowed}_k\\ 0 & \text{otherwise}\end{cases} \tag{2}$$

where, $p_{ij}^{k}(t)$ denotes the transition probability in which the ant *k* will traverse from grid *i* to *j* at time *t*; $\tau_{ij}(t)$ is the intensity of the pheromone between grid *i* and *j* at time *t*; $\eta_{ij}(t)$ represents the heuristic function between grid *i* and *j*; α is information inspiration factor and β is hope inspiration factor; allowed$_k$ denotes a set of grids the ant *k* selecting in next step.

The process of robot path planning based on ant colony algorithm can be divided into two stages: optimization stage and stagnation stage. In optimization stage, the algorithm should have stronger ability of global search and can rapidly converge; and during stagnation stage, the algorithm should can automatically jump out of local optimal solution and continue to search the global optimal solution. To avoid premature to fall into local optimal solution and blind searching, the algorithm requires randomness of search and accuracy of solution. But randomness and accuracy are usually both contradictory and interrelation. In order to overcome defects of ACO, this paper presents an improved ant colony algorithm (IACO).

4. The Improved Ant Colony Algorithm

4.1. Optimize the Initial Pheromone

For the basic ant colony algorithm, the searching time is too long. One of the reasons is that the direction of the initial moment when the ants begin to search is uncertain. To improve the problem, and speed up the search, we initialize pheromone decreasing with distance so that the ant has a clear motion direction during the initial search, the method is as follows:

$$\tau_{ig}(0)=\frac{C_1}{d_{ig}} \tag{3}$$

$$d_{ig}=\sqrt{(x_i-x_g)^2+(y_i-y_g)^2} \tag{4}$$

where, $\tau_{ig}(0)$ denotes the initial pheromone level of a free grid i (x_i, y_i); C_1 is a pheromone constant; d_{ig} represents the distance between the grid i (x_i, y_i) and the target position G (x_g, y_g).

4.2. Adjust Pheromone Evaporation Rate Dynamically

In ACO, pheromone evaporation rate ρ is unchanged and $\rho \in (0,1)$, which is directly related to the global search ability and convergence speed. If ρ is too small, ACO is prone to local convergence. On the contrary, if ρ is too large, although it can improved ACO random search performance and global search capability, the convergence rate of the algorithm will be reduced. So the paper adjusts ρ dynamically with the change of the number of cycles. In order to reduce the amount of calculation, ρ does not adjust in each cycle. In the early stage of algorithms, ρ has a large value to enhance the global search ability of the algorithm. With the number of cycles increases, ρ reduced appropriately, and the algorithm can converge to the optimal solution quickly. Specific rules are as follows:

$$\rho=1-\rho_0\cdot\frac{ceil(k/F)}{ceil(k_{\max}/F)} \tag{5}$$

where, k represents the current number of cycle; $k_{\max}$ represents the maximum number of cycle; ρ_0 is the initial value of the pheromone evaporation rate, F is the frequency factor, the bigger F is , the slower ρ will be adjusted.

4.3. Improved Heuristic Function

In ACO, the heuristic function is:

$$\eta_{ij}(t)=\frac{1}{d_{ij}} \tag{6}$$

where, d_{ij} represents the distance between grid i and the next grid j. But in grid map, the difference of value between adjacent grids is not obvious. In order to improve search efficiency of the algorithm, we use d_{jg} instead of d_{ij}, and d_{jg} represents the distance the next grid j to the target grid g.

In grids map, when the target position is known, a ant can calculate distances from its surrounding eight grids to the target point. The smaller the distance, the larger the value of η_{ij}. So the heuristic function can be expressed as

$$\eta_{ij}(t)=\frac{C_2}{d_{jg}} \tag{7}$$

where, C_2 is heuristic factor, and it is a constant.

4.4. Modified Pheromone Update Strategy

Scholars have proposed several different pheromone update models, there are three mainly models: Ant-Cycle model, Ant-Quantity model and Ant-Density model. The paper chooses Ant-Cycle model, which can be expressed as follows:

$$\Delta\tau_{ij}^k=\begin{cases}\dfrac{Q}{L_k}, & \text{if ant} k \text{ use edge}(i,j) \text{ in its tour}\\ 0, & \text{otherwise}\end{cases} \tag{8}$$

where Q is a constant of pheromone, L_k is the length of path passed by the ant k, and $\Delta\tau_{ij}^k(0)=0$.

All ants move one step called once iteration, after n iterations, all ants finish a cycle. In a cycle, there is a current optimal solution L_{best} and a worst solution L_{worst}. Using these two values to select some good solutions and update pheromone quantity, the rules are as follows:

$$L \le \frac{L_{best}+L_{worst}}{2} \tag{9}$$

$$\tau_{ij}(t+1)=(1-\rho)\tau_{ij}(t)+\Delta\tau_{ij} \tag{10}$$

$$\Delta\tau_{ij}^k=\begin{cases}\frac{Q}{L}, & \text{if ant}k \text{ use edge}(i,j) \text{ in its tour}\\ 0, & \text{otherwise}\end{cases} \tag{11}$$

4.5. Limited Pheromone Quantity

When the pheromone quantity is too high, the algorithm is prone to premature, and it will reduce the optimization capability. The paper introduces Max-Min ant method to pheromone update strategy, which can be described as follows:

$$\tau_{ij}=\begin{cases}\tau_{\min}, & \text{if } \tau_{ij}\le\tau_{\min}\\ \tau_{ij}, & \text{if } \tau_{\max}<\tau_{ij}<\tau_{\min}\\ \tau_{\max}, & \text{if } \tau_{ij}\ge\tau_{\max}\end{cases} \tag{12}$$

In Formula (12), the pheromone quantity is limited between $[\tau_{\min},\tau_{\max}]$, which can make the difference of the pheromone quantity between the worst path and the optimal path is not too large.

4.6. Algorithm Steps

Step 1: Initialization parameter: setting the start and target point, the maximum number of cycles $k_{\max}$, the number of ant colony, and other parameters such as α,β,ρ, etc.;

Step 2: Put all ants on the start point, and each ant determines which path it will select according to the transition probability equation;

Step 3: Save the path and the path length of each ant in each cycle;

Step 4: When one cycle finish, update pheromone quantity according to pheromone strategies;

Step 5: Loop execution Step 2 to 4 until the optimal solution is got or reach the maximum number of cycles.

5. Simulation and Analysis

In order to verify the effectiveness of the improved ant colony algorithm, the paper simulates on Matlab 7.1 platform. The main parameters as follows: $\rho = 0.7$; $\alpha = 1$; $\beta = 5$; Q = 1000.

To compare with the basic ant colony algorithm (ACO), we design the five different grid maps as **Figures 3-7**, which are the results of the simulation. **Figure 1** is a 10 × 10 size and the simple and small map. **Figure 2** is a 15 × 15 size and more complex than **Figure 1**. **Figures 3-5** is a 20 × 20 size, 25 × 25 size and 30 × 30 size respectively. In different figures, there are two parameter were changed during simulation, they are the maximum number of cycles and the number of ants. In each Figure, (a) and (b) show the optimal path with IACO and ACO; (c) and (d) are the convergence curve about the average length of path and the number of cycle.

In the different five maps, each map simulates 20 times independently. **Table 1** shows the results of comparison between IACO and ACO. Form **Table 1**, we can see that the average number of cycles with IACO is much less than ACO, which nearly reduce one-third of the number of cycles. Also, the average length of the optimal path with IACO is shorter than ACO. It's because IACO improves the transition probability, and ants are prone to find the optimal path. In other words, the results show that global optimization ability and searching speed are improved obviously with IACO.

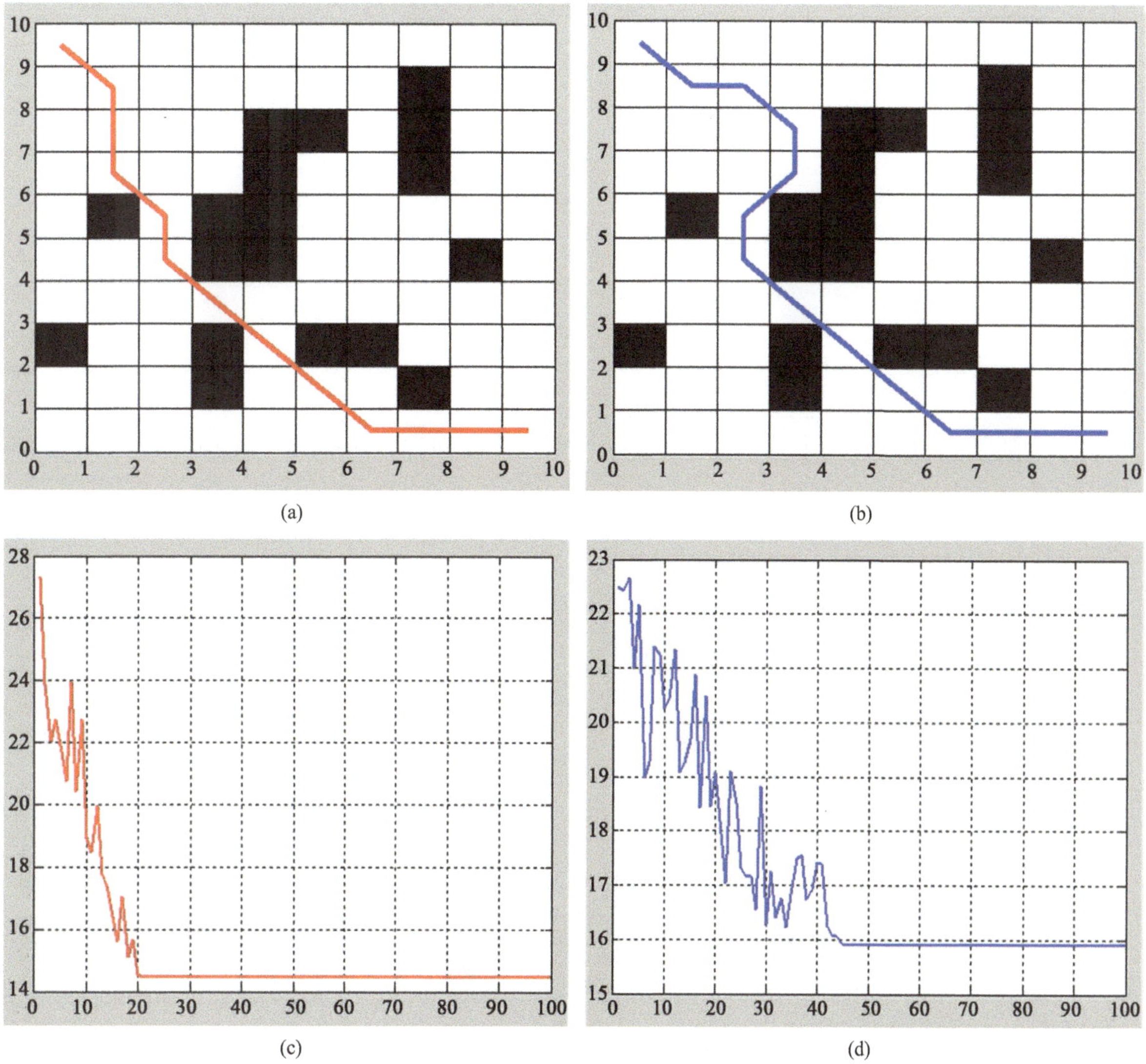

Figure 3. (a) IACO; (b) ACO; (c) IACO; (d) ACO.

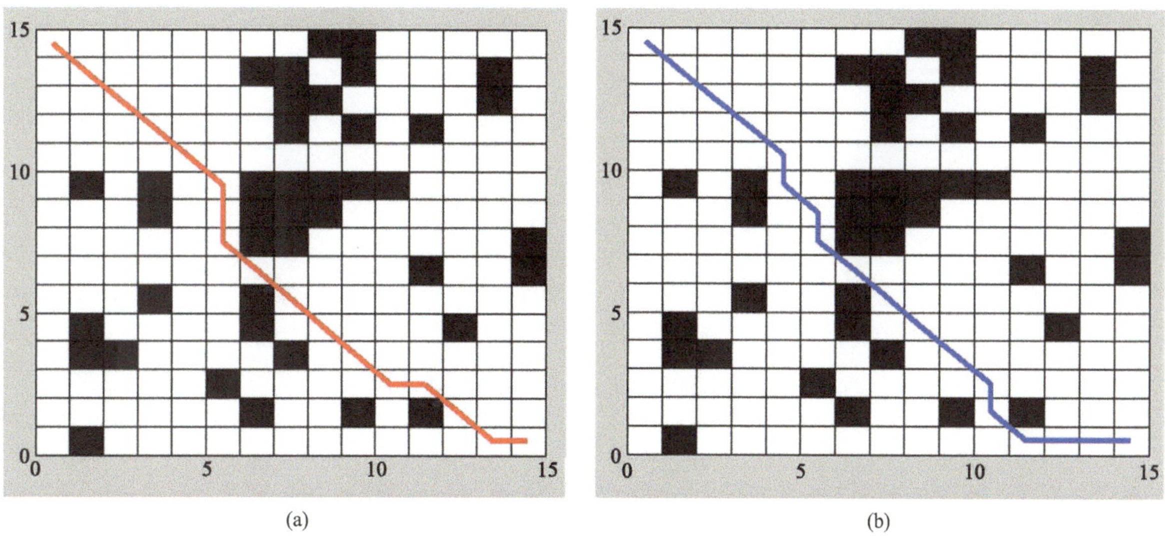

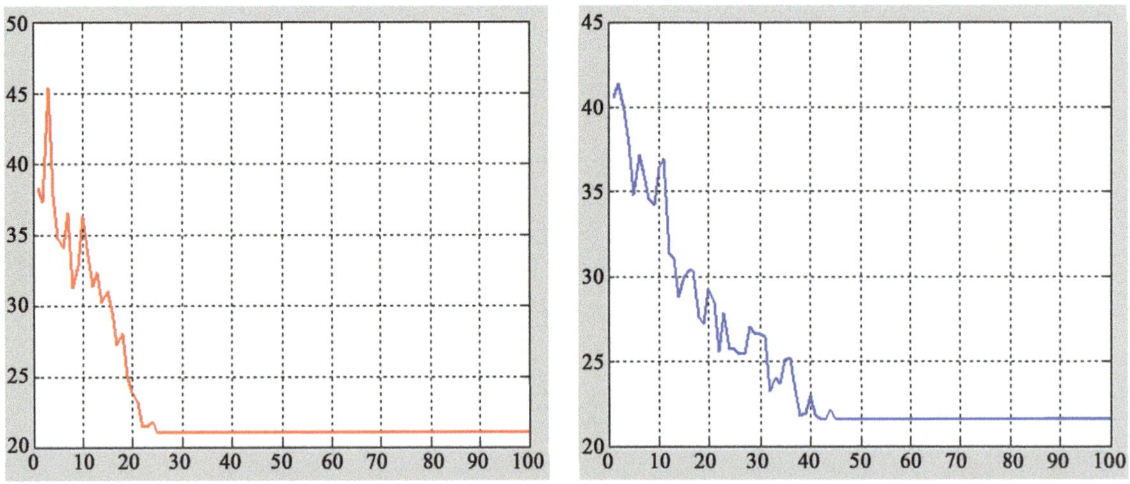

(c) (d)

Figure 4. (a) IACO; (b) ACO; (c) IACO; (d) ACO.

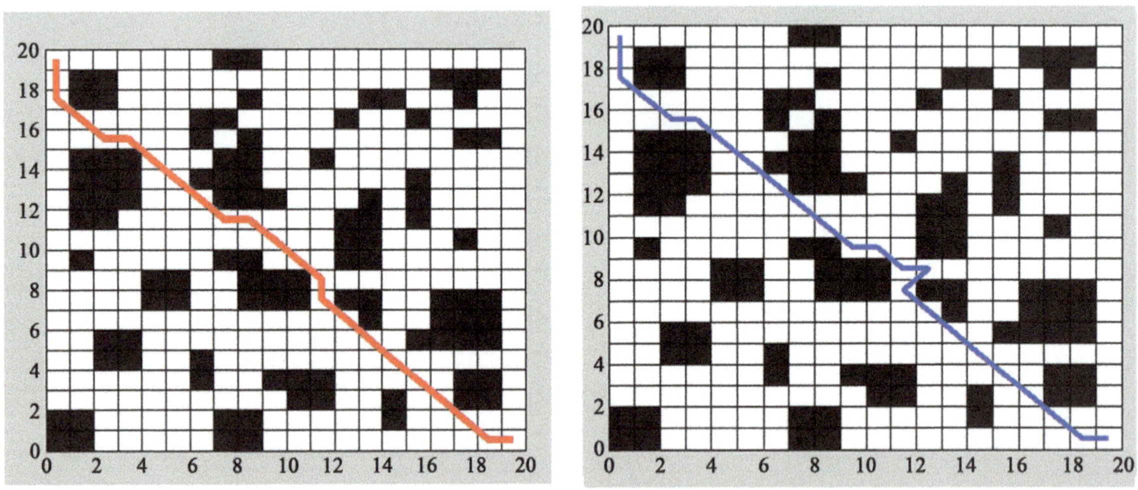

(a) (b)

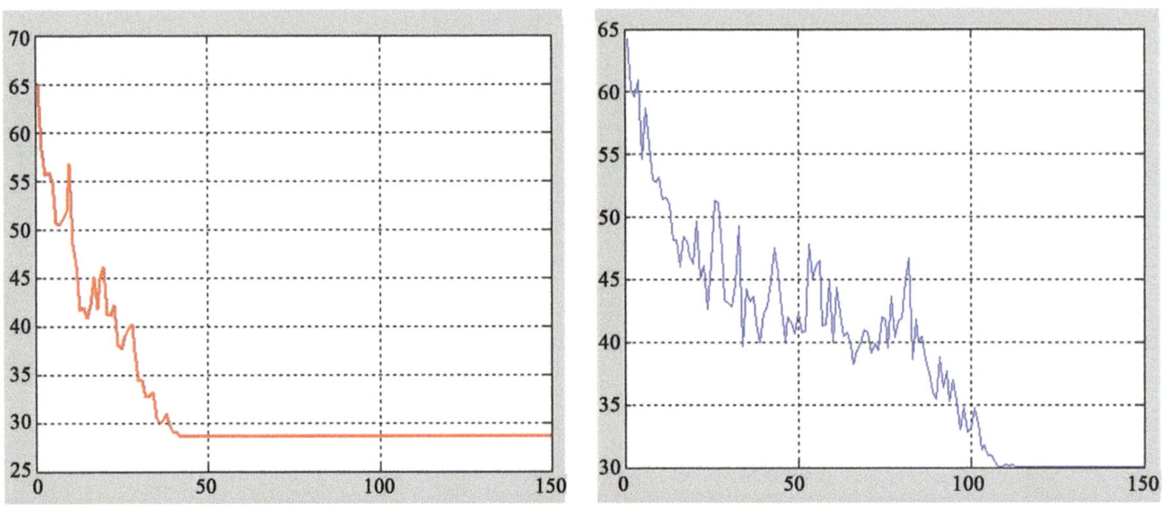

(c) (d)

Figure 5. (a) IACO; (b) ACO; (c) IACO; (d) ACO.

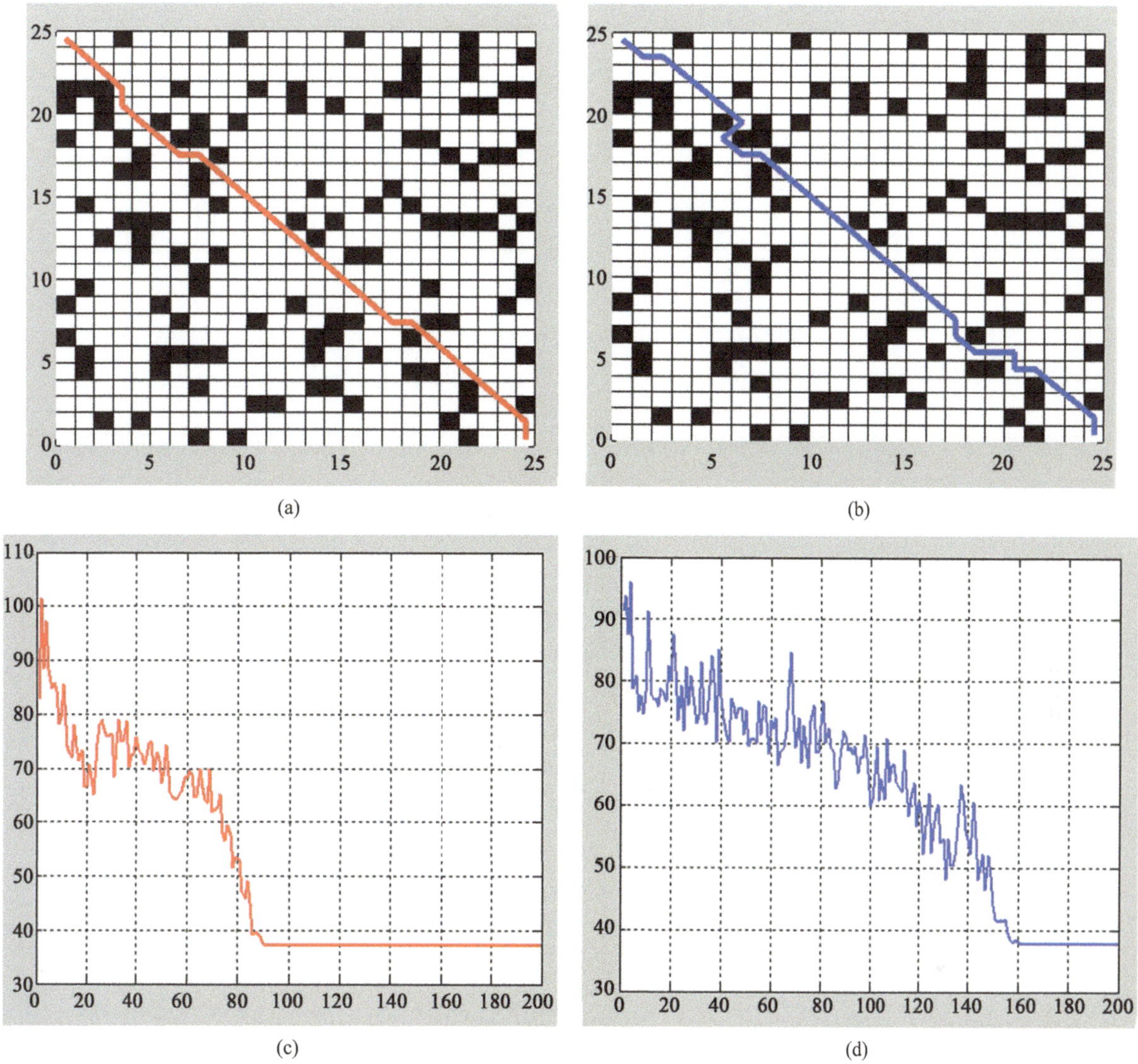

Figure 6. (a) IACO; (b) ACO; (c) IACO; (d) ACO.

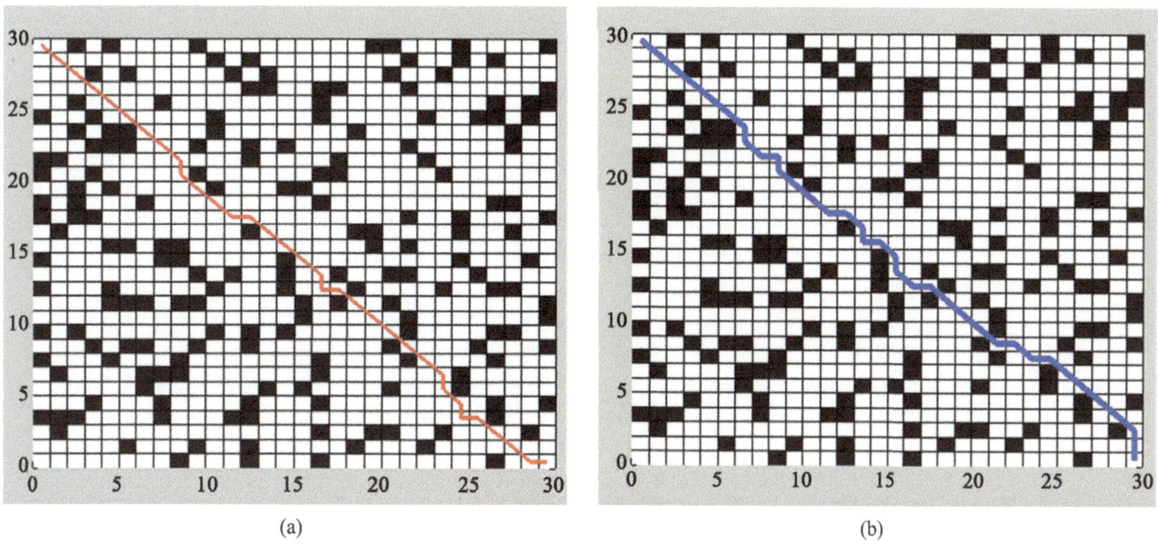

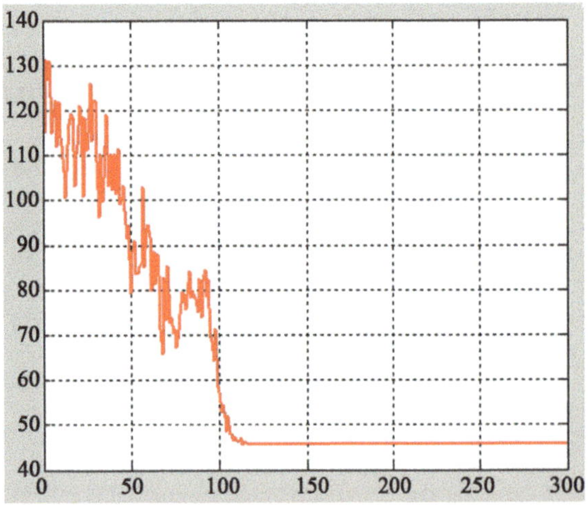

(c)

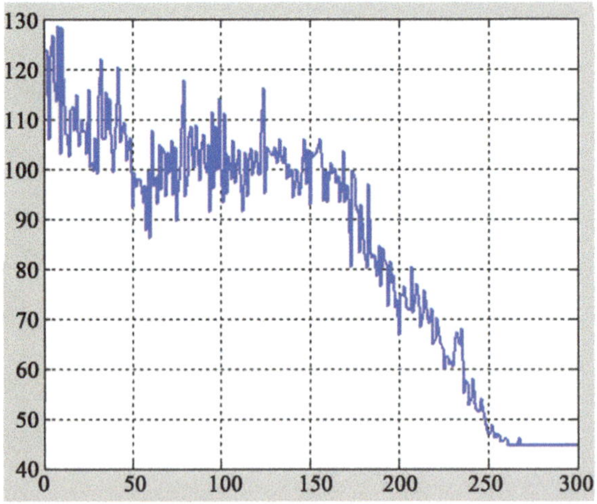

(d)

Figure 7. (a) IACO; (b) ACO; (c) IACO; (d) ACO.

Table 1. The results of the comparison.

Type of map	The average number of cycles		The average length of the optimal path	
	ACO	IACO	ACO	IACO
Figure 3	31	21	15.7279	14.8995
Figure 4	43	34	22.799	21.3848
Figure 5	99	50	30.2668	29.1153
Figure 6	153	95	37.0692	36.2885
Figure 7	212	147	44.5269	43.1838

6. Conclusion

This paper proposes an improved ant colony algorithm and applies it to the robot path planning. In the improved algorithm, pheromone quantity is reinforced in some short paths of each cycle, and pheromone evaporation rate is adjusted dynamically and so the transition probability is improved. The simulation results show that these measures are effective and can enhance global optimization ability, searching speed, and can avoid premature.

References

[1] Sedighi, K.H., Ashenayi, K., Manikas, T.W., Wainwright, R.L. and Tai, H.M. (2004) Autonomous Local Path Planning for a Mobile Robot Using a Genetic Algorithm. *Proceedings of the* 2004 *IEEE Congress on Evolutionary Computation*, **2**, 1338-1345. http://dx.doi.org/10.1109/CEC.2004.1331052

[2] Yee, Z.C. and Ponnambalam, S.G. (2009) Mobile Robot Path Planning Using Ant Colony Optimization. *IEEE/ASME International Conference on Advanced Intelligent Mechatronics*, *AIM* 2009, Singapore, 14-17 July 2009, 851-856.

[3] Zhou, L.F. and Hong, B.R. (2006) A Knowledge Based Genetic Algorithm for Path Planning of a Mobile Robot. *Chinese Journal of Acta Electronica Sinica*, **34**, 911-914.

[4] Zhang, W.X., Zhang, X.L. and Li, Y. (2014) Path Planning for Intelligent Robots Based on Improved Particle Swarm Optimization Algorithm. *Chinese Journal of Computer Applications*, **34**, 510-513.

[5] Deng, W.Y., Zheng Q.H., Chen, L. and Xu, X.B. (2010) Research on Extreme Learning of Neural Networks. *Chinese Journal of Computers*, **33**, 279-287. http://dx.doi.org/10.3724/SP.J.1016.2010.00279

[6] Dorigo, M., Maniezzo, V. and Colorni, A. (1996) Ant System: Optimization by a Colony of Cooperating Agents. *IEEE Transactions on Systems*, *Men and Cybernetics—Part B*, **26**, 29-41. http://dx.doi.org/10.1109/3477.484436

[7] Wang, H.H. and Liu, Q. (2006) Convergence Analysis of a MAX-MIN ant Colony Algorithm. *Dynamics of Continuous Discrete and Impulsive Systems*, *Series B—Applications & Algorithms*, **13E**, 3901-3904.

24

Integrate Liveness Detection with Iris Verification to Construct Support Biometric System

Hanaa Mohsin Ahmad, Bushra Jabbar Abdulkareem

Computer Science, University of Technology, Baghdad, Iraq
Email: uot_info@uotechnology.edu.iq

Abstract

The probabilities for the technology to be spoofed are widely acknowledged in biometric verification system. Important efforts have been conducted to study such threats and to develop countermeasures to direct attacks to the biometric verification system to ensure the security of these systems against spoof attacks and reduce this risk, by using another module that is added to the biometric verification system called the "liveness detection" which uses different anatomical properties to distinguish between real and fake traits. Thus, the robustness of the system against direct attacks can be improved through increasing the security level offered to the final user. This paper is an attempt to construct support biometric security system to protect the iris biometric verification system from spoof attacks, through integrating the iris verification system with addition module called liveness detection which composed of two sub-modules (static and dynamic). A test has been performed, for iris verification phase performed on two types of database (MMU DB) for 180 samples and (CASIA DB) for 90 samples, and gave accuracy (99.44%) with FAR of (0.0277) and FRR (0.0055) for MMU DB, and accuracy (97.77%) with FAR of (0.0333) and FRR (0.0222) for CASIA DB.

Keywords

Liveness Detection, Iris Verification, Countermeasures, Direct Attacks, Fingerprint Verification

1. Introduction

The biometric based verification system is essentially a pattern-recognition system that recognizes a person based on a feature vector extracted from physical or behavioral traits. This type of systems has evolved to play a critical role in personal, national and global security [1], which includes a variety of applications that require reliable verification techniques, such as passports, border control, and used in computer systems, cellular phones, medical records management, banking. The systems are also used in important and sensitive government departments, which include the traffic offices.

The probabilities for the technology to be spoofed are widely acknowledged. Absolute security does not really exist, but many researchers study such threats and develop countermeasures to direct attacks to the biometric system attempted to ensure the security of these systems against spoof attacks and reduce this risk [2]: one of the main points focused on it by researchers is the way in which another module is added to the biometric verification system called the "liveness detection" which uses different anatomical properties to distinguish between real and fake traits. Thus, improve the robustness of the system against direct attacks through increasing the security level offered to the final user [3] [4]. These methods for liveness assessment represent a challenging problem because they have to satisfy certain requirements [5]:

- Non-invasive: the technique should in no case penetrate the body or present and excessive contact with the user,
- Fast: results should be produced in very few seconds as the user cannot be asked to interact with the sensor for a long period of time,
- User-friendly: people should not be reluctant to use it,
- Low cost: a wide use cannot be expected if the cost is very high,
- Performance: it should not degrade the recognition performance of the biometric system.

On the other hand, Iris is an important feature of the human body and very stable throughout a lifetime of a person. Iris is used as a biometric trait and offers many advantages over other human biometric features. It is only an internal human body organ that is visible from the outside and is well protected from external modifiers. For example, voice patterns may be altered due to vocal diseases, a fingerprint may suffer transformations due to harm or aging. The iris is a thin circular diaphragm, which lies between the cornea and the lens of the human eye [6].

Iris's image is characterized with richness of its texture details such as cornea, crypts, filaments, flecks, radial furrows, stripes, arching ligaments, etc. Therefore, the iris region in the eye image of an individual contains completely independent patterns, and minute details of the iris are so randomly distributed patterns, which make the human iris be one of the most important biometric characteristics [7].

This paper is an attempt to address the biometric security issue through integrating the iris verification system with addition module called liveness detection which composed of two sub-modules (static and dynamic) to protect the iris biometric system from spoof attacks, the system consists of two basic phases: liveness detection phase and iris verification phase and will be implemented successively.

The rest sections of the paper are structured as follows. Section 2 presents a collection of previous studies related to the paper theme. Section 3 explains algorithm for proposed system and results and discussion are done in Section 4, and Section 5 deals with the conclusions.

2. Literature Survey

Many researchers and designers of biometric systems studied the possibility of exposure these systems to attempts to attack it by attackers, these attempts affected on data security through alteration in order to the system becomes inactive. In spite of the human iris considered as one of the most important biometric characteristics. There are many fake techniques evolved to cheat the security of these systems, here is a collection of previous studies related to the paper theme:

In 2002, Daouk *et al.* [8] present a new technique to create the new biometric system for iris recognition. Canny edge detection and circular Hough transform used to detect iris region from input eye image by detect inner boundary and the outer boundary of the iris, And then using the Haar Wavelets to extract the Patterns of an iris in the form of a feature vector. In the last step, the matching score has been produced using Hamming Distance operator as a method to compare feature vector for submitted iris with the feature vector stored in a database (template) to find the similarity between two irises. Finally, the decision that the person is genuine or imposter has been made according to the similarity degree. This work results in successfully and simple and not cost Biometric Recognition system used iris trait. The ratio of error in the accuracy of this system can be easily overcome by the use of constant equipment.

In 2012, Amel Saeed Tuama [9] presented a typical iris recognition method which performed by: first, eye image capturing, second, iris segmentation through pupil localization by dividing eye image into 8 × 8 regions and compute the mean intensity for each region, lowest mean used as threshold to detect pupil region, and then iris localized by detect the outer boundary from horizontal line starting from the pupil to iris boundary, then normalize the detected iris region from Cartesian to polar, the third step is feature extracted, by filtering the normalized iris region by convolution with a pair of Gabor filters, finally using Hamming distance approach to producing the similarity between two irises to decide whether these two eye images belong to the same person or not. This algorithm enhances the performance of iris recognition system by detected and segment the iris region in a fast and effective manner and use statistical features for iris recognition and using Hamming distance to perform the comparison of two iris patterns.

In 2011, Shashi Kumar D R, *et al.* [7] New algorithm for iris Recognition has been proposed. In this algorithm using a morphological process to remove the noise is caused by eyelids and eyelashes from upper and lower portion of the iris. Forty-five pixels to left and right of the pupil boundary detected as iris template, and then enhanced the image using Histogram Equalization to get high contrast. Discrete Wavelet Transform (DWT) is applied on histogram equalized iris template to get DWT coefficients. And then extract the features from the DWT coefficients using Principal Component Analysis (PCA). In matching module, multiple classifiers are used such as K-Nearest Neighbors (KNN), Random forest RF and SVM. The proposed algorithm has better performance parameters compared to existing algorithm. And the success rate is better in the case of KNN classifier compared to RF and SVM.

In 2012, Ashish kumar, and Majid Ahmad [10] developed an iris recognition system to verify the uniqueness of the human iris through used it as a biometric. The iris recognition starting with automatic iris segmentation, here use circular Hough transform technique to determine the parameters of circles(radius and centre coordinates of the pupil and iris regions) and then, remapping of the iris region from Cartesian coordinates to the polar coordinates to get rectangular block representation with constant dimensions to account for imaging inconsistencies. Feature vector has extracted using Log-Gabor filter. Finally, for matching, the Hamming distance was chosen as a metric for recognition. The performance of the system has been perfect and produces good recognition results.

In 2012 Gil Santos, Edmundo Hoyle [11] The recognition techniques used in this work performed by five steps: first, iris boundary detection, by following procedure: construct binary for the iris information, A contour is extracted from the white region of binary image, Hough transform is applied to obtain the circle for the iris region. Convert the eye image to grayscale and enhance it using Canny edge detection which applied inside the circular region finally, a Hough transform is used on the resulting to obtain the circle of the pupil. The second step is Iris normalization, then the third step feature extracted through five recognition techniques: Scale-Invariant Feature Transform, Local Binary Patterns, 1-D Wavelet Zero-Crossing, 2-D Dyadic Wavelet Zero-Crossing, Comparison Maps. In the fourth step is matching performed by five techniques: Distance-Ratio Based Scheme, Euclidean Distance, Dissimilarity Using Correlation Coefficient, Dissimilarity Using Correlation Coefficient, and Spatial and Frequency Analysis; The Decision is taken in the last step. In this system several different autonomous approaches were tested; the performances of each individual were evaluated in identification and verification modes and then the methods were fused and caused to improved accuracy, and showed that combining features extracted from the iris region with particular information improve the performance in both recognition modalities.

3. The Main Approach of Proposed Algorithm

This algorithm represents new try to construct a biometric system using iris trait and integrates this system by adding another module called liveness detection to protect it from spoof attack. The architecture of the proposed algorithm composed of two main phases' liveness detection phase and iris verification phase. The liveness detection performance by two sub-modules: (static sub-module and dynamic sub-module). And then iris verification phase is performed to produce matching scores for iris biometric. Finally, the decision is taken if a person is declared as genuine or an imposter as following:

3.1. Iris Liveness Detection

The existing techniques for liveness detection, depicted in **Figure 1**, can broadly be divided into two classes:

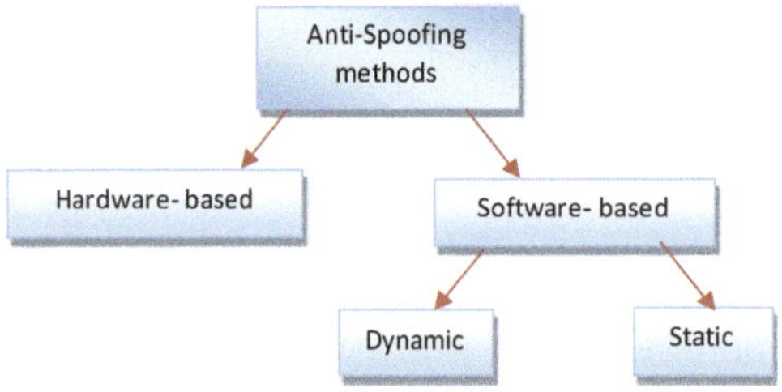

Figure 1. The existing anti-spoofing approaches.

[12] [13]:

- Hardware-based techniques: exploit characteristics of vitality from the available biometric at the acquisition stage, by adding an extra device to the sensor in order to acquire life signs from the presented biometric sample such as the blood pressure, skin distortion, or the odor.
- Software-based techniques: in this case fake traits are detected once the sample has been acquired with a standard sensor during the processing stage. (*i.e.*, feature used to distinguish between real and fake trait), as it used in this proposed algorithm.

Software-based approaches can extract any one peculiarity of live signs from the acquired sample using static techniques (using single sample), or dynamic techniques (using multiple samples) as showed in **Figure 1** [12].

Iris liveness detection module aims to ensure that an input eye image is from a live subject instead of a counterfeited eye image. In this proposed system focused on the establishment of countermeasure to iris photograph spoof attack, by using static and dynamic techniques as in subsection.

3.1.1. Static Technique

Considering the degree of focus property in acquired eye image, the real eye image is 3D volume object, while fake eye image (printed eye image), is a 2D surface. Thus the focus in 2D less than in 3D image, as in **Figure 2**, also defocus primarily suppresses high spatial frequencies which reduce the sharpening of the image, so that, the focus of a fake iris will differ from that of a genuine sample [4].

High pass filter used for this purpose. Which is accomplished using a kernel containing a mixture of positive and negative coefficients such as (Sobel filter) to compute the gradient of the image which represent the change in intensity level. Since an image $f(x,y)$ is a two-dimensional function, its gradient is a vector:

$$\begin{bmatrix} G_x \\ G_y \end{bmatrix} = \begin{bmatrix} \frac{\mathrm{d}f}{\mathrm{d}x} \\ \frac{\mathrm{d}f}{\mathrm{d}y} \end{bmatrix}. \tag{1}$$

The magnitude of the gradient may be computed by any of these two ways:

$$G[f(x,y)] = \sqrt{G_x^2 + G_y^2}. \tag{2}$$

$$G[f(x,y)] = |G_x| + |G_y|. \tag{3}$$

3.1.2. Dynamic Technique

Conceding Pupil variation in size with change in illumination, by comparing the size of pupils of two eye image samples of the same person acquired in different illumination, the difference in size of two pupils is measured (**Figure 3**). If a variation is in the range of 5% - 15%, then it is considered as real eye image, else fake eye image [14]. The percentage variation in size can be computed by this formula [15]:

$$\left| \frac{\text{First size} - \text{Second size}}{\text{First siz} + \text{Second size}/2} \right| \times 100\%. \tag{4}$$

3.2. Iris Verification Phase

This phase executed after liveness detection phase and if the output from this phase detected that the eye image

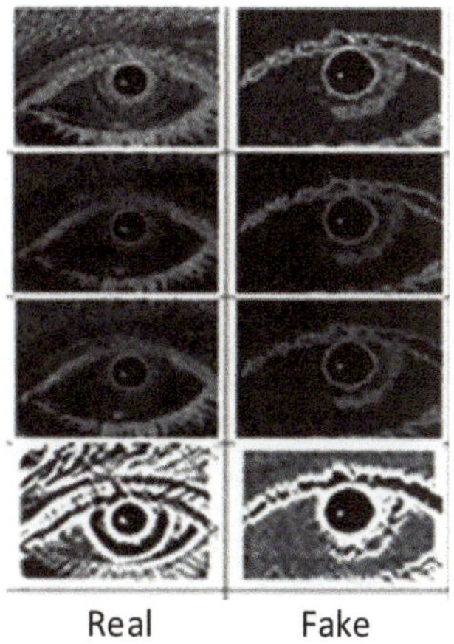

Figure 2. Different in focus quality features for a real and fake iris [4].

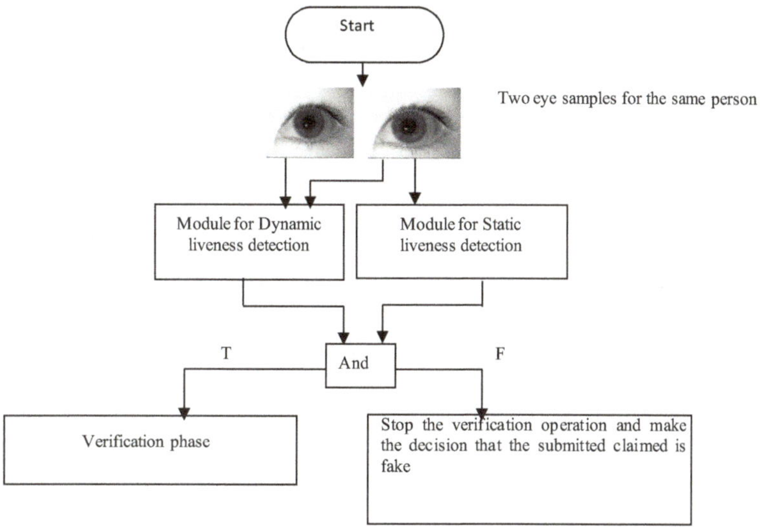

Figure 3. The flowchart of liveness detection phase algorithm.

samples is not spoofed. This phase contains two stage, Enrolled (training) stage and Verification (testing) stage:

- Enrolled stage: This stage deals with how to create reference set (template) for iris biometric system which composed of feature vector of iris biometric traits and an identifier (ID) for each authorized individual based on his input samples and stored in DBs to be used in matching operation in the verification (testing) module.
- Verification (testing) stage: In this stage Person submits the requested biometric traits to the sensors, and claims the identity of a client. This sample detected if it is fake or real sample by liveness detection module, and then pass through sequins of steps to extract feature vector and comparing this submitted feature vector with previously enrolled (template) for this claimed person, using a matching function. The output is matching scores for iris sample. The threshold used to accept the testing sample for the submitted person is (85%).

Iris Verification Steps

The iris verification phase composed of four steps performed in sequential mode, these are: Initialization, Iris Segmentation, Feature Extraction and Matching. As described below and depicted in **Figure 4**:

Step-1: Initialization: Input the eye image to extract iris feature set from it. The eye dataset used in this proposed system (MMU database and CASIA database) is grayscale eye images; therefore there is no need to convert from RGB to gray image

Step-2: Iris Segmentation: The main objective here is to isolate iris region from other regions of the eye image, by detecting papillary (inner) and limbic (outer) boundaries. It is the most important step in iris recognition

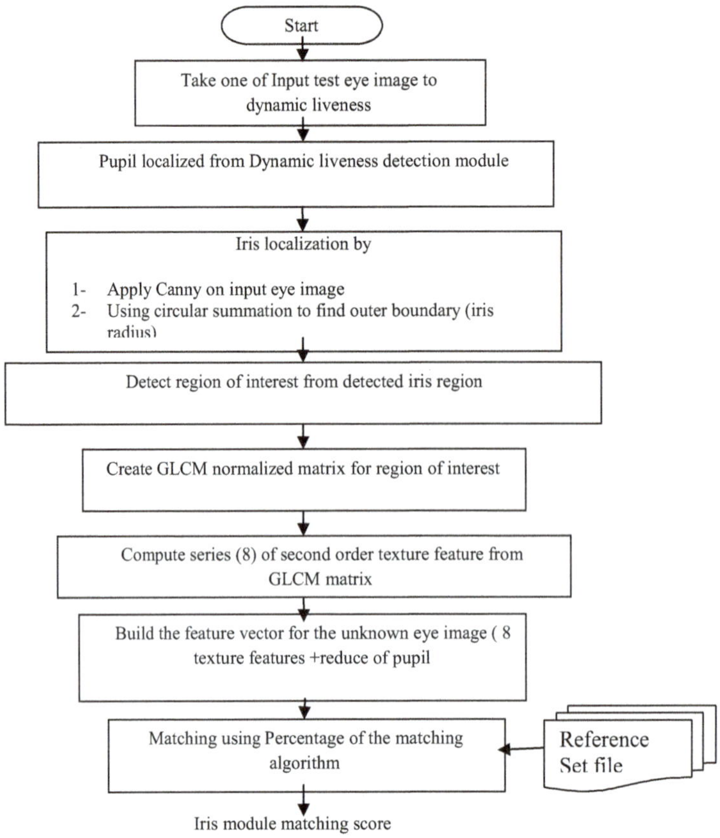

Figure 4. Block diagram for the eye verification using second order texture feature from GLCM, AndPercentage of the matching algorithm.

systems because all the subsequent steps depend on its accuracy. And detect the region of interest to extract a feature from it. This step is performed in three stages:

Stage 1: Pupil localization: The assumption that the Pupil is considered as the darkest region in an eye image. And the pupil considers a very compact and connected region, that the difference between standard deviations for the coordinates (x) and (y) of the pixel inside of the pupil region tends to be the minimum value in eye image. On this basis, a new method for the pupil localization, which is based on the morphology of the binary image, constructed from the original eye image as following: the region with lowest gray scale take white color other regions take black color. The morphology operation was performed by using structuring element which represented by a disk of radius 2.5 shown in **Figure 5**. And then computes the difference between standard deviation for the coordinate of white region of binary image result from the morphological process (Sdt_{Diff}).

$$Sdt_{Diff} = \left|Std_x - Std_y\right|. \tag{5}$$

$$Std_x = \sqrt{\frac{1}{n}\sum_{i=1}^{n}(x_i - \bar{x})^2}. \tag{6}$$

$$Std_y = \sqrt{\frac{1}{n}\sum_{i=1}^{n}(y_i - \bar{y})^2}.$$

$$\bar{y} = \frac{\sum_{i=1}^{n}(y_i)}{n}. \tag{7}$$

$$\bar{x} = \frac{\sum_{i=1}^{n}(x_i)}{n}.$$

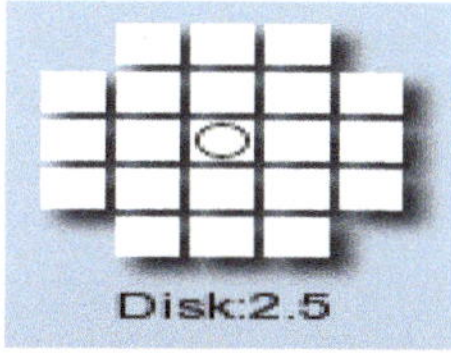

Figure 5. Structure element used in morphological operation.

After detecting the true pupil, next step is to find its radius R_p and center coordinates (C_{px}, C_{py}). Using this equation:

$$R_p = \frac{(y_{\max} - y_{\min})}{2}. \quad (8)$$

$$C_{py} = \frac{(y_{\max} + y_{\min})}{2}. \quad (9)$$

$$C_{px} = x_{\max} - R_p. \quad (10)$$

Stage 2: Iris localization

The outer boundary of an iris (the boundary between the sclera and the iris) has been found based on Intensity variation between iris and sclera. This operation can be conducted by using the canny operator with default two threshold values which have detected by the training data, to obtain the gradient image. And exploit the center coordinate and radius of the pupil which have been detected in the pupil localization stage and apply a Circular summation which consists of summing the intensities over all circles, pass over all possible radii starting from pupil's radius +15 to the pupil's radius +50 and center coordinates of the pupil. The circle with the highest summation corresponds to the outer boundary.

Stage 3: Detect the Region Of Interest (ROI): In the proposed system the conventional methods by using Daugman's rubber sheet model for iris normalization, is avoided as number of researchers do so, such as Debnath *et al.* [16] who Extracting a 8 × 12 Iris Pattern from Edge Detected IRIS Image, and Shashi Kumar D R, *et al.* [7] who select the iris regions to the left and right of the pupil From the localized image. Here, the region of interest is detected by selecting the part of iris region to the left and right and to the bottom of the pupil and then make all gray level values in pupil region equal to zero to isolate it from detected region to extract features from this detected region as templates. Selection of the region in this way reduces the dimension as the non-useful information is cropped out.

The training data turned out to be that radius of the pupil of the MMU database ranges from 20 to 30. The region that will be taken is 84 horizontally and 65 vertically around the Centre of the pupil.

Step-3: Feature Extraction: Once the segmentation has been performed and the region of interest (ROI) has been detected. The next step is to extract the feature from it to create the reference template. This operation performed by two stages:

Stage 1: Create a GLCM Normalized matrix for the detected interest region. The Grey Level Co-occurrence Matrix (GLCM) is a matrix fill with how many times different neighbor combinations of pixel's values occur in an image.

Stage 2: Find a series of "second order" texture as features. These texture measures consider the relationship between groups of two neighboring pixels in the original image, these (8) feature calculated from normalized GLCM matrix result from the previous stage and adding to them a radius of the pupil to represent feature vector consist of (9) features.

Step-4: Matching: this operation carry out by comparing the submitted feature vector with the template feature vector for the claimed person stored in the database using a matching algorithm, in order to compute a similarity degree between these two feature vectors, this similarity degree represented by a value called (matching score).

Many algorithms have been proposed to match two biometric trait representations such as Euclidean distance, City block distance, Hamming distance. However, it is still trying to design a matching algorithm suitable of satisfying the right requirements of many real applications in terms of verification errors. The (Percentage of the matching) is metric used in proposed system for measuring the matching score between two feature vectors and is performed by the (algebraic sum of the percentages of the matching achieved by each element in the test vector set with the corresponding element in the template vector set). Finally, the decision has been taken if the submitted biometric trait belongs to the genuine or impostor person.

4. Results

4.1. Iris Liveness Detection Results

Database that is set up to test the robustness of the proposed algorithm through iris liveness detection process through static and dynamic modules consists of 15 folders of original (MMU database) each folder contain tow eye images sample represent live tries. And 15 folders of(MMU database) eye images printed using the scanner as a devise and recapture using a specific camera and resaved in the computer to represent 15 attempted spoof attack against the system each folder contained two samples of fake eye image. **Table 1** shows experiment results for dynamic iris liveness module and **Table 2** show the results of static iris liveness module.

4.2. Iris Verification Experimental Results

Two types of iris database used to training and testing the proposed system: MMU Iris database and CASIA-IrisV1database. The training of iris verification algorithm consists of three experiments as follows:

The first experiment conducted to test the proposed iris verification algorithm by applied on 180 eye image of (MMU database) for 30 persons for left and right eyes three samples for the left eye and three sample for the right eye for each person. The second experiment testing of the proposed iris verification algorithm phase by applied on 90 images of (CASIA-IrisV1 database) for 30 people's three samples of eye image for each person. In the third experiment, the eye images database collected for testing the proposed biometric verification algorithm as completed system. Started from liveness detection phase and ending with verification phase consists of 15

Table 1. The experiment results for dynamic iris liveness module of 15 real tries and 15 fake tries.

Person No.	Results by appling on original eye images			Results by appling on recaptured eye images		
	No. of pixel in pupil for two samples with different eliminations	Percentage difference in pupil size	Decision	No. of pixel in pupil for two samples with different eliminations	Percentage difference in pupil size	Decision
1	291210 325125	11.005	Live	694620 695895	0.183	Fake
2	482460 540090	11.272	=	682380 691050	1.613	Fake
3	457980 525300	13.693	=	1418310 1419330	0.072	Fake
4	330735 372300	11.824	=	961605 974355	1.317	=
5	699210 759900	8.319	=	512295 772395	40.492	=
6	410550 474300	14.409	=	930495 759900	20.184	=
7	408000 431460	5.589	=	1266075 1327530	4.739	=
8	337875 368220	8.595	=	542640 548760	1.121	=
9	248625 288150	11.116	=	482715 685695	34.745	=
10	514845 566355	9.528	=	675750 907800	29.308	=
11	585735 627555	6.894	=	1564425 1195695	26.718	=
12	469455 536010	13.239	=	828750 562785	38.226	=
13	439110 469710	6.734	=	1036065 1171980	12.311	Faulty live
14	336345 336345	5.529	=	818805 842265	2.825	Fake
15	719865 762450	5.746	=	1002915 860880	15.241	Fake

Table 2. Show the experiment results of applying static module on original eye images and recaptured eye images for 15 person enter to the system.

Person No.	Results by appling static module on original eye images		Results by appling static module on recaptured eye images	
	Mean of gradient for input original MMU eye sample images	Decision	Mean of gradient for input recaptured MMU eye sample images(spoof image)	Decision
1	45.867 38.969	Live =	28.808 32.486	Fake =
2	36..203 58.622	Live =	29.355 39.577	Fake Faulty Live
3	42.561 37.927	Live =	34.55 30.977	Faulty Live Fake
4	35.306 35.998	Live	29.382 28.72	Fake =
5	55.638 49.592	Live =	32.273 32.884	= =
6	36.935 41.593	Live =	26.155 28.721	= =
7	36.489 36.454	Live	27.464 29.281	= =
8	35.789 45.489	Live	26.947 32.051	= =
9	35.49 35.186	Live =	26.22 25.554	= =
10	35.511 34.5	Live =	27.708 23.303	= =
11	35.122 35.052	Live =	25.392 28.361	= =
12	39.363 37.276	Live =	28.55 28.323	= =
13	36.702 38.173	Live =	23.235 28.416	= =
14	34.363 35.389	Live =	23.112 27.843	= =
15	35.55 33.55	Live =	26.295 24.792	= =

folders of (MMU database) for 15 persons each folder contained four samples for left eye image. **Table 3** clarify the results of testing operation for verification phase of iris model and the accuracy computed and The False Accepted Rate (FAR) and False Rejected Rate (FRR) which computed by following equations:

$$\text{FAR} = \frac{\text{Number of times different member matching} \times 100}{\text{Number of comparison between difference members}}. \quad (11)$$

$$\text{FRR} = \frac{\text{Number of times same member rejected} \times 100}{\text{Number of comparison between same members}}. \quad (12)$$

5. Conclusions

By extensive and hard work in the design of the proposed system which is represented by iris verification to improve the security and accuracy of it through adding another module (liveness detection) to it, we reach to a number of conclusions.

- The dynamic method which used to detect liveness is more effective, more accuracy and more successful

Table 3. Experiment results of the iris verification phase.

Eye image database	No. of people	No. of samples	No of samples successfully verification	No. of samples faulty accepted	No. of samples faulty rejected	Tame of verification	Time for matching	FRR %	FAR%	Average accuracy
MMU DB	30	180	179	5	1	20 second	0.8 second	0.0055	0.0277	99.44%
MMU DB	15	60	59	2	1	20 second	0.8 second	0.0166	0.0333	98.33%
CASIA (Version 1.0)	30	90	88	3	2	24 second	0.8 second	0.0222	0.0333	97.77%

than static method, as shown in **Table 1** for iris dynamic liveness detection. The Static method needs to detect fixed threshold, and is extracted from original properties of the biometric trait to accept image sample as the real sample. Choosing this threshold will be influenced by the variance of these original properties from person to another, such as the degree of sharpening of original eye image and the way used to collect the original image database and manufacture spoof database, which causes the ratio of error in liveness decision.

- The sequence of morphological operations has been followed to isolate a pupil object from eye images and the limitation used to extract the region of interest from detected iris region is not fixed for all types of eye images database with any conditions. That means the sequence applied in (MMU database) is closing first and then opening. In CASIA database, different sequence of morphological operations is applied, and the radius of the pupil in MMU DB which it is in range (20 to 30) is different from radius in CASIA DB which it is in the range (30 to 45)
- For matching module of this proposed system, there is a specific percentage rate for similarity between each element in test feature vector with the corresponding element in template feature vector allowed to it to enter into the comparison process to produce the final matching score. If the percentage rate of similarity for any element in the feature vector is less than the specific percentage rate, it will be not take part in the matching process to avoid FAR in this system.
- Due to the richness of the texture details in the iris, we find that the calculating texture features from the normalized GLCM (series of second order texture features) is the most convenient and very successful method and satisfies good accuracy in calculating the features of this region.

The experiment results show that the performance of the proposed system in MMU DB is better than performance in CASIA DB. This is shown from the results in **Table 3** which showed that the verification operation is faster in applying on MMU DB than in CASIA DB, and the accuracy in MMU DB is 99.44%, while in CASIA DB is 97.77%.

References

[1] Akhtar, Z. (2012) Security of Multimodal Biometric Systems against Spoof Attacks. Thesis of Ph.D. in Electronic and Computer Engineering, University of Cagliari, Cagliari.

[2] Galbally, J., Alonso-Fernandez, F., Fierrez, J. and Ortega-Garcia, J. (2009) Fingerprint Liveness Detection Based on Quality Measures. 2009 *International Conference on Biometrics, Identity and Security*, Tampa, 22-23 September 2009, 1-8. http://dx.doi.org/10.1109/bids.2009.5507534

[3] Marasco, E. and Ross, A. (2014) A Survey on Anti-Spoofing Schemes for Fingerprint Recognition Systems. *Journal ACM Computing Surveys* (CSUR), **47**, Article A.

[4] Galbally, J., Alonso-Fernandez, F., Fierrez, J. and Ortega-Garcia, J. (2012) Iris Liveness Detection Based on Quality Related Features. 2012 *5th IAPR International Conference on Biometrics* (*ICB*), New Delhi, March 29 2012-April 1 2012, 271-276. http://dx.doi.org/10.1109/icb.2012.6199819

[5] Herrero, J.G. (2009) Vulnerabilities and Attack Protection in Security Systems Based on Biometric Recognition. Thesis of Ph.D., Universidad Autonoma de Madrid, Madrid.

[6] Patil, B.G. and Subbaraman, S. (2011) SVD-EBP Algorithm for Iris Patten Recognition. *International Journal of Advanced Computer Science and Applications*, **2**, 115-119.

[7] Shashi Kumar, D.R., Raja, K.B., Chhootaray, R.K. and Pattnaik, S. (2011) PCA Based Iris Recognition Using DWT. *International Journal of Computer Technology and Applications*, **2**, 884-893.

[8] Daouk, C.H., El-Esber, L.A., Kammoun, F.D. and Al Alaoui, M.A. (2002) Iris Recognition. *IEEE ISSPIT*, Marrakesh, 1.

[9] Tuama, A.S. (2012) Iris Image Segmentation and Recognition. *International Journal of Computer Science & Emerging*

Technologies, **3**, 60-65.

[10] Dewangan, A.K. and Siddhiqui, M.A. (2012) Human Identification and Verification Using Iris Recognition by Calculating Hamming Distance. *International Journal of Soft Computing and Engineering* (*IJSCE*), **2**, 334-338.

[11] Santos, G. and Hoyle, E. (2012) A Fusion Approach to Unconstrained Iris Recognition. *Pattern Recognition Letters*, **33**, 984-990. http://dx.doi.org/10.1016/j.patrec.2011.08.017

[12] Singh, Y.N. and Singh, S.K. (2011) Vitality Detection from Biometrics: State-of-the-Art. *IEEE World Congress on Information and Communication Technologies*, Mumbai, 11-14 December 2011, 106-111.

[13] Anonymous (2014) Fingerprint Liveness Detection Using Convolutional Networks. *IJCB*, *Confidential Review*, 1-2.

[14] Rajesh M. Bodade, Sanjay N. Talbar (2014) Iris Analysis for Biometric Recognition Systems. Springer, New Delhi, Heidelberg, New York, Dordrecht and London. http://dx.doi.org/10.1007/978-81-322-1853-1

[15] (2014) Percentage Difference. https://www.mathsisfun.com/percentage-difference.html

[16] Bhattacharyya, D., Das1, P., Bandyopadhyay, S.K. and Kim, T.-H. (2008) IRIS Texture Analysis and Feature Extraction for Biometric Pattern Recognition. *International Journal of Database Theory and Application*, **Vol.**, 53-60.

25

Investigation and Analysis on DRS Condition of Agricultural University Libraries in China

Linlin Zhang

Heilongjiang Bayi Agricultural University Library, Heilongjiang Bayi Agricultural University, Daqing, China
Email: liblin@126.com

Abstract

31 investigated higher agricultural college libraries all provide the reference consulting service. The service mode includes asynchronous reference consulting, synchronous reference consulting, synchronous and asynchronous reference consulting, joint reference consulting service and traditional reference consulting service. Service items include interlibrary borrowing and document delivery, reader training, special database, resource navigation and so on. But there are some problems to be solved.

Keywords

Agricultural University Library, Digital Reference Consulting, Problems, Countermeasures

1. Introduction

Extensive use of electronic information resources and extensive propagation of network information will make the library unavoidably encounter austere challenges. It is obvious that the traditional service mode of the libraries cannot fully satisfy the requirements of readers, so the library should explore the digital services and avoid edge effect in digital age by using the strength in self-resource. Now the digital libraries are emerging consecutively. Digital libraries can transfer the information resources to the reader's desktop via the network. Generally readers do not know how to effectively utilize these resources, so readers request the libraries to provide related fitting services. Digital reference consulting service (called as DRS) emerges [1]. It symbolizes that libraries are exploring their professional services by using the network platform. The digital reference consulting service becomes the new trend of the reference consulting service development of libraries under the network information environment.

Digital libraries are the important reference information guarantee base for teaching and scientific research, which have plentiful opportunities in digital reference consulting services by depending on strengths in the hardware devices and talent of the universities. Although some well-known university libraries lead in China by referring to successful experiences in the digital reference consulting service and the gap with the foreign libraries reduces, the fact is not satisfactory. To improve comprehensive recognition on the digital reference consulting service of the university libraries, the author investigates the library DRS conditions of the agricultural universities in China in order to find the problems via the empirical research and decide purposeful countermeasures, which is practical for promoting research on the library DRS of agricultural universities.

2. Survey Approaches and Methods

The author visited CERNET, carried on the detailed investigation and research on the web pages of libraries of 31 ordinary higher agricultural colleges and universities in "Education Resources—China's University" Column, extracted related content of reference consulting service of libraries of 22 colleges and universities, including service form and service projects, and entered them into Excel and Word documents for detailed screening, classification, sorting, statistics and analysis.

3. Network Investigation and Analysis of DRS Condition of Agricultural Universities in China

Libraries provide DRS service later in China. Except a minority of well-known university libraries in China, which have stronger comprehensive strength and actual operation experiences, most university libraries have different problems in DRS service. The author investigates 31 agricultural university libraries in China via the network on 14 January 2015, including the state-affiliated key universities and general local agricultural universities. The invested universities basically cover most areas in China.

In 31 investigated agricultural university libraries, the homepages of 5 libraries fail to be linked. The homepages of 2 libraries include DRS services, namely 22% invested libraries have network problems and cannot be further investigated, namely 6% university libraries do not provide DRS service.

The investigation results indicate that although 22 agricultural university libraries implement different DRS, the DRS pattern is asynchronously interactive. Few DRS implements interactive DRS. No university implements the cooperative DRS. Specific research projects and survey result see **Table 1**. Survey results show, 32% investigated libraries have no FAQ. 82% libraries perform DRS via emails or form. 77% libraries perform DRS inside the libraries. 50% libraries perform real-time reference consulting (referring to **Table 2**). It indicates that the digital reference consulting services implemented by the agricultural academies in China is under preliminary phase. DRS depth and extensiveness should be further explored. The service influence should be further improved, the service quality should be further perfected, and the recognition on DRS should be further strengthened.

4. Problems in DRS of Agricultural University Libraries in China

4.1. Non-Perfect Service Function and Worse Whole Level

The investigation indicates that the whole level of agricultural university libraries is low in DRS service. 60% investigated libraries perform DRS service, but the DRS service features obviously insufficient depth and extensiveness and is mainly implemented via E-mail, FAQ and Web form and its combined manner. It belongs to first-generation DRS service. The real-time DRS service indicates that the experts (or reference librarians) directly "face" users under virtual environment, namely answer the questions proposed by users, which is also called as the real-time interactive reference service (Interactive Reference Services, IRS). Now the software such as chartroom or video conference is mainly used for implementation [2]. **Table 1** indicates that the agricultural university libraries are based on the digital reference service using the real-time interactive technologies. The DRS features slow development and popularization speed due to technology reasons. Such worse conditions are caused by many factors, including insufficient human resources, insufficient resources, lack of digital reference consulting service experiences and adaptation software, no network or insufficient network service and no attention from leaders, so the DRS of the university libraries should be further improved and perfected.

Table 1. 22 agricultural university libraries reference form.

Library Name	FAQ	E-Mail	Forms Advisory	Visitors Book	Blog	QQ Online Consultation	Online Service Timely Answers	Mobile Library	Cooperative Reference
Beijing Forestry University	Y		Y	Y		Y			
China Agricultural University	Y	Y		Y		Y	Y		Y
Agricultural University of Hebei Province		Y							Y
Shanxi Agricultural University	Y		Y			Y	Y		
Inner Mongolia Agricultural University		Y	Y				Y		
Shenyang Agricultural University		Y							
Jilin Agricultural University	Y	Y				Y		Y	Y
Northeast Agricultural University		Y		Y					Y
Northeast Forestry University		Y						Y	Y
Nanjing Agricultural University	Y	Y		Y		Y		Y	
Nanjing Forestry University	Y			Y		Y			
Anhui Agricultural University	Y			Y				Y	
Shandong Agricultural University	Y	Y				Y			Y
Zhejiang A & F University		Y		Y		Y			
Jiangxi Agricultural University	Y	Y				Y			Y
Fujian Agriculture and Forestry University	Y	Y		Y			Y		
Huazhong Agricultural University	Y	Y							Y
Hunan Agricultural University	Y	Y	Y			Y			Y
Henan Agricultural University	Y	Y	Y	Y	Y		Y		Y
South China Agricultural University	Y	Y	Y			Y	Y		Y
Sichuan Agricultural University		Y		Y					
Northwest A & F University	Y	Y	Y					Y	
Total	15	18	7	9	1	11	6	5	11

Table 2. Implementation conditions of reference consulting of 22 agricultural universities.

Reference Consulting Form	University Quantity	Percent
FAQ service	15	68
E-mail service	18	82
Consulting inside library	17	77
Real-time reference on cooperative reference consulting	11	50
User education	11	50

4.2. DRS Position Is Not Highlighted on the Library Homepage and the Pattern Is Not Standardized

The investigation indicates that the percent of the digital reference consulting service is insufficient in the whole work of libraries, e.g. the identifier and names of the DRS are chaotic. As a symbolic service, it should be located at a highlighted position on the library homepage and can show its importance. The investigation results indicate that some libraries locate the DRS on the L2 link. Readers cannot easily find this link to use these services. Some universities mix the reference consulting with the reader guide, so readers cannot easily find it. Generally the DRS is called as the consulting expert, virtual reference consulting, electronic reference consulting, remote reference consulting, network reference consulting, but some libraries name it as the "online word leaving" and "word leaving plate". Although the service pattern of the digital reference consulting includes the word leaving board as the asynchronous interaction manner, the service names should be standardized to correctly guide the users to use the service. The above problems are generated due to insufficient recognition on importance of the digital reference consulting in the digital reference information service.

4.3. Lack of Cooperative DRS

The reference consulting services of the agricultural university libraries are performed separately. No cooperative reference consulting service is available, so it is unfavorable to the emerging DRS service and the research foundation for cooperation is not solid in DRS. It will affect DRS much. On the contrary, it will restrict evolvement of DRS and form malicious circulation. The DRS quality cannot be ensured. The DRS of libraries are under the preliminary phase. Quality of persons and owned resources are not perfect. All libraries should be united to carry out the cooperative reference consulting. All libraries should exert their strengths, offset weaknesses, and provide higher quality.

4.4. Lack of Professional DRS Talents

The consulting experts control quality of the digital reference consulting. Their information quality and occupational morality will directly affect quality of the information service and DRS. The consulting experts not only have better information quality, rich knowledge, higher foreign language level, and certain professional knowledge, collect, arrange, analyze and use information via different means, but also should have better occupational morality and dedication spirits and can carefully deal with and handle each consulting requests of users. Now most DRS consultors are acted by librarians of a department (e.g. information department) of the libraries in China. Most consultors are not from the book intelligence specialty, so they are lack of the knowledge extensiveness and information integration and application capability.

As the "processor of reference consulting", the Samuel·Greekn, director of the Wooster public library in America, highlighted "importance of the individual help for readers seeking information" and pointed out function information, guidance and reader education of the reference consulting [3]. It launches theory of the reference consulting service. Now some agricultural university libraries cannot better explain three functions of the reference consulting in the digital reference consulting service. No library can perform the reference consulting service completely in three fields, so the service quality should be further improved and the meaning should be further mined. The service means should be further perfected.

5. DRS Development Countermeasures and Recommendations of Agricultural University Library

5.1. Strengthen Advertisement and Improve DRS Idea of Users

As the new form of the information service in the digital age, DRS changes the service idea, service pattern, and service effect of the reference consulting as a new service mode. Different agricultural university libraries pay different attention. The author thinks that the link of the DRS service should be located at the highlighted position on the library homepage. The readers should be reminded at proper time to cultivate DRS service consciousness of the users and facilitate use [4]. The advertisement should be strengthened to enhance the information quality of the users and make them truly understand potential value. The user requirements are the main force to drive digital reference consulting service.

5.2. Strengthen Standardization of DRS

DRS requires cooperation from different parties, especially cooperative DRS implementation, so the DRS standardization is very important and urgent. DRS standards involve many aspects, including technology standard, service standard, information exchange format standard, information sharing standard, metadata standard and quality control standard. If no standard is available, DRS implementation only keeps at a lower level, will not reach true purpose of the DRS, cannot meet user requirements, and further affect DRS development. How to refer to DRS experiences of other libraries and enrich our standardization issues should be emphasized by the agricultural university library.

5.3. Carry out Cooperation by Participating in and Simulating Domestic Well-Known Cooperative DRS Projects

Now the domestic well-known cooperative reference consulting service in China includes the state library information consulting and cooperation network and united network knowledge navigation website. We can invite the agricultural university library to join in it, so it can improve quality of the agricultural university DRS and further drive the cooperative DRS of the agricultural university libraries in China. This means we need not start with beginning, so it can save investment of some manpower and finance in establishment period of cooperative DRS and increase implementation possibility of cooperative DRS in an intangible manner. For issues such as separate development and no uniform plan, the state large-scale library and related governmental departments should have a whole plan for system construction, integrate resources of different organizations, coordinate relation among members, break current separate layout, establish uniform reference consulting platform, and provide comprehensive high-quality service, especially cooperative reference consulting and distributed reference consulting not only integrates limited resources of libraries and meets different user requirements as much as possible, but also solves the lack of consulting experts, so it should be encouraged.

5.4. Found "Virtual Team" to Improve DRS Quality

The "virtual team" is different from traditional entity team. Members do not depend on tangible office site. They will work in a virtual space and mainly depend on modern communication and information technologies to realize the remote communication and coordination [5]. The virtual team mainly engages the high-knowledge work. Members should be composed of high-level professional experts, who can cross geographical position specialty and function department. They should contact and exchange with each via advanced communication technologies such as electronic means, including email, network chat, telephone, and video conference. Their work and service is mainly implemented via a remote manner. The members of the virtual team will cooperate with each other to solve complicated problems to meet the requirements of the users and organizations via their knowledge.

Now DRS has acted as the main role in the library service. Its service quality will directly affect whole image of the library. If the service quality is not guaranteed and no corresponding control measures of the consulting quality, DRS cannot be recognized and extensively utilized by extensive users, so we should strengthen deep research on the DRS. By combining the actual DRS experiences of the agricultural university library, we can launch personalized, proactive, multi-level and multi-view whole-process digital reference consulting service, and provide beneficial help for the university libraries to carry out DRS.

Fund

Heilongjiang Philosophy and Social Science Plan research project "information service system building of university libraries in big data age" (project number: 14E045).

Heilongjiang Art and Science Plan co-building project "research on library service innovation development in new media age" (project number: 2014D014).

References

[1] Liu, L. and Li, X.J. (2012) University Library Marketing and Reference with Network Embedded Community: Exploration of Library Service Based on Sina Microblogging. *Library Journal*, No. 7.

[2] Shang, F.-Y. (2015) Research on the Mode of Digital Reference Services Based on WeChat Public Plat Form. *Journal of Academic Library and Information Science*, No. 4.

[3] Yang, F. (2014) Exploring New Field of Reference Consulting Service in the Big Data Age. *Journal of Library and Information Sciences in Agriculture*, No. 12.

[4] Yuan, H.-J. (2015) Research on the Ecological Mechanism for Knowledge Transfer in Collaborative Digital Reference Service. *Information Science*, No. 1.

[5] Liu, H.H. (2008) DRS Development Trend Analysis of Domestic Libraries. *Journal of Xiangtan Normal College*, No. 7, 55-58.

26

Enhancing Amharic Information Retrieval System Based on Statistical Co-Occurrence Technique

Abey Bruck, Tulu Tilahun

Department of Computer Science & IT, AMiT, Arba Minch University, Arba Minch, Ethiopia
Email: bruckabey@gmail.com, tuttilacs@gmail.com

Abstract

Information retrieval (IR) systems are designed to help information seekers retrieving relevant information from vast document. The need for relevant information from a vast amount of document gave birth to IR systems. Even though different IR systems exist, they cannot meet all users' expectations. A different level of users' knowledge makes queries to be expressed in different ways. As a result, the system may miss the core meaning of users query and retrieve dissatisfactory results. This happens mainly because of the ambiguities of words involved in the natural languages and expression mismatch among users and authors. The existing ambiguities in Amharic language have negative impacts on the performance of Amharic IR system. Some of the ambiguities for this type of problem are: spelling variants of the same word, polysemous and synonymous terms. If users are not fully knowledgeable about the information domain area, they will mostly formulate weak queries to retrieve documents. Thus, they end up frustrated with the results found from an IR system. This research has been conducted, aiming at augmenting the recall of previous work. Statistical co-occurrence technique has been used in order to expand query terms. The main reason for performing query expansion is to provide relevant documents as per users' query that can satisfy their information need. Statistical co-occurrence method considers, frequently appearing terms with the query term, regardless of their position. The efficiency of proposed technique has been tested on the prototype system and the result found compared with the result of previous study. Accordingly, 6% recall and 2% f-measure improvement has been made. Hence, the statistical co-occurrence method outperformed the bi-gram based IR system.

Keywords

Statistical, Co-Occurrence, Information Retrieval, Query Expansion, Amharic

1. Introduction

As early as 1945 [1], Bush talked about a device in which an individual stores all his books, records, and communications, which has been mechanized so that it may be consulted with exceeding speed and flexibility. Bush's ideal device has become a reality, that the term "Information Retrieval" was coined as its name in 1952 [2]. IR is one of the ingenious solutions mankind invented to solve the obvious problem of searching for information. The idea of using computers to search for relevant pieces of information was popularized by Vannevar Bush in 1945 [1]. Bush [1] stated, "Instruments are at hand which, if properly developed, will give man access to and command over the inherited knowledge of the ages". This system predefined by Bush, is an IR system, which is used in all aspects of information retrieval now a days. The need for relevant information from a vast amount of document gave birth to information retrieval (IR). IR is the art and science of retrieving relevant documents by searching and measuring similarity between users query and documents in a certain corpora.

Information seekers thus, should implement an IR system to satisfy their information need [3]. The main goal of an IR system is, to support and nourish information-seeking behaviors of users [4]. Even though the task of information retrieval systems is retrieving relevant information, there is no one system, which is capable of retrieving relevant and only relevant documents as per users' query.

To investigate the underlying principles/theories, philosophies, techniques and tools of the various methods that can be adopted, manipulated or simply selected we have reviewed literatures and then collected data for testing purposes. And then, queries that have polysemous and synonymous terms are formulated in order to challenge the Amharic IR system and measure the extent to which it can discriminate their various meanings.

Nowadays, the IR research focuses on, the problem of making existing IR systems retrieve better results. Regarding this problem, different mechanisms have been introduced by many researches, in order to achieve a solution [5]. Some of these proposed techniques, as listed in [6], are: User-oriented search result organization; Incorporating user negative feedback; supporting effective browsing and Effective query reformulation. Among the listed mechanisms, effective query reformulation or query expansion based on semantic thesaurus has been found to be better [6].

After receiving a query, an IR system goes through a series of steps, in order to provide seemingly relevant information to users. First, it clusters documents as relevant and non-relevant as per the query, then ranks the relevant documents and displays them according to a certain similarity measurement [7]. Other than implementing such steps, a good IR system's designer should need to have knowledge of the relationship between a query, its user and the environment Stubinz and Whighli [8]. According to [8], before considering the experiments in designing a good information retrieval, impact of language and text needs to be considered. Researches introduced various methods in information retrieval to satisfy users' information need. Since the initiation of automated information retrieval, Boolean queries were an accurate reflection of user's information needs [9]. In time, language and needs outmoded this approach, leading to the need for approximated similarity measures, such as the inner product, the cosine similarity measure and probabilistic retrieval [10]. For many years these were the methods of choice, but were based on studies that neglected the user, and now the possibilities have been completely explored [8]. Therefore, the frontier of information retrieval research is, to exploit new methods which can take the exact users information need in to consideration. In this research, it is suggested to achieve this objective, through pre-calculated query operation.

IR system designers may face challenges on query reformulation issues, besides challenges that occur from the user's perspective. For example, when synonyms or polysemous terms, in queries are encountered, it can be difficult to relate and identify other terms for expansion. That is because, when such terms are used in queries, they seem to convey a certain meaning, while the users' intention was another [7] [11] [12]. Thus, this problem apparently becomes a major challenge, for IR designers to specify a solution to. Altogether, ambiguity that arises between users and authors, context similarity in written documents and existence of polysemous words in the natural language, exaggerates the problem further. Therefore, IR systems should comprise a more sophisticated technique and strategy that enables it, to cope up with the problem and have more of user-centered approach. Such techniques and strategies would be, ontology based query expansion method, meaning oriented thesaurus usage, expanding the whole query with common expanding terms and strategies which seek meaning out of users' queries.

Different methods can be used for query expansion [7] [13] [14]; such as word space nearest-neighbor and thesauruses method, cluster-based query expansion, latent semantic indexing, concept-based query expansion.

Even though many kinds of query expansion methods exist, their goal is the same; making users search task easy as much as possible, aiming at delivering information that satisfies users' information need.

The purpose of this research is to design statistical co-occurrence based query expansion technique that can overcome the limitations of the previous research aiming at delivering relevant documents as close as professionals' do. In turn, research result introduces a technique to develop a fully-fledged IR system that can behave like professionals so that users can be satisfied with results retrieved. Further, to enhances the precision of the system so that more relevant documents can be found among the retrieved ones.

2. Amharic Writing System and Ambiguities

Amharic is the official working language of Federal Democratic Republic of Ethiopia. Its alphabet is grouped into either consonantal or syllabary writing system. A consonantal system represents symbols separately as consonants and vowels, while the syllabary writing system has individual signs for syllables (*i.e.* consonant + vowel combinations). There is a dispute as to whether or not the Amharic writing system is a syllabary [15]. It could be argued that Amharic to some extent resembles other Semitic scripts such as the Arabic and Hebrew consonantal systems, which basically indicate consonants but, for teaching purposes, etc, have developed optional diacritics to signify vowels [15]. Unlike the Arabic and Hebrew, the Amharic writing system is written from left to right.

Amharic writing system has no upper and lower case latter variations and has no conventional cursive (*i.e.* written in a connected letters) form. Unlike Hebrew and Arabic, there are no systematic variations in the form of the symbol according to its position in the word [15].

When Ge'ez became the spoken and written language in common use in northern Ethiopia, it took only 24 of the 29 Sabaean symbols, modify most of them and add two new symbols to represent sounds of Greek and Latin loanwords not found in Ge'ez [7]. The two extra symbols are added to cope with two new sounds (/p/ as in "ፐ" and /p'/ as in "ጰ"), originally required for use in ecclesiastical Greek and Latin borrowings and names (e.g. Poulos, Police) [15]. Amharic inherited symbols from Ge'ez. It took all of the symbols and added eight new ones ("ቸ", "ጨ", "ጀ", "ኘ", "ሸ", "ሸ", "ኸ", and "ዠ") that represent sounds not found in Ge'ez [7]. Alphabetic system of the Amharic writing system consists of 34 base symbols with seven orders, representing seven vowels. For example the consonant /b/ has seven different symbols (*i.e.* በ, ቡ, ቢ, ባ, ቤ, ብ, ቦ read as bə, bu, bi, ba, be, bɨ, bo, respectively) representing seven vowels combined with the consonant /b/. Taking another consonant like /z/, the seven symbols ዘ, ዙ, ዚ, ዛ, ዜ, ዝ and ዞ read as zə, zu, zi, za, ze, zɨ and zo respectively can be found.

From the construction of the seven consonant-vowel combinations of the /b/ and /z/ consonants, one can see that there is a consistent pattern in the shapes, except the 1st symbols pronounced as (consonant + "ə") and the 6th order which are less systematic. Considering all the 34 consonant symbols, there is considerable regularity of letter shapes, but some orders are more regular than others. The shapes are most consistent in the 5th order (e.g. ሄ, ሌ, ሜ, ሴ, ሬ having a ring on their right legs except ጬ), slightly less in the 3rd (e.g. ሂ, ሊ, ሚ, ሲ, ቢ having a small hyphen like extension on the bottom of their right legs, except ሪ and ዪ), slightly less again in the 2nd (e.g. ሁ, ሉ, ሱ, ቡ having a small hyphen like extension on their right legs except ሩ, ፉ and ው), still less in the 4th (e.g. ላ, ሳሻ, ባ, ካ, ዛ, ዣ, ዳ, ጃ, ጳ have their left leg shortened, on the other hand ጣ and ጫ has there two left legs shortened, others like ቃ, ታ, ቻ, ጋ, ፓ have bow like legs) and even less in the 7th (e.g. ሶ, ሾ, ቦ, ኮ, ዞ, ዦ, ዶ have their right legs shortened, on the other hand ጮ and ጦ have their two right legs shortened, others like ሆ, ሎ, ኖ, ኞ, ኖ, ሮ have a ring somewhere on their base letter or have some kind of modification like ሞ and ሦ) and the 6th (e.g. ህ, ል, ም, ስ, ድ, ቅ, ት, ፕ, ጭ having completely distinctive and pattern less structures) order, with the greatest number of patterns, [15]. Therefore the system is composed of largely unpredictable patterns. For example the set of syllables with /g/ starts off regularly enough except the 4th "ጋ", 6th "ግ" and 7th "ጎ" orders. The /w/ set is even ambiguous having the 2nd "ዉ" read as "wu" and 6th "ው" read as "wɨ" symbols highly alike. The system has regularity in writing which makes it easy for a person to learn the language. For example, the symbol "ሰ" read as "sə" and "ሸ" read as "shə" have relatively the same kind of structure as their accent. Therefore, a person experiences any difficulty to learn a symbol, given that he/she knows the other (for some of the symbols). The Amharic writing system has 34 basic characters and their seven orders give 238 distinct symbols. In addition, there are forty others that contain a special feature usually representing labialization e.g ቿ, ቋ [7].

Amharic writing system adopted all the symbols in Ge'ez and added 8 other symbols and the other 44 symbols. The result is that there is a considerable systemic redundancy of several consonant sounds which lacks in the phonology of Ge'ez [15]. Ambiguities in Amharic writing system arise mainly due to symbol redundancy [7]. Thus, 4 distinct sets of 7 can represent the sound /h/ + vowel: ("ሀ", "ሐ", "ኃ", "ኸ"), 2 sets represent /s/: ("ሰ", "ሠ") and 2 /s'/ ("ጸ", "ፀ") [15]. A similar problem is observed in usage of some letters interchangeably, like "ቆ" vs "ቈ" [7]. In addition to the symbolic redundancy of characters, Amharic writing system suffers slightly from visual similarity or different character, such as ፐ and ፕ, ፖ and ፓ, ዪ and ዩ, ጕ and ጐ [7]. The different forms of spelling variants of the same word are shown in **Table 1**. Because of those symbols having the same accent in speaking, they can be used interchangeably in the various words of Amharic language, thus, forming different forms of spelling for the same word. Using these symbols interchangeably in words doesn't make reading, or forwarding ideas difficult for human beings. But unlike humans it is difficult for systems to consider them having the same meaning. Because IR systems only match the symbols in words to check weather a word from a document has the same meaning as in the query (*i.e.* if the words are a match then they are the same and have same meaning) encountering the different interchangeable symbols in words forces it to consider them as different *i.e.* the system considers ፀሀይ and ጸሃይ, as different words with different meaning.

The other challenge Amharic IR systems face is the combination of two words. There is no convention as to which words should be combined as one word or separately during writing. For example, the word "megneta bet" which means "bed room" can possibly be written as መኝታቤት (without space) and መኝታ ቤት (with space) and also the word "betemekides" which means "temple" can be written as "ቤተመቅደስ" (without space) and "ቤተ መቅደስ" (with space) which makes it difficult for the IR system to differentiate between them [7]. These kinds of situations makes IR systems' task difficult.

According to 1998 statistical census [16], Amharic language has 17.4 million speakers as a mother tongue and 5.1 million speakers as their second tongue. Many recent researches have indicated that, electronic documents in Amharic keep on growing every year [7] [17]. This apparently makes it difficult for IR systems, to find relevant document from the vast amount of electronic documents available nowadays.

There are information retrieval systems designed for Amharic language [17]-[22] that attempts to retrieve relevant documents as per information need of users. Even though such IR systems exist, they cannot meet all users' expectations. In order to satisfy the need of extracting relevant information, data mining and text processing techniques have been applied so far [23] [24].

Different levels of users' knowledge makes queries to be expressed in different ways, as different authors express ideas using different terms. As a result, systems most of the time, misses the core meaning of users query and retrieve dissatisfactory results. This happens mainly because of the ambiguities of words involved in the natural languages and expression mismatch among users and authors.

The prevalence of synonyms query terms tends to decrease precision at higher recall levels [12]. One of the solutions suggested to solve such a problem is query operation. A recent research paper done by Alemayehu [7] is worth mentioning here, which attempts to apply query expansion to control synonyms words using thesaurus. Alemayehu tried to enhance the system's recall at the expense of its precision. The performance analyses show that there is an enhancement of recall from an average of 0.29% to 0.65%. On the other hand, because of a tradeoff between recall and precision and because of polysemous query terms existence, his proposed system decreased the overall precision from 0.91% to 0.57% on average. This happens because some synonymous query terms can also be polysemous.

Words in the natural language can be regarded as polysemous or synonymous according to the context they are used in [25]. A word is polysemous if it has different contextual meaning. On the other hand, if a couple or more words refer to one meaning, they are termed as synonymous words. For example in these two sentence

Table 1. Spelling variants of the same word.

Canonical Amharic	Common Amharic	Possible but improbable Amharic
ዓለም	አለም	ዐለም ፣ አለም
ፀሐይ	ጸሃይ ፣ ፀሃይ ፣ ጸሐይ	ጸሀይ ፣ ጸሓይ ፣ ጸኀይ ፣ ጸኃይ ፣ጸኻይ ፣ ፀሀይ ፣ ፀሓይ ፣ ፀኀይ ፣ ፀኃይ ፣ ፀኻይ
ኃይለ ሥላሴ	ኀይለ ሥላሴ ፣ ኃይለ ስላሴ ፣ ኀይለ ስላሴ ፣ ኀይለ ስላሤ ፣ ሀይለ ስላሤ ፣ ሀይለ ሥላሴ ፣ ሃይለ ስላሴ ፣ ሃይለ ሥላሴ ፣ ሃይለ ስላሤ ፣ ኀይለ ሥላሤ ፣ ሀይለ ሥላ ሤ ፣ ሃይለ ሥላ ሤ ፣ ሐይለ ስላሴ ፣ ሐይለ ስላሤ ፣ ሐይለ ሥላሴ	ሓይለ ሥላሤ ፣ ሓይለ ሥላሤ ፣ ሓይለ ሥላሴ ፣ ሓይለ ስላሤ ፣ ሓይለ ስላሴ ፣ ኻኀይለ ሥላሴ ፣ ኻይለ ሥላሴ ፣ ኻይለ ስላሤ ፣ ኻይለ ስላሴ

"አበበ ካራ እየሳለ ነው" and "አበበ ቢላ እየሳላ ነው", which their equivalent meaning in English is "Abebe is sharpening a knife", the two words "ካራ" and "ቢላ" refers to a "knife" and thus, they are synonyms to each other. In another two sentences "አበበ ካራ እየሳለ ነው" and "አበበ ስእል እየሳለ ነው" the word "እየሳለ" refers to "sharpening a knife" in the first sentence and "drawing a picture" in the later. Therefore, this word "እየሳለ" changes its meaning according to the context it is used and thus it is polysemous. In a sentence "አበበ ጠላ ይወዳል" which means "Abebe likes tela", "tela" or "ጠላ" refers to an Ethiopian traditional drink. In this context, the term "ጠላ" is a polysemous term which means "a traditional drink" or "hatred", which can also be a synonym for other terms that bare one of its meanings. If the term "ጠላ" is used in a query, an Amharic query expansion technique may try to expand it using terms such as, "መጠጥ" and "ባህላዊ" for the meaning "traditional drink" or "መጥላት" and "አለመዋደድ" for "hatred". Thus, expanding the query with all of the expansion terms mentioned above, may make the system to retrieve all the relevant documents, but with many non-relevant ones in between. In addition to the example given, there are many polysemous words in Amharic language. To mention some of them; "በቅሎ" means "a mule" or "to grow", "ዋና" means "main" or "to swim", "አንቀላፋ" means "to sleep" or "to die", "ሳለ" means "to cough", "sharpen a knife", "to draw a picture" or it can even express a "period" in time, according to a context it is used. There are also phrases which can be equally ambiguous. The phrase "አበባው በቀለ" can be understood as a person's name or "the flower has grown". Therefore, it is necessary to consider polysemous nature of terms in addition to synonym terms for query expansion. It is therefore the purpose of the current research to investigate the possibility of building statistical co-occurrence based query expansion that can control polysemous and synonymous words to enhance the performance of the system by considering users' information need.

3. Query Expansion Based on Statistical Co-Occurrence

An IR system takes a string query, and retrieves documents based on a certain similarity measurement technique. In addition, an IR system has its own performance level as measured in terms of recall and precision. As any other systems' performance measurement, it is most unlikely to score 100% on both precision and recall in the case of information retrieval systems. But good systems are designed to enhance both precision and recall to the possible limit. Thus the aim of this research is, to design a good system, which can enhance the recall of the system without affecting its precision. This can be achieved by integrating a query expansion model with the system, so that it finds words having similar meanings with users' query and retrieves relevant documents that satisfies users information need.

This method generates synonymous expanding terms for a query term based on index terms co-occurrence information. It analyzes the presence of a term with another term to decide whether they have same or different meaning. The overall architecture of process and data flows involved for the statistical co-occurrence technique is presented in **Figure 1**.

The set of top 10 documents retrieved using the refined query is the pseudo relevance feedback for statistical co-occurrence method. These 10 relevant retrieved documents are indexed in a separate inverted index to extract expansion terms easily. This index is over written every time a query is given to the system. That is because; different relevant documents are retrieved for each different users query. The reformulation process then, based on the statistical co-occurrence technique, selects expansion terms for each query term, and finally selects common expansion terms suitable for the whole query. The reformulation process involves query expansion and term selection sub-processes. Finally, the reformulated query is fed to the original IR system to retrieve re-ranked documents.

On expansion terms selection, co-occurrence frequency values of terms with the query term are analyzed, among the refined retrieved document set index. These found expansion terms for each query term is then saved separately, so that common terms found in all of the query terms can be extracted. Expanding users query using common expanding terms for the whole query has been hypothesized in this research. **Figure 2** shows a java written code for expanding term selection for each query terms.

As shown in **Figure 2**, a term is selected as an expansion term if it co-occurred with a query term even once, among the ten top retrieved documents set. Then these terms are ordered in descending order. The final task is to select those terms which are common for the whole query in order to eliminate polysemous terms ambiguity characteristics.

For statistical co-occurrence technique, the query expansion task is carried out based on the refined document set. For relevance feedback the first k out of the refined documents are selected. That is because; it is assumed

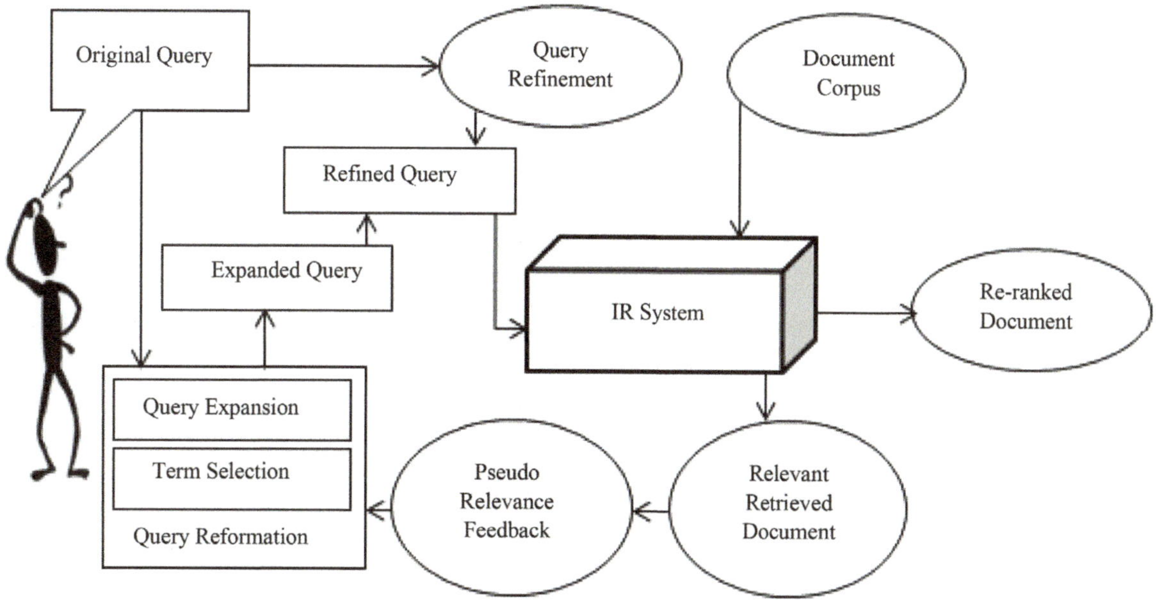

Figure 1. Statistical co-occurrence based query expansion architecture.

```
for(int i = 0; i<termsToSelectFromArray.length;i++)
  if(!termsToSelectFromArray[i].isEmpty())
  int docCount = hitCoutnt(queryTerm + "AND" + termsToSelectFromArray[i]);
  if(docCount >= 1)
  expandingTerms = expandingTerms + termsToSelectFromArray[i] + "-" + docCount
return expandingTerms
```

Figure 2. Statistical co-occurrence based query expansion.

that query refinement process ought to populate more relevant documents at the top retrieved documents set. Hence, these k documents are responsible for generating the expansion words. To overcome the polysemous characteristics of query terms, expansion terms are selected which represent every query terms. This is done by selecting those terms which appear in every expansion terms set of every query term. This technique assumes each query term as a polysemous term or having the potential of possessing polysemous characteristic as per the document collection. The second assumption in this technique is that, query terms other than the polysemous one have the potential to indicate the meaning the user intended. This means if a query term has two different meanings, other query terms aside from that polysemous query term has the potential to select from the two meanings.

For example let there be three query terms, q_1, q_2 and q_3. Let x, y and z be three meanings or collection of expansion terms for q_1, q_2 and q_3 respectively. A certain set of terms C which have common meaning with x, y and z is selected for expansion. **Figure 3** presents a pictorial representation this example given.

Here is a brief discussion of this technique. First the pseudo-relevance feedback of k top ranked relevant retrieved documents are taken and indexed in a separate inverted index. Then all the terms in this index are taken and each term is searched in that same index with each query term. Next the top n terms are selected, which frequently co-occurred with for each query term. The frequency is calculated by simply adding one every time a query term and a potential expansion term are found together in one of the k documents. Given q and t as query term and expansion term respectively and k amount of documents as pseudo-relevance feedback in which Equation (1) is used for frequency calculation.

$$Sim(q,t) = \sum_{j=1}^{k} x = \begin{cases} x = 1, & (q \in d_j \text{ AND } t \in d_j) \\ x = 0, & (q \notin d_j \text{ OR } t \notin d_j) \end{cases} \tag{1}$$

At last common expansion terms are selected from all of the query terms, which are also good expansion terms for the overall query rather than for each query term. Given query terms $q_1, q_2, \cdots, q_n$, index terms $t_1, t_2, \cdots, t_m$ and a constant v. **Figure 4** shows, how technique1 selects expansion terms.

E is a special variable and deliverable of the program which holds the total expansion terms of each query term. The $search$ (q_k, t_j) function compares the co-occurrence value for q_k and t_j, and returns the term t_j if it has a hit count greater than a certain value v, where v is less than k. k is the number of relevant retrieved documents gained from the pseudo-relevance feedback. The process left the variable E with the total expansion terms of the query. E is passed to the function SCT which Selects Common Terms from E and assigns it on itself. The last step returns common expansion terms for the whole query.

Given the query Q and the total expansion terms E **Figure 5** shows the algorithm; how the set of common expansion terms C is selected.

L is the number of query terms. An expansion term that found L times in E is finally taken as a common term.

4. Performance Evaluation

The experimentation phase holds implementation, testing and discussion of the challenges and findings that are recorded for each proposed techniques. Data preparation and selection, testing procedure through empirical testing and threshold selection are also discussed. A prototype has been built using Java NetBean.

In order to evaluate performance of this prototype system we have used recall, precision and F-measure. Determination of greater value for F-measure can be interpreted as an attempt to find the best possible compromise between recall and precision.

Precision and recall are the basic measures used in evaluating relevant document retrieval strategies. Recall is the ratio of the number of relevant records retrieved (*Ra*) to the total number of relevant records in the corpus (*R*), Equation (2). Precision is the ratio of the number of relevant records retrieved (*Ra*) to the total number of

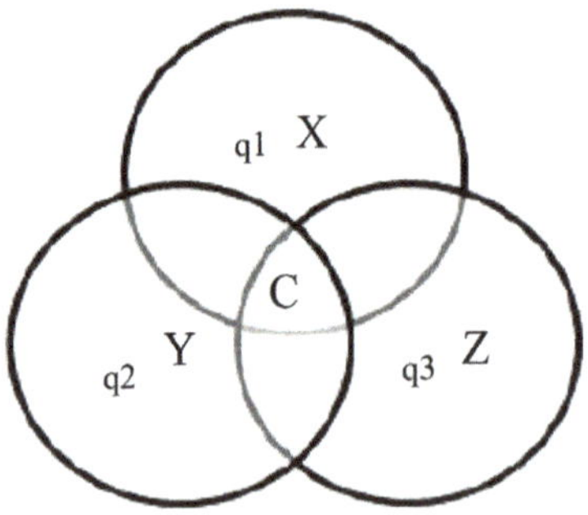

Figure 3. Pictorial representation of common meaning assumption.

```
for k = 0 to n do
    for j=0 to doc
   search(qk, tj)
          if(hitcount > V) then
      add(tj, E)
   E= SCT(E)
   Return E
```

Figure 4. Pseudo code for statistical co-occurrence method.

```
SCT(Q, E)
L = length (Q)
for i=0 to length (E) do
      if count (Ei, E) = L
           C = C + Ei
return C
```

Figure 5. Pseudo code for common terms selection.

irrelevant and relevant records retrieved (*A*), Equation (3). Unless either recall or precision is needed it is good to show the result in F-measure (*F*) which is calculated from recall and precision. These evaluation metrics are usually expressed as a percentage, Equation (4).

Prototype system centric testing is carried out to evaluate the statistical co-occurrence technique's performance. The statistical co-occurrence's performance is recorded in terms of recall, precision and f-measure as shown on **Table 2**.

$$\text{Recall} = Ra/R \tag{2}$$

$$\text{Precision} = Ra/A \tag{3}$$

F-measure (*F*) is a harmonic mean evaluation measurement, which combines both recall (*Re*) and precision (*Pr*), Equation (4).

$$F = 2 * Re * Pr/Re + Pr \tag{4}$$

The intension of this study is to improve the recall of previous study which has been done based on bi-gram method [26]. Result difference shown in **Table 3** and **Figure 6**. Even though, precision of retrieving relevant

Table 2. Statistical co-occurrence's performance.

Query	REL	Using statistical co-occurrence method				
		RET	RETREL	R	P	F
ቀና እና ቅን ስራዎች	140	652	129	0.19	0.82	0.32
ሰው የተባለ አለቀ	10	28	7	0.7	0.25	0.36
የጌታዬ ባሪያ አለቀ	100	226	72	0.72	0.31	0.44
ዘይት ቀባው	69	121	45	0.65	0.37	0.47
መጥፎ ስራውን ገሰጸ	18	6	6	0.33	1.0	0.5
እጥፍ ሰጠው	26	9	9	0.34	1.0	0.51
አመድ እና ማቅ ለብሶ	35	60	27	0.77	0.45	0.56
አመታት ተቀመጠ	326	244	209	0.64	0.85	0.73
ሲጨልም አንቀላፋ	10	14	10	1	0.71	0.83
			Average	0.73	0.53	0.63

Table 3. Comparing performance of bi-gram based with statistical co-occurrence based.

	Average Recall	Average Precision	Average F-measure
Original query	0.83	0.39	0.53
The Bi-gram method	0.67	0.57	0.61
Statistical co-occurrence	0.73	0.53	0.63

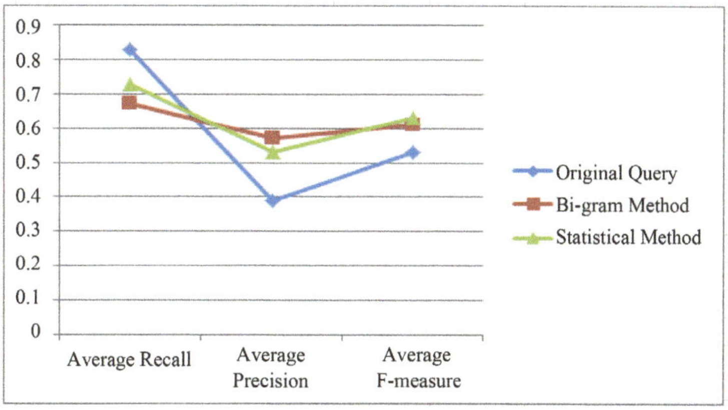

Figure 6. Performance graph of original query, bi-gram and statistical methods.

document down by 4%, the recall of retrieving relevant document improved by 6%. However, F-measure increased by 2%. Accordingly, statistical based information retrieval system performs more than bi-gram based information retrieval system.

5. Conclusion

Based on the frequency, statistical co-occurrence method selects common words which are expected to be synonyms for the polysemous word or words as per the user's intention. Then, these common words are expected to be synonyms for the polysemous word or words as per the user's intention. It is finally concluded that the statistical methods outperformed bi-gram based information retrieval system and scored 6% recall and 2% F-measure. This is an encouraging result to design an applicable search engine for Amharic language information retrieval system. The performance of the system can further be improved by designing hybrid, bi-gram and statistical co-occurrence, based query expansion.

References

[1] Bush, V. (1945) As We May Think. *The Atlantic Monthly*, **176**, 101-108.

[2] Baeza-Yates, R. and Ribeiro-Neto, B. (1999) Modern Information Retrieval. 2nd Edition, Addison-Wesley-Longman Publishers, England.

[3] Spink, A. and Wilson, T.D. (2002) Towards a Theoretical Framework for Information Retrieval (IR) Evaluation in an Information Seeking Context. *Journal of the American Society for Information Science*, **51**, 841-857.

[4] Belkin, N.J. (1993) Interaction with Texts: Information Retrieval as Information Seeking Behavior. Universität Regensburg and Universitätsverlag Konstanz, SchriftenzurInformationswissenschaft Band 12, 55-66.

[5] Wang, X. (2009) Improving Web Search for Difficult Queries. Unpublished paper available at University of Illinois at Urbana-Champaign.

[6] Greenberg, J. (2001) Optimal Query Expansion (QE) Processing Methods with Semantically Encoded Structured Thesauri Terminology. *Journal of the American Society for Information Science and Technology*, **52**, 487-498. http://dx.doi.org/10.1002/asi.1093

[7] Alemayehu, N. (2002) Application of Query Expansion for Amharic Information Retrieval System. M.Sc. Thesis, Addis Ababa University, Addis Ababa.

[8] Stubinz, J. and Whighli, S. (1998) Information Retrieval System Design for Very High Effectiveness. Division of Computer Science, Endeavour Research and Development (BVI), 12 October 1998, 1-7.

[9] Ashington, B. (1956) An Automatic System for Retrieval of Electronic Documents. *Third British Colloquium on Electronic Computing*, Manchester, 118-121.

[10] Dagnachew, A. and Worku, A. (1986) Yeamaregna Felitoch. Kuraz Asatami Dereget.

[11] Germann, D.C., Villavicencio, A. and Siqueira M. (2010) An Investigation on Polysemy and Lexical Organization of verbs. *Proceedings of the NAACL HLT* 2010 *First Workshop on Computational Neurolinguistics*, Los Angeles, June 2010, 52-60.

[12] Billhardt, H., Borrajo, D. and Maojo, V. (2002) A Context Vector Model for Information Retrieval. *Journal of American Society for Information Science and Technology*, **53**, 236-249. http://dx.doi.org/10.1002/asi.10032

[13] Kalmanovich, I.G. and Kurland, O. (2009) Cluster-Based Query Expansion. In: *Proceedings of the* 32*nd International ACM SIGIR Conference on Research and Development in Information Retrieval*, ACM, New York, 646-647. http://dx.doi.org/10.1145/1571941.1572058

[14] Li, L. (2007) A Query Expansion Method Based on Semantic Element. *Eighth ACIS International Conference on Software Engineering, Artificial Intelligence, Networking, and Parallel/Distributed Computing*, Qingdao, 30 July-1 August 2007, 587-590. http://dx.doi.org/10.1109/SNPD.2007.249

[15] Bloor, T. (1995) The Ethiopic Writing System: A Profile. *Journal of the Simplified Spelling Society*, **19**, 30-36.

[16] Gizaw, S. (2009) Multiple Pronunciation Model for Amharic Speech Recognition System. M.Sc. Thesis, Addis Ababa University, Addis Ababa.

[17] Redwan, H. and Atnafu, S, (1995) Design and Implementation-Algorithms of Amharic Search Engine System for Amharic Web Contents. M.Sc. Thesis, Addis Ababa University, Addis Ababa.

[18] Mengistu, T.M. (2007) Design and Implementation of Amharic Search Engine. M.Sc. Thesis, Addis Ababa University, Addis Ababa.

[19] Hailemeskel, T. (2003) Amharic Text Retrieval: An Experiment Using Latent Semantic Indexing (LSI) with Singular Value Decomposition (SVD). M.Sc. Thesis, Addis Ababa University, Addis Ababa.

[20] Saba, A. (2001) The Application of Information Retrieval Techniques to Amharic Documents on the Web. M.Sc. Thesis, Addis Ababa University, Addis Ababa.

[21] Gezmu, A.M. (2009) Automatic Thesaurus Construction for Amharic Text Retrieval. M.Sc. Thesis, Addis Ababa University, Addis Ababa.

[22] Bethlehem, M.A. (2002) The Application n-Gram-Based Indexing in Amharic Text Retrieval. M.Sc. Thesis, Addis Ababa University, Addis Ababa.

[23] Tilahun, T. (2014) Linguistic Localization of Opinion Mining from Amharic Blogs. *International Journal of Information Technology & Computer Sciences Perspectives*, **3**, 890.

[24] Tilahun, T. and Sharma, D. (2015) Design and Development of E-Governance Model for Service Quality Enhancement. *Journal of Data Analysis and Information Processing*, **3**, 55-62. http://dx.doi.org/10.4236/jdaip.2015.33007

[25] Jing, H and Tzoukermann, E. (1999) Information Retrieval Based on Context Distance and Morphology. In: *Proceedings of the 22nd Annual International Conference on Research and Development in Information Retrieval* (*SIGIR* "99), ACM, New York, 90-96. http://dx.doi.org/10.1145/312624.312661

[26] Bruck, A. and Tilahun, T. (2015) Bi-Gram Based Query Expansion Technique for Amharic Information Retrieval System. *IJIEEB*, **7**, 1-7.

27

Modeling and Stability Analysis of a Communication Network System

Zvi Retchkiman Königsberg

Centro de Investigacion en Computacion, Instituto Politecnico Nacional, Mexico City, Mexico
Email: mzvi@cic.ipn.mx

Abstract

In this work, the modeling and stability problem for a communication network system is addressed. The communication network system consists of a transmitter which sends messages to a receiver. The proposed model considers two possibilities. The first one, that messages are successfully received, while in the second one, during the sending process the transmitter breaks down and as a result the message does not reach the receiver. Timed Petrinets is the mathematical and graphical modeling technique utilized. Lyapunov stability theory provides the required tools needed to aboard the stability problem. Employing Lyapunov methods, a sufficient condition for stabilization is obtained. It is shown that it is possible to restrict the communication network system state space in such a way that boundedness is guaranteed. However, this restriction results to be vague. This inconvenience is overcome by considering a specific recurrence equation, in the max-plus algebra, which is assigned to the timed Petri net graphical model.

Keywords

Communication Network System, Transmitter Breakdown, Discrete Event Dynamical Systems, Max-Plus Algebra, Lyapunov Method, Timed Petri Nets

1. Introduction

In this work, the modeling and stability problem for a communication network system is addressed. The communication network system consists of a transmitter which sends messages to a receiver. The proposed model considers two possibilities. The first one, that messages are successfully received, while in the second one, during the sending process the transmitter breaks down and as a result the message does not reach the receiver. A communication network system can be considered as a dynamical system whose state evolves in time by the occurrence of events at possibly irregular time intervals, (therefore belong to the class of dynamical systems known as discrete event systems). Place-transitions Petri nets (commonly called Petri nets) are a graphical and mathematical modeling tool that can be applied to the communication network system in order to represent its states evolution. Petri nets are known to be useful for analyzing the systems properties in addition of being a paradigm for describing and studying information processing systems. Timed Petri nets are an extension of Petri nets, where now the timing at which the state changes are taken into consideration. This is of critical importance since it allows considering useful measures of performance as for example: how long does the communication network system spends at a given state etc. For a detailed discussion of Petri net theory see [1] and the references quoted therein. One of the most important performance issues to be considered in a communication network system is its stability. Lyapunov stability theory provides the required tools needed to aboard the stability problem for communication network systems modeled with timed Petri nets whose mathematical model is given in terms of difference equation. By proving practical stability one is allowed to preassigned the bound on the communication network systems dynamics performance. Moreover, employing Lyapunov methods, a sufficient condition for the stabilization problem is also obtained. It is shown that it is possible to restrict the communication network systems state space in such a way that boundedness is guaranteed. However, this restriction results to be vague. This inconvenience is overcome by considering a specific recurrence equation, in the max-plus algebra, which is assigned to the timed Petri net graphical model. The paper is organized as follows. In Section 2, Lyapunov theory for discrete event systems modeled with Petri nets is given. Section 3 presents max-plus algebra. In Section 4, the solution to the stability problem for discrete event systems modeled with timed Petri nets is considered. Finally, in Section 5 the modeling and stability analysis for communication network systems is addressed. Some conclusion remarks are also provided.

2. Lyapunov Stability and Stabilization of Discrete Event Systems Modeled with Petri Nets [2] [3]

NOTATION: $N=\{0,1,2,\ldots\}$, $R_+=[0,\infty)$, $N_{n0}^+=\{n_0,n_0+1,\ldots,n_0+k,\ldots\},n_0\geq 0$. Given $x,y\in R^n$, $x\leq y$ is equivalent to $x_i\leq y_i,\forall i$. A function $f(n,x)$, $f:N_{n0}^+\times R^n\to R^n$ is called nondecreasing in x if given $x,y\in R^n$ such that $x\geq y$ and $n\in N_{n0}^+$ then, $f(n,x)\geq f(n,y)$. Consider systems of first ordinary difference equations given by

$$x(n+1)=f[n,x(n)],x(n_o)=x_0,n\in N_{n0}^+ \tag{1}$$

where $n\in N_{n0}^+$, $x(n)\in R^n$ and $f:N_{n0}^+\times R^n\to R^n$ is continuous in $x(n)$.

Definition 1. *The n vector valued function* $\Phi(n,n_0,x_0)$ *is said to be a solution of* (1) *if* $\Phi(n_0,n_0,x_0)=x_0$ *and* $\Phi(n+1,n_0,x_0)=f(n,\Phi(n,n_0,x_0))$ *for all* $n\in N_{n0}^+$.

Definition 2. *The system* (1) *is said to be practically stable, if given* (λ,A) *with* $0<\lambda<A$, *then* $|x_0|<\lambda\Rightarrow|x(n,n_0,x_0)|<A,\forall n\in N_{n0}^+,n_0\geq 0$.

Definition 3. *A continuous function* $\alpha:[0,\infty)\to[0,\infty)$ *is said to belong to class* $\mathcal{K}$ *if* $\alpha(0)=0$ *and it is strictly increasing.*

Consider a vector Lyapunov function $v(n,x(n))$, $v:N_{n0}^+\times R^n\to R_+^p$ and define the variation of v relative to (1) by

$$\Delta v=v(n+1,x(n+1))-v(n,x(n)) \tag{2}$$

Then, the following result concerns the practical stability of (1).

Theorem 4. *Let* $v:N_{n0}^+\times R^n\to R_+^p$ *be a continuous function in x, define the function* $v_0(n,x(n))=\sum_{i=1}^p v_i(n,x(n))$ *such that satisfies the estimates*

$$b(|x|)\leq v_0\left(n,x\left(n\right)\right)\leq a(|x|);a,b\in\mathcal{K},\Delta v(n,x(n))\leq w(n,v(n,x(n))) \tag{3}$$

for $n\in N_{n0}^+$, $x(n)\in R^n$, *where* $w:N_{n0}^+\times R_+^p\to R^p$ *is a continuous function in the second argument. Assume that:* $g(n,e)\triangleq e+w(n,e)$ *is nondecreasing in e,* $0<\lambda<A$ *are given and finally that* $a(\lambda)<b(A)$ *is satisfied. Then, the practical stability properties of*

$$e(n+1)=g(n,e(n)),e(n_0)=e_0\geq 0. \tag{4}$$

imply the practical stability properties of system (1).

Corollary 5. *In Theorem* (4): *if* $w(n,e)\equiv 0$ *we get uniform practical stability of* (1) *which implies structural stability.*

Definition 6. *A Petri net is a 5-tuple,* $PN=\{P,T,F,W,M_0\}$ *where:* $P=\{p_1,p_2,\ldots,p_m\}$ *is a finite set of places,* $T=\{t_1,t_2,\ldots,t_n\}$ *is a finite set of transitions,* $F\subset(P\times T)\cup(T\times P)$ *is a set of arcs,* $W:F\to N_1^+$ *is a* *weight function,* $M_0: P \to N$ *is the initial marking,* $P \cap T = \varnothing$ *and* $P \cup T \neq \varnothing$.

Definition 7. *The clock structure associated with a place* $p_i \in P$ *is a set* $\mathbf{V} = \{V_i : p_i \in P\}$ *of clock sequences* $V_i = \{v_{i,1}, v_{i,2}, ...\}$, $v_{i,k} \in R^+, k = 1, 2, ...$

The positive number $v_{i,k}$, associated to $p_i \in P$, called holding time, represents the time that a token must spend in this place until its outputs enabled transitions $t_{i,1}, t_{i,2}, ...,$ fire. We partition *P* into subsets P_0 and P_h, where P_0 is the set of places with zero holding time, and P_h is the set of places that have some holding time.

Definition 8. *A timed Petri net is a 6-tuple* $TPN = \{P, T, F, W, M_0, \mathbf{V}\}$ *where* $\{P, T, F, W, M_0\}$ *are as before, and* $\mathbf{V} = \{V_i : p_i \in P\}$ *is a clock structure. A timed Petri net is a timed event Petri net when every* $p_i \in P$ *has one input and one output transition, in which case the associated clock structure set of a place* $p_i \in P$ *reduces to one element* $V_i = \{v_i\}$.

Notice that if $W(p,t) = \alpha$ (or $W(t,p) = \beta$) then, this is often represented graphically by α, (β) arcs from *p* to *t* (*t* to *p*) each with no numeric label.

Let $M_k(p_i)$ denote the marking (*i.e.*, the number of tokens) at place $p_i \in P$ at time *k* and let $M_k = [M_k(p_1), ..., M_k(p_m)]^T$ denote the marking (state) of *PN* at time *k*. A transition $t_j \in T$ is said to be enabled at time *k* if $M_k(p_i) \geq W(p_i, t_j)$ for all $p_i \in P$ such that $(p_i, t_j) \in F$. It is assumed that at each time *k* there exists at least one transition to fire. If a transition is enabled then, it can fire. If an enabled transition $t_j \in T$ fires at time *k* then, the next marking for $p_i \in P$ is given by

$$M_{k+1}(p_i) = M_k(p_i) + W(t_j, p_i) - W(p_i, t_j). \tag{5}$$

Let $A = [a_{ij}]$ denote an $n \times m$ matrix of integers (the incidence matrix) where $a_{ij} = a_{ij}^+ - a_{ij}^-$ with $a_{ij}^+ = W(t_i, p_j)$ and $a_{ij}^- = W(p_j, t_i)$. Let $u_k \in \{0,1\}^n$ denote a firing vector where if $t_j \in T$ is fired then, its corresponding firing vector is $u_k = [0, ..., 0, 1, 0, ..., 0]^T$ with the one in the jth position in the vector and zeros everywhere else. The nonlinear difference matrix equation describing the dynamical behavior represented by a *PN* is:

$$M_{k+1} = M_k + A^T u_k \tag{6}$$

where if at step *k*, $a_{ij}^- < M_k(p_j)$ for all $p_i \in P$ then, $t_i \in T$ is enabled and if this $t_i \in T$ fires then, its corresponding firing vector u_k is utilized in the difference equation to generate the next step. Notice that if M' can be reached from some other marking *M* and, if we fire some sequence of *d* transitions with corresponding firing vectors $u_0, u_1, ..., u_{d-1}$ we obtain that

$$M' = M + A^T u, u = \sum_{k=0}^{d-1} u_k. \tag{7}$$

Let (N_{n0}^m, d) be a metric space where $d: N_{n0}^m \times N_{n0}^m \to R_+$ is defined by

$d(M_1, M_2) = \sum_{i=1}^{m} \zeta_i \,|\, M_1(p_i) - M_2(p_i) \,|; \zeta_i > 0$ and consider the matrix difference equation which describes the dynamical behavior of the discrete event system modeled by a *PN*, see (7).

Proposition 9. *Let PN be a Petri net. PN is uniform practical stable if there exists a* Φ *strictly positive m vector such that*

$$\Delta v = u^T A \Phi \leq 0 \tag{8}$$

Moreover, PN is uniform practical asymptotic stable if the following equation holds

$$\Delta v = u^T A \Phi \leq -c(e), c \in K \tag{9}$$

Lemma 10. *Let suppose that Proposition* (9) *holds then,*

$$\Delta v = u^T A \Phi \leq 0 \Leftrightarrow A \Phi \leq 0 \tag{10}$$

Remark 11. *Notice that since the state space of a TPN is contained in the state space of the same now not timed PN, stability of PN implies stability of the TPN.*

Lyapunov Stabilization

Definition 12. *Let PN be a Petri net. PN is said to be stabilizable if there exists a firing transition sequence* *with transition count vector u such that system (7) remains bounded.*

Proposition 13. *Let PN be a Petri net. PN is stabilizable if there exists a firing transition sequence with transition count vector u such that the following equation holds*

$$\Delta v = A^T u \le 0 \tag{11}$$

Remark 14. *By fixing a particular u, which satisfies* (11), *the state space is restricted to those markings that are finite.*

3. Max-Plus Algebra [4] [5]

3.1. Basic Definitions

NOTATION: $\epsilon = -\infty$, $e = 0$, $\mathbb{R}_{\max} = \mathbb{R} \cup \{\epsilon\}$, $\underline{n} = 1,2,\ldots,n$. Let $a,b \in \mathbb{R}_{\max}$ and define the operations $\oplus$ and $\otimes$ by: $a \oplus b = \max(a,b)$ and $a \otimes b = a + b$.

Definition 15. *The set* $\mathbb{R}_{\max}$ *with the two operations* $\oplus$ *and* $\otimes$ *is called a max-plus algebra and is denoted by* $\mathfrak{R}_{\max} = (\mathbb{R}_{\max}, \oplus, \otimes, \epsilon, e)$.

Definition 16. *A semiring is a nonempty set R endowed with two operations* $\oplus_R$, $\otimes_R$, *and two elements* ϵ_R *and* e_R *such that:* $\oplus_R$ *is associative and commutative with zero element* ϵ_R, $\otimes_R$ *is associative, distributes over* $\oplus_R$, *and has unit element* e_R, $\in_R$ *is absorbing for* $\otimes_R$ *i.e.,* $a \otimes_R \epsilon = \epsilon_R \otimes a = a$, $\forall a \in R$. *In addition if* $\otimes_R$ *is commutative then R is called a commutative semiring, and if* $\oplus_R$ *is such that* $a \oplus_R a = a$, $\forall a \in R$ *then it is called idempotent.*

Theorem 17. *The max-plus algebra* $\mathfrak{R}_{\max} = (\mathbb{R}_{\max}, \oplus, \otimes, \epsilon, e)$ *has the algebraic structure of a commutative and idempotent semiring.*

3.2. Matrices and Graphs

Let $\mathbb{R}_{\max}^{n \times n}$ be the set of $n \times n$ matrices with coefficients in $\mathbb{R}_{\max}$ with the following operations: The sum of matrices $A, B \in \mathbb{R}_{\max}^{n \times n}$, denoted $A \oplus B$ is defined by: $(A \oplus B)_{ij} = a_{ij} \oplus b_{ij} = \max(a_{ij}, b_{ij})$, for *i* and $j \in \underline{n}$. The product of matrices $A \in \mathbb{R}_{\max}^{n \times l}, B \in \mathbb{R}_{\max}^{l \times n}$, denoted $A \otimes B$ is defined by: $(A \otimes B)_{ik} = \bigotimes(a_{ij} \otimes b_{jk})$ for *i* and $k \in \underline{n}$. Let $\mathcal{E} \in \mathbb{R}_{\max}^{n \times n}$ denote the matrix with all its elements equal to ϵ and denote by$^{j=1}E \in \mathbb{R}_{\max}^{n \times n}$ the matrix which has its diagonal elements equal to *e* and all the other elements equal to ϵ. Then, the following result can be stated.

Theorem 18. *The* 5-*tuple* $\mathfrak{R}_{\max}^{n \times n} = (\mathbb{R}_{\max}^{n \times n}, \oplus, \otimes, \mathcal{E}, E)$ *has the algebraic structure of a noncommutative idempotent semiring.*

Definition 19. *Let* $A \in \mathbb{R}_{\max}^{n \times n}$ *and* $k \in \mathbb{N}$ *then the k-th power of A denoted by* $A^{\otimes k}$ *is defined by:* $A^{\otimes k} = A \otimes A \otimes \cdots \otimes A$, *(k times), where* $A^{\otimes 0}$ *is set equal to E.*

Definition 20. *A matrix* $A \in \mathbb{R}_{\max}^{n \times n}$ *is said to be regular if A contains at least one element distinct from* ϵ *in each row.*

Definition 21. *Let* $\mathcal{N}$ *be a finite and non-empty set and consider* $\mathcal{D} \subseteq \mathcal{N} \times \mathcal{N}$. *The pair* $G = (\mathcal{N}, \mathcal{D})$ *is called a directed graph, where* $\mathcal{N}$ *is the set of elements called nodes and* $\mathcal{D}$ *is the set of ordered pairs of nodes called arcs. A directed graph* $G = (\mathcal{N}, \mathcal{D})$ *is called a weighted graph if a weight* $w(i,j) \in \mathbb{R}$ *is associated with any arc* $(i,j) \in \mathcal{D}$.

Let $A \in \mathbb{R}_{\max}^{n \times n}$ be any matrix, a graph $\mathcal{G}(A)$, called the communication graph of *A*, can be associated as follows. Define $N(A) = \underline{n}$ and a pair $(i,j) \in \underline{n} \times \underline{n}$ will be a member of $\mathcal{D}(A) \Leftrightarrow a_{ji} \ne \epsilon$, where $\mathcal{D}(A)$ denotes the set of arcs of $\mathcal{G}(A)$.

Definition 22. *A path from node i to node j is a sequence of arcs* $p = \{(i_k, j_k) \in \mathcal{D}(A)\}_{k \in \underline{m}}$ *such that* $i = i_1, j_k = i_{k+1}$, *for* $k < m$ *and* $j_m = j$. *The path p consists of the nodes* $i = i_1, i_2, \ldots, i_m, j_m = j$ *with length m denoted by* $|p|_1 = m$. *In the case when* $i = j$ *the path is said to be a circuit. A circuit is said to be elementary if nodes* i_k *and* i_l *are different for* $k \ne l$. *A circuit consisting of one arc is called a self-loop.*

Let us denote by $P(i,j;m)$ the set of all paths from node *i* to node *j* of length $m \ge 1$ and for any arc $(i,j) \in \mathcal{D}(A)$ let its weight be given by a_{ij} then the weight of a path $p \in P(i,j;m)$ denoted by $|p|_w$ is defined to be the sum of the weights of all the arcs that belong to the path. The average weight of a path *p* is given by $|p|_w / |p|_1$. Given two paths, as for example, $p = ((i_1,i_2),(i_2,i_3))$ and $q = ((i_3,i_4),((i_4,i_5)$ in $\mathcal{G}(A)$ the concatenation of paths $\circ : \mathcal{G}(A) \times \mathcal{G}(A) \to \mathcal{G}(A)$ is defined as $p \circ q = ((i_1,i_2),(i_2,i_3),(i_3,i_4),(i_4,i_5))$. The communication graph $\mathcal{G}(A)$ and powers of matrix A are closely related as it is shown in the next theorem.

Theorem 23. *Let* $A \in \mathbb{R}_{\max}^{n\times n}$, *then* $\forall k \geq 1$: $[A^{\otimes k}]_{ji} = \max\{|p|_w : p \in P(i,j;k)\}$, *where* $[A^{\otimes k}]_{ji} = \epsilon$ *in the case when* $P(i,j;k)$ *is empty i.e., no path of length k from node i to node j exists in* $\mathcal{G}(A)$.

Definition 24. *Let* $A \in \mathbb{R}_{\max}^{n\times n}$ *then define the matrix* $A^{+} \in \mathbb{R}_{\max}^{n\times n}$ *as:* $A^{+} = \bigoplus_{k=1}^{\infty} A^{\otimes k}$. *Where the element* $[A^{+}]_{ji}$ *gives the maximal weight of any path from j to i. If in addition one wants to add the possibility of staying at a node then one must include matrix E in the definition of matrix* A^{+} *giving rise to its Kleene star representation defined by:* $A^{*} = \bigoplus_{k=0}^{\infty} A^{\otimes k}$.

Lemma 25. *Let* $A \in \mathbb{R}_{\max}^{n\times n}$ *be such that any circuit in* $\mathcal{G}(A)$ *has average circuit weight less than or equal to* ϵ. *Then it holds that:* $A^{*} = \bigoplus_{k=0}^{n-1} A^{\otimes k}$.

Definition 26. *Let* $G = (\mathcal{N}, \mathcal{D})$ *be a graph and* $i, j \in \mathcal{N}$, *node j is reachable from node i, denoted as* $i\mathcal{R}j$, *if there exists a path from i to j. A graph G is said to be strongly connected if* $\forall i, j \in \mathcal{N}, j\mathcal{R}i$. *A matrix* $A \in \mathbb{R}_{\max}^{n\times n}$ *is called irreducible if its communication graph is strongly connected, when this is not the case matrix A is called reducible.*

Remark 27. *In this paper irreducible matrices are just considered. It is possible to treat the reducible case by transforming it into its normal form and computing its generalized eigenmode see* [4].

3.2.1. Spectral Theory and Linear Equations

Definition 28. *Let* $A \in \mathbb{R}_{\max}^{n\times n}$ *be a matrix. If* $\mu \in R_{\max}$ *is a scalar and* $v \in R_{\max}^{n}$ *is a vector that contains at least one finite element such that:* $A \otimes v = \mu \otimes v$ *then,* μ *is called an eigenvalue and v an eigenvector.*

Let $\mathcal{C}(A)$ denote the set of all elementary circuits in $\mathcal{G}(A)$ and write: $\lambda = \max_{p\in\mathcal{C}(A)} \frac{|p|_w}{|p|_1}$ for the maximal average circuit weight. Notice that since $\mathcal{C}(A)$ is a finite set, the maximum is attained (which is always the case when matrix A is irreducible). In case $\mathcal{C}(A) = \varnothing$ define $\lambda = \epsilon$.

Definition 29. *A circuit* $p \in G(A)$ *is said to be critical if its average weight is maximal. The critical graph of A, denoted by* $G^{c}(A) = (\mathcal{N}^{c}(A), \mathcal{D}^{c}(A))$, *is the graph consisting of those nodes and arcs that belong to critical circuits in* $G(A)$.

Theorem 30. *If* $A \in \mathbb{R}_{\max}^{n\times n}$ *is irreducible, then there exists one and only one finite eigenvalue (with possible several eigenvectors). This eigenvalue is equal to the maximal average weight of circuits in* $G(A)$ $\lambda(A) = \max_{p\in\mathcal{C}(A)} \frac{|p|_w}{|p|_1}$.

Theorem 31. *Let* $A \in \mathbb{R}_{\max}^{n\times n}$ *and* $b \in \mathbb{R}_{\max}^{n}$. *If the communication graph* $G(A)$ *has maximal average circuit weight less than or equal to e, then* $x = A^{*} \otimes b$ *solves the equation* $x = (A \otimes x) \oplus b$. *Moreover, if the circuit weights in* $G(a)$ *are negative then, the solution is unique.*

3.3. Max-Plus Recurrence Equations for Timed Event Petri Nets

Definition 32. *Let* $A_m \in \mathbb{R}_{\max}^{n\times n}$ *for* $0 \leq m \leq M$ *and* $x(m) \in \mathbb{R}_{\max}^{n}$ *for* $-M \leq m \leq -1$; $M \geq 0$. *Then, the recurrence equation:* $x(k) = \bigoplus_{m=0}^{M} A_m \otimes x(k-m); k \geq 0$ *is called an Mth order recurrence equation.*

Theorem 33. *The Mth order recurrence equation, given by equation* $x(k) = \bigoplus_{m=0}^{M} A_m \otimes x(k-m); k \geq 0$, *can be transformed into a first order recurrence equation* $x(k+1) = A \otimes x(k)$; $k \geq 0$ *provided that* A_0 *has circuit weights less than or equal to zero.*

With any timed event Petri net, matrices $A_0, A_1, ..., A_M \in \mathbb{N}^n \times \mathbb{N}^n$ can be defined by setting $[A_m]_{jl} = a_{jl}$, where a_{jl} is the largest of the holding times with respect to all places between transitions t_l and t_j with m tokens, for $m = 0,1,...,M$, with M equal to the maximum number of tokens with respect to all places. Let $x_i(k)$ denote the kth time that transition t_i fires, then the vector $x(k) = (x_1(k), x_2(k),...x_m(k))^T$, called the state of the system, satisfies the Mth order recurrence equation: $x(k) = \bigoplus_{m=0}^{M} A_m \otimes x(k-m); k \geq 0$ Now, assuming that all the hypothesis of theorem (33) are satisfied, and setting $\hat{x}(k) = (x^T(k), x^T(k-1),...,x^T(k-M+1))^T$, equation $x(k) = \bigoplus_{m=0}^{M} A_m \otimes x(k-m); k \geq 0$ can be expressed as: $\hat{x}(k+1) = \hat{A} \otimes \hat{x}(k); k \geq 0$, which is known as the standard autonomous equation.

4. The Solution to the Stability Problem for Discrete Event Dynamical Systems Modeled with Timed Petri Nets

Definition 34. *A TPN is said to be stable if all the transitions fire with the same proportion i.e., if there exists* $q \in \mathbb{N}$ *such that*

$$\lim_{k\to\infty} \frac{x_i(k)}{k} = q, \forall i = 1,...,n \tag{12}$$

This means that in order to obtain a stable *TPN* all the transitions have to be fired q times. It will be desirable to be more precise and know exactly how many times. The answer to this question is given next.

Lemma 35. *Consider the recurrence relation* $x(k+1) = A \otimes x(k), k \geq 0$, $x(0) = x_0 \in \mathbb{R}^n$ *arbitrary. A an irreducible matrix and* $\lambda \in \mathbb{R}$ *its eigenvalue then,*

$$\lim_{k\to\infty} \frac{x_i(k)}{k} = \lambda, \forall i = 1,...,n \tag{13}$$

Proof. Let v be an eigenvector of A such that $x_0 = v$ then,

$$x(k) = \lambda^{\otimes k} \otimes v \Rightarrow x(k) = k\lambda + v \Rightarrow \frac{x(k)}{k} = \lambda + \frac{v}{k} \Rightarrow \lim_{k\to\infty} \frac{x_i(k)}{k} = \lambda .$$

Now starting with an unstable *TPN*, collecting the results given by: proposition (13), what has just been discussed about recurrence equations for *TPN* at the end of subsection (3.3) and the previous lemma (35) plus theorem (30), the solution to the problem is obtained.

5. Modeling and Stability Analysis of a Communication Network System

In this section, the modeling and stability analysis for a communication network system is addressed. The communication network system consists of a transmitter which sends messages through a communication channel to a receiver. The proposed model considers two possibilities. The first one, that messages are successfully received, while in the second one, during the sending process the transmitter breaks down and as a result the message does not reach the receiver. Consider a communication network system whose *TPN* model is depicted in **Figure 1**.

Where the events (transitions) that drive the system are: q: receivers connect to the communication network, s: messages are sent, b: the transmitter breaks, r: the transmission is restored, d: the message has been successfully received. The places (that represent the states of the queue) are: A: receivers concentrating, P: the receiver is waiting for a message to be sent, B: the message is being received, D: transmitter breaks down, I: the transmitter is idle. The holding times associated to the places A and I are Ca and Cd respectively (with $Ca > Cd$). The incidence matrix that represents the PN model is

$$A = \begin{bmatrix} 0 & 1 & 0 & 0 & 0 \\ 0 & -1 & 1 & -1 & 0 \\ 0 & 0 & -1 & 1 & 0 \\ 0 & 0 & -1 & 0 & 1 \\ 0 & 0 & 0 & 1 & -1 \end{bmatrix}$$

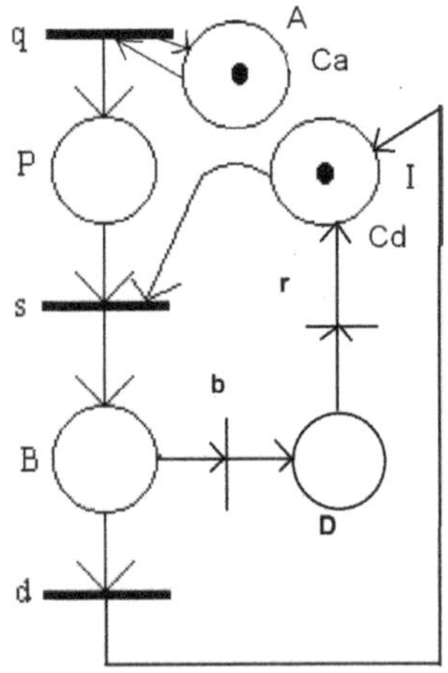

Figure 1. Transmitter breakdown timed Petri net model.

Therefore since there does not exists a Φ strictly positive m vector such that $A\Phi \le 0$ the sufficient condition for stability is not satisfied, (moreover, the PN (TPN) is unbounded since by the repeated firing of q, the marking in P grows indefinitely). However, by taking $u=[k,k,k/2,k/2,k/2];k>0$ (but unknown) we get that $A^T u \le 0$. Therefore, the PN is stabilizable which implies that the TPN is stable. Now, let us proceed to determine the exact value of k. From the TPN model we obtain that:

$$A_0 = \begin{pmatrix} \varepsilon & \varepsilon & \varepsilon & \varepsilon & \varepsilon \\ 0 & \varepsilon & \varepsilon & \varepsilon & \varepsilon \\ \varepsilon & 0 & \varepsilon & \varepsilon & \varepsilon \\ \varepsilon & \varepsilon & \varepsilon & \varepsilon & \varepsilon \\ \varepsilon & \varepsilon & \varepsilon & 0 & \varepsilon \end{pmatrix} \text{ and } A_1 = \begin{pmatrix} Ca & \varepsilon & \varepsilon & \varepsilon & \varepsilon \\ \varepsilon & \varepsilon & Cd & \varepsilon & Cd \\ \varepsilon & \varepsilon & \varepsilon & \varepsilon & \varepsilon \\ \varepsilon & \varepsilon & \varepsilon & \varepsilon & \varepsilon \\ \varepsilon & \varepsilon & \varepsilon & \varepsilon & \varepsilon \end{pmatrix} \text{ which implies } A_0^* = \begin{pmatrix} 0 & \varepsilon & \varepsilon & \varepsilon & \varepsilon \\ 0 & 0 & \varepsilon & \varepsilon & \varepsilon \\ 0 & 0 & 0 & \varepsilon & \varepsilon \\ \varepsilon & \varepsilon & \varepsilon & 0 & \varepsilon \\ \varepsilon & \varepsilon & \varepsilon & 0 & 0 \end{pmatrix},$$

leading to:

$$\hat{A} = A_0^* \otimes A_1 = \begin{pmatrix} Ca & \varepsilon & \varepsilon & \varepsilon & \varepsilon \\ Ca & \varepsilon & Cd & \varepsilon & Cd \\ Ca & \varepsilon & Cd & \varepsilon & Cd \\ \varepsilon & \varepsilon & \varepsilon & \varepsilon & \varepsilon \\ \varepsilon & \varepsilon & \varepsilon & \varepsilon & \varepsilon \end{pmatrix}$$

Therefore, $\lambda(A) = \max_{p \in C(A)} \frac{|p|_w}{|p|_1} = \max\{Ca, Cd\} = Ca$. This means that in order for the TPN to be stable and work properly the speed at which the receivers concentrate has to be equal to Ca (the firing speed of transition q) which is attained by taking $k = Ca$.

6. Conclusion

This paper studies the modeling and stability problem for communication network systems using timed Petri nets, Lyapunov methods and max-plus algebra. The results obtained are consistent to what was expected.

References

[1] Murata, T. (1989) Petri nets: Properties, Analysis, and Applications. *Proc. IEEE*, **77**. http://dx.doi.org/10.1109/5.24143

[2] Retchkiman, Z. (2005) Stability Theory for a Class of Dynamical Systems Modeled with Petri Nets. *International Journal of Hybrid Systems*, **4**.

[3] Retchkiman, Z. (1999) From Stability to the Stabilization Problem of Discrete Event Systems Modeled by Petri Nets.

American Control Conference'99, San Diego, June 1999.

[4] Heidergott, B., Olsder, G.J. and van der Woude, J. (2006) Max Plus at Work. Princeton University Press.

[5] Baccelli, F., Cohen, G., Olsder, G.J. and Quadrat, J.P. (2001) Synchronization and Linearity. Web-Edition.

28

Analysis and Design of 4 × 4 MIMO-Antenna Systems in Mobile Phone

Deshun Sun, Chongyu Wei

Qingdao University of Science & Technology, Qingdao, China
Email: 811093787@qq.com

Abstract

MIMO antenna systems can greatly improve the capacity and transmission rate of wireless communication. In this letter, two four-port multiple-input-multiple-output (MIMO) antenna systems are proposed in mobile devices. An algorithm of calculating envelope correlation (ρ_e) between antenna units is also proposed. By this algorithm the ρ_e of the PIFA-MIMO antenna system has been calculated. First, a PIFA-MIMO antenna system is designed for ISM (2.4 GHZ) application. The slots are etched on the ground and the antenna units are placed in such a way that the polarization diversity is exploited to enhance the impedance matching and isolation. Based on this, a printed wideband MIMO antenna system is proposed for portable 3G and 4G applications by means of multi-branch and the defect of ground technology, increasing the bandwidth and reducing mutual coupling between the antennas. The simulation results show that the return loss and isolation of the two antenna systems can meet the performance requirements of the antennas in the mobile phone.

Keywords

Isolation, Wideband, Multiple-Input-Multiple-Output (MIMO), Mutual Coupling, Defect Ground, Correlation Coefficient

1. Introduction

With the rapid development of wireless communication technology and the increasing demand of multimedia service, it is extremely urgent to realize the data communication with high speed and large capacity. As the core technology of wireless communication, MIMO technology has not only achieved a great increase in the capacity of communication, but also greatly improved the communication quality. By the literature [1], the capacity of the MIMO systems can be increased by the MIMO systems when channels are independent and identically distributed. The number and spacing of antennas are the key parameters of the MIMO antenna. Theoretically, the more the number of antennas are, the better the performance is. However, in the mobile phone and other small mobile terminals, with multiple antennas, antenna spacing between the antennas is not enough to ensure to reach high isolation between antennas. Actually, there are technical challenges in implementing mobile antennas with both a low correlation property and a sufficient antenna gain in mobile devices. If the isolation within the multi-antenna system is insufficient, system performance is reduced in terms of the gain and correlation due to the coupling between antennas [2].

In recent years, multiple-input-multiple-output (2 × 2 MIMO) communication systems use the multiple antennas that are equipped at receiver and transmitter ends to enhance not only the available data rate but the capacity in multipath environments. In order to improve the isolation between the two ports of the MIMO antenna, many designs have adopted the defect ground structure to meet the requirements of high isolation in MIMO antenna [3] [4]. [3] designs an ultra wideband (UWB) antenna. T he seam gap and vertical polarization are combined to improve the isolation between two antennas in band of 3 - 4.5 GHZ. [4] shows that the MIMO system is composed of two symmetrical printed monopole antenna. A high isolation between antenna elements can be obtained with two inverted L-branch and a rectangular slots embedded in the floor. The letter [5] investigates a method to use the DGS to the dual-band antenna with high port-isolation. Study [6] provides a design of two-port UWB MIMO antenna system with high isolation using a fork-shaped structure. An ultra wideband (UWB) MIMO antenna is designed in [7], which uses the floor gap and the vertical polarization mode to effectively improve the isolation of the two antennas in the 3 - 4.5 GHZ band. Antenna designs of 4 × 4 MIMO system have been realized in the literature [8] and [9]. [8] is a compact printed ultra wideband antenna and the antenna uses the EBG structure. In order to reduce the coupling between antenna elements, the four antenna elements are arranged vertically to achieve vertical polarization effect; in the design of [9], UWB antenna is a printed dipole antenna, and antenna element is also arranged vertically to achieve high isolation between the antennas. Based on these, in this paper, I design two new 4 × 4 MIMO antenna model. One works in a single frequency 2.4 GHZ (ISM), and the other can cover large bandwidth in 3G and 4G-LTE.

2. Correlation Analysis of MIMO Antennas Systems

2.1. Key Parameters of Diversity Antenna

Beyond the existing performance standards for a single-receiver mobile device, there are two antenna parameters that impact mobile receive diversity (MRD) performance: 1) the gain difference between the two antennas, and 2) the fading (envelope) correlation coefficient, ρ_e, between the two antennas.

2.2. Envelope Correlation Coefficient (ρ_e)

ρ_e is the most relevant to overall diversity performance. It is directly related to diversity gain (fading mitigation), and is statistically related to interference rejection gain.

2.2.1. Envelope Correlation Coefficient (ρ_e)

The fading (envelope) correlation coefficient ρ_e is a key indicator of MRD performance, and is strongly influenced by the incident EM field characteristics. The wireless multipath channel—due to its significant impact on the incident EM field must be considered as well as the antenna array and each element within the array.

Let me identify one of two mobile antenna ports. According to [10], the signal observed at each antenna port due to the *k*th multipath is a function of the antenna field pattern $E_m(\theta,\phi)$ and the incident field $F_m^k(\theta,\phi)$.

$$\begin{aligned} V_{Lm}^k &= \int_0^{2\pi}\int_0^{\pi} E_m(\theta,\phi)\cdot F_m^k(\theta,\phi)\cdot\sin\theta\cdot \mathrm{d}\theta\cdot \mathrm{d}\phi \\ &= \int_0^{2\pi}\int_0^{\pi}\left[E_{\theta,m}(\theta,\phi)\cdot F_{\theta,m}^k(\theta,\phi)+E_{\varphi,m}(\theta,\phi)\cdot F_{\varphi,m}^k(\theta,\phi)\right]\cdot\sin\theta\cdot \mathrm{d}\theta\cdot \mathrm{d}\phi \end{aligned} \tag{1}$$

$E_{\theta,m}(\theta,\phi)$ and $E_{\varphi,m}(\theta,\phi)$ are the vertical and horizontal (respectively) polarization patterns of antenna m. It is assumed that the incident fields from different directions and polarizations are independent from one another.

The variance of the total received signal can be expressed as follows:

$$E\left[\left|V_{Lm}^{k}\right|^{2}\right]=P_{V}\int_{0}^{2\pi}\int_{0}^{\pi}\left|E_{\theta,m}(\theta,\phi)\right|^{2}\cdot P_{\theta}(\theta,\phi)\cdot\sin\theta\cdot d\theta\cdot d\phi+P_{H}\int_{0}^{2\pi}\int_{0}^{\pi}\left|E_{\phi,m}(\theta,\phi)\right|^{2}\cdot P_{\phi}(\theta,\phi)\cdot\sin\theta\cdot d\theta\cdot d\phi \quad (2)$$

$P_\theta(\theta,\phi)$ and $P_\phi(\theta,\phi)$ are the angular density functions of the vertically and horizontally polarized components of the incident field, both contributing to the total incident field. P_V and P_H are the mean powers that would be received by isotropic vertical and horizontal polarized antennas.

Similarly, the cross-correlation between the complex signal-envelopes received at the two antennas can be expressed as follows:

$$\begin{aligned}E\left[V_{L_1}^{k}\cdot V_{L_2}^{k\,*}\right]&=E\left[V_{L\theta,1}^{k}\cdot V_{L\theta,2}^{k}{}^{*}\right]+E\left[V_{L\varphi,1}^{k}\cdot V_{L\varphi,2}^{k}{}^{*}\right]\\&=P_H\cdot\int_0^{2\pi}\int_0^{\pi}\Big[XPR\cdot E_{\theta,1}(\theta,\varphi)\cdot E_{\theta,2}^{*}(\theta,\phi)\cdot P_\theta(\theta,\phi)+E_{\varphi,1}(\theta,\phi)\\&\quad\cdot E_{\theta,2}^{*}(\theta,\phi)\cdot P_\theta(\theta,\phi)+E_{\varphi,1}(\theta,\phi)\cdot\sin\theta\cdot d\theta\cdot d\phi\end{aligned} \quad (3)$$

XPR is the cross-polarization power ratio, $XPR = P_V/P_H$. Using the two signal variances and the cross-correlation, the complex correlation coefficient can be calculated as follows:

$$\rho=\frac{E\left[V_{L1}^{k}\cdot V_{L2}^{k\,*}\right]}{\sqrt{E\left[V_{L1}^{k}\cdot V_{L1}^{k*}\right]\cdot E\left[V_{L2}^{k}\cdot V_{L2}^{k\,*}\right]}} \quad (4)$$

The envelope correlation ρ_e is defined as the correlation coefficient of the signal envelopes received at the two antenna ports. It is approximately equal to the magnitude-square of the complex correlation coefficient:

$$\rho_e\cong|\rho|^2=\frac{\left|E\left[V_{L1}^{k}\cdot V_{L2}^{k\,*}\right]\right|^2}{E\left[V_{L1}^{k}\cdot V_{L1}^{k*}\right]\cdot E\left[V_{L2}^{k}\cdot V_{L2}^{k\,*}\right]} \quad (5)$$

2.2.2. Incident Field Distribution Models

In order to compute the variance and correlation, the wireless channel is modeled to specify the incident field angular density function (or incident field distribution). The incident field distribution depends upon the scattering properties of the wireless environment. The most common assumption is that there are numerous scatterers present around the mobile, causing a uniform incident field in azimuth. Gaussian and Laplacian distributions are also used for cases when the incident wave is more concentrated from certain directions (smaller angular spread).

Several example incident field distribution expressions are provided below; they are all applicable to both polarizations and are all linear functions (not logarithmic).

1. Uniform in both azimuth and elevation:

$$P(\theta,\phi)=\frac{1}{4\pi}\quad 0<\theta<\pi \text{ and } 0<\phi<2\pi \quad (6)$$

2. Uniform in azimuth, Gaussian in elevation:

$$P(\theta,\phi)=A1\cdot\exp\left\{-\frac{[\theta-m_\theta]^2}{2\sigma_\theta^2}\right\}\quad 0<\theta<\pi \text{ and } 0<\phi<2\pi \quad (7)$$

3. Gaussian in both azimuth and elevation:

$$P(\theta,\phi)=A2\cdot\exp\left\{-\frac{[\phi-m_\phi]^2}{2\sigma_\phi^2}\right\}\cdot\exp\left\{-\frac{[\theta-m_\theta]^2}{2\sigma_\theta^2}\right\} \quad (8)$$

$$0<\theta<\pi \text{ and } 0<\phi<2\pi$$

m_ϕ and σ_ϕ are the mean and the standard deviation of the azimuth angle of arrivals. *A2* is a scaling constant that forces the density function to unity when integrated over the surface of a unit sphere.

2.2.3. Calculating ρ_e at Discrete Angles

In engineering, the envelope correlation coefficient is calculated from the measured complex quantities $E_{\theta,1}(\theta,\phi)$, $E_{\phi,1}(\theta,\phi)$ and $E_{\theta,2}(\theta,\phi)$, $E_{\phi,2}(\theta,\phi)$. Recognizing that measurements occur at discrete angles, the expressions for ρ_e variables can be rewritten as follows:

$$R_{12} = \sum_{j=1}^{N\varnothing}\sum_{i=1}^{N\theta}\left(XPR \cdot E\theta 1_{i,j} \cdot E\theta 2^{*}_{i,j} \cdot P\theta_i + E\varnothing 1_{i,j} \cdot E\varnothing 2^{*}_{i,j} \cdot P\varnothing_j\right) \cdot \sin\theta_i \cdot \Delta\theta \cdot \Delta\varnothing$$

$$\sigma_1 = \sum_{j=1}^{N\varnothing}\sum_{i=1}^{N\theta}\left(XPR \cdot E\theta 1_{i,j} \cdot E\theta 1^{*}_{i,j} \cdot P\theta_i + E\varnothing 1_{i,j} \cdot E\varnothing 1^{*}_{i,j} \cdot P\varnothing_j\right) \cdot \sin\theta_i \cdot \Delta\theta \cdot \Delta\varnothing$$

$$\sigma_2 = \sum_{j=1}^{N\varnothing}\sum_{i=1}^{N\theta}\left(XPR \cdot E\theta 2_{i,j} \cdot E\theta 2^{*}_{i,j} \cdot P\theta_i + E\varnothing 2_{i,j} \cdot E\varnothing 2^{*}_{i,j} \cdot P\varnothing_j\right) \cdot \sin\theta_i \cdot \Delta\theta \cdot \Delta\varnothing$$

$XPR = P_V/P_H$; $p\theta$ and $p\phi$ are the incident field power densities defined in the previous section. $\Delta\theta$ and and $\Delta\phi$ are the angular increments for the measurement, and $N\theta$ and $N\phi$ are the number of angular measurement points in θ and ϕ assuming the measurement is done over a sphere of angles(θ varies from 0 to π and ϕ varies from 0 to 2π).

Finally, the envelope correlation coefficient is calculated from:

$$\rho_e = \frac{|R_{12}|^2}{\sigma_1 \cdot \sigma_2} \tag{9}$$

2.2.4. Other ρ_e Techniques

The complex antenna pattern correlation method recommended above takes into account many factors that can impact the correlation between signals from dual antennas on handsets in a mobile environment. The designer may prefer to use more simple and rapid method to estimate the envelope correlation coefficient. For example, S Blanch *et al.* describes a calculation of envelope correlation coefficient using S formula:

$$\rho_e = \frac{|S_{11} * S_{12} + S_{21} * S_{22}|^2}{\left(1-\left(|S_{11}|^2+|S_{21}|^2\right)\right)\left(1-\left(|S_{21}|^2+|S_{12}|^2\right)\right)} \tag{10}$$

Although attractive in its simplicity and ease of measurement, this expression only considers the direct contribution of mutual coupling to the antenna correlation, the properties of the incident RF field and its interaction with the antennas' gain patterns are not considered in this expression.

3. PIFA-MIMO Antenna System

Due to the low profile and easy to match of the PIFA antennas, Mobile phone built-in antenna always adopts PIFA (Planar inverted F antenna) antennas. This part presents a design of 4 × 4 MIMO antenna system model, operating in 2.4 GHZ (ISM band). The antenna system uses air as the medium. Because of the four antenna elements sharing one floor, the current of one antenna element will be coupled to another antenna element through the floor, which leads to mutual coupling and reduces the isolation between antenna elements. The antenna system adopts defect ground structure to extending the current length, as a result, effectively reducing the mutual coupling between the antenna elements. In addition, the antenna space-coupling by space radiation field will influence the performance of the MIMO antenna system. In this chapter, the four port antenna elements of this antenna system are placed vertically with each other, reducing the space-coupling between antenna units, increasing the isolation. Simulation results show that the antenna can meet the performance requirements of the MIMO antenna systems.

3.1. Antenna Structure

Figure 1 shows a planar and solid structure. The antenna system is composed of four parts, the radiation patch, the feed structure, the short circuit unit and the floor. In the design, the medium filled with air is between antenna radiation patch and the floor. Four identical PIFA antennas are located at the four corners of the floor. The

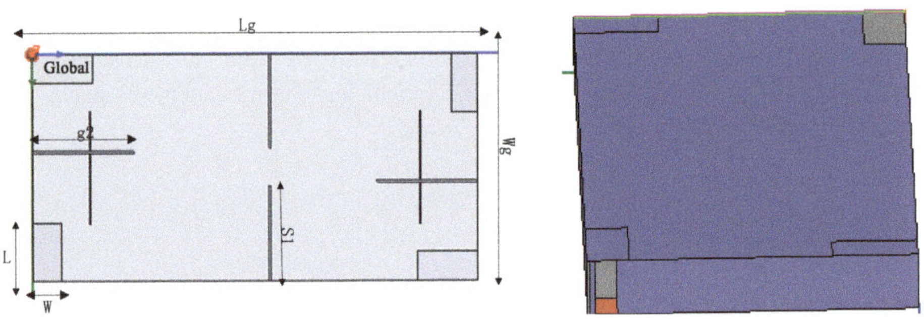

Figure 1. Antenna structure.

overall size of the machine is 120 × 60 × 10 mm^3, and the size of each PIFA antenna element is 16 × 7.5 × 10 mm^3.

The specific size of antenna elements: Lg = 120 mm, Wg = 60 mm, L = 16 mm, W = 7.5 mm, S1 = 27 mm, g2 = 30 mm.

As shown in **Figure 1**, two "+" shaped slots and two line-shaped slots are etched on the ground, as well as, four antenna units are arranged vertically between each other to reduce the space-coupling.

3.2. Antenna Simulation Results

Using the simulation software (HFSS) to simulate the antenna model, as shown in **Figure 2(a)**, the operating frequency of the four PIFA antennas (−15 dB) all cover 2360 - 2420 MHZ, meeting the performance requirements of the mobile phone MIMO antenna in 3G and 4G-LTE band; **Figure 2(b)** is the mutual correlation coefficient between two antennas, the correlation coefficient of the antenna system in operating band is basically below −15 dB. Using the simulation software (HFSS), the antenna radiation pattern data can be obtained, and then importing the data into the software of Matlab, according to the Equation (9) shown above, envelope correlation coefficient curve between adjacent antenna elements is shown in **Figure 3**. In the calculation, we always assume that the incident field in horizontal and vertical direction was uniformly distributed, XPR is 6 dB. The ρ_e in center frequency is that, $\rho_e = 0.0054$. Hence, the isolation between antennas elements meets well the requirements of mobile antenna systems for isolation.

4. Monopole-MIMO Antenna with Large Bandwidth

4.1. Antenna Structure

Printed monopole antenna has been widely used in many wireless devices due to its low cost, light weight and easy processing. As a new type of antenna, high dielectric antenna has the advantages of small size, simple structure and low cost. With the application of 4G wireless communication, the MIMO antenna systems operating in 3G and 4G-LTE frequency band have a good application prospect and demand. In order to achieve the large bandwidth of the antenna system, this paper mainly uses the multi-branch technology to achieve the principle of multi-resonant, through connecting the two or more adjacent resonant frequency band together, hence, obtaining the expansion of the bandwidth.

In this paper, the design of a broadband monopole-MIMO antenna system is composed of four monopole printed antennas, floor and dielectric substrate. The overall structure of the antenna system is shown in **Figure 4**. Four monopole antennas are printed on the four sides of the FR4 substrate, and the floor is located at the bottom of the substrate. In order to achieve the large bandwidth of the antenna, the multi branch structure is adopted to realize the multi resonance. The sizes of the antenna system are shown as: p1 = 21.25 mm, p2 = 9.5 mm, p3 = 38 mm, p5 = 21.5 mm, Lg = 60 mm, Wg = 46 mm.

4.2. Antenna Simulation Results

By simulation, the return loss of the four single pole printed antennas is shown in **Figure 5**. The operating frequency of the four antenna elements (S11 < −6 dB) covers the 3G and 4GLTE bands, and the large bandwidth of the MIMO antenna system is realized.

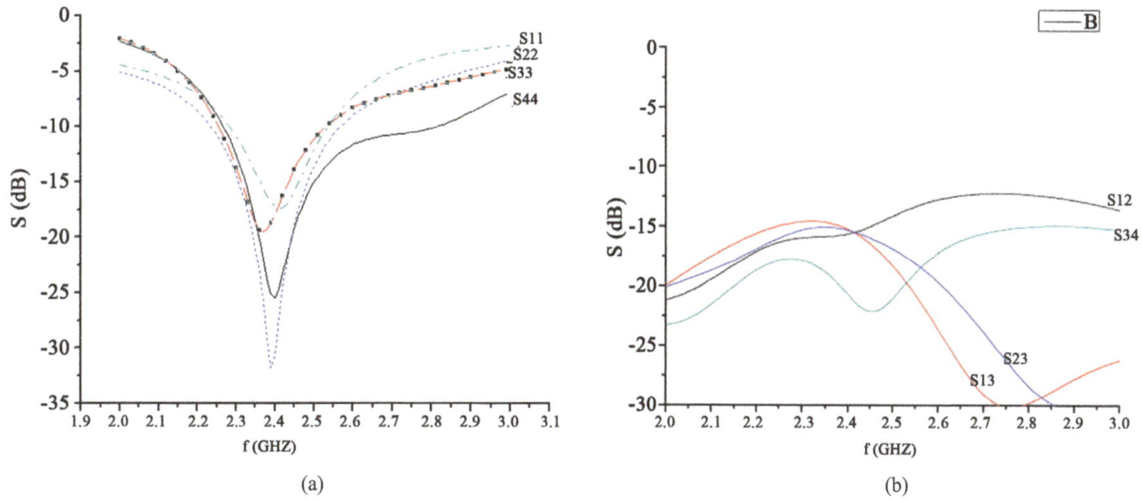

Figure 2. (a) The curves of S22, S33, S11 and S44; (b) The curves of S12, S13, S23 and S34.

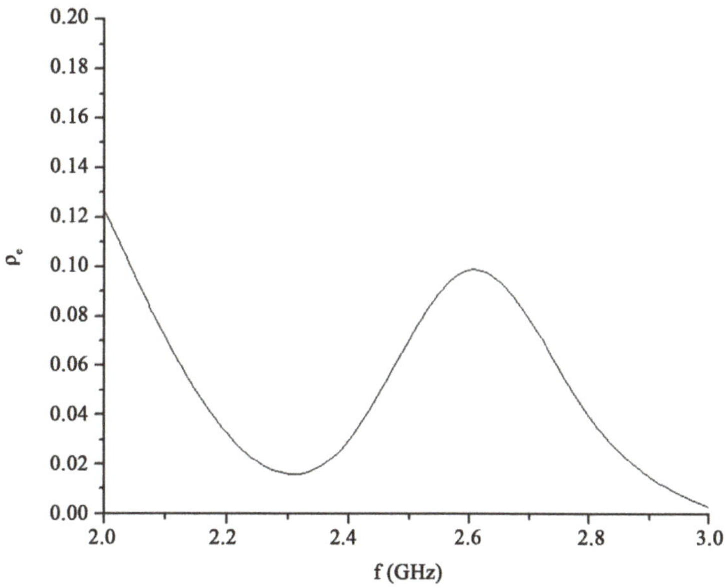

Figure 3. The curve of ρ_e.

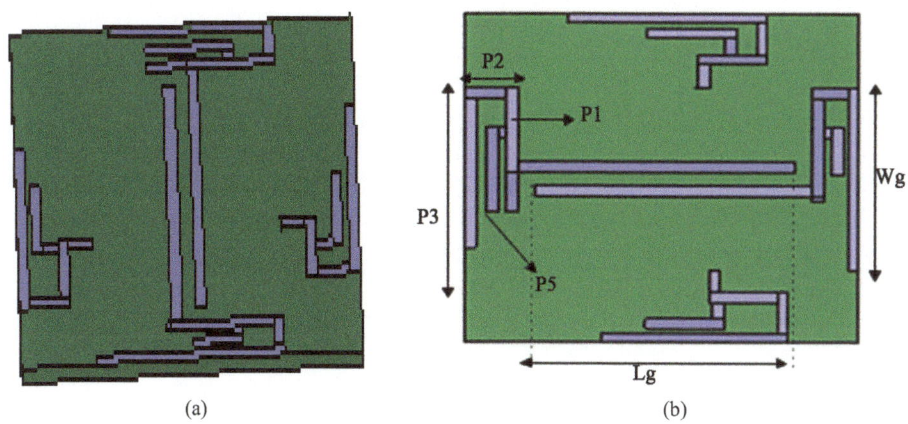

Figure 4. Top structure of antenna.

The MIMO antenna system designed in this paper is composed of four printed monopole antennas with nearly identical multi branch. The four antenna elements are located on the four side of the dielectric substrate respectively, without floor under the antenna elements. In order to obtain a better impedance matching and increase the isolation between adjacent antenna units, the floor under the medium substrate is slotted; the slotted size is 46 mm × 1 mm, so that the floor is divided into two parts, as shown in **Figure 6**. Using Equation (9), the ρ_e between the antenna units can reach the magnitude of 0.01, far less than 1. According the result above, the antenna system can work well in operating frequency band.

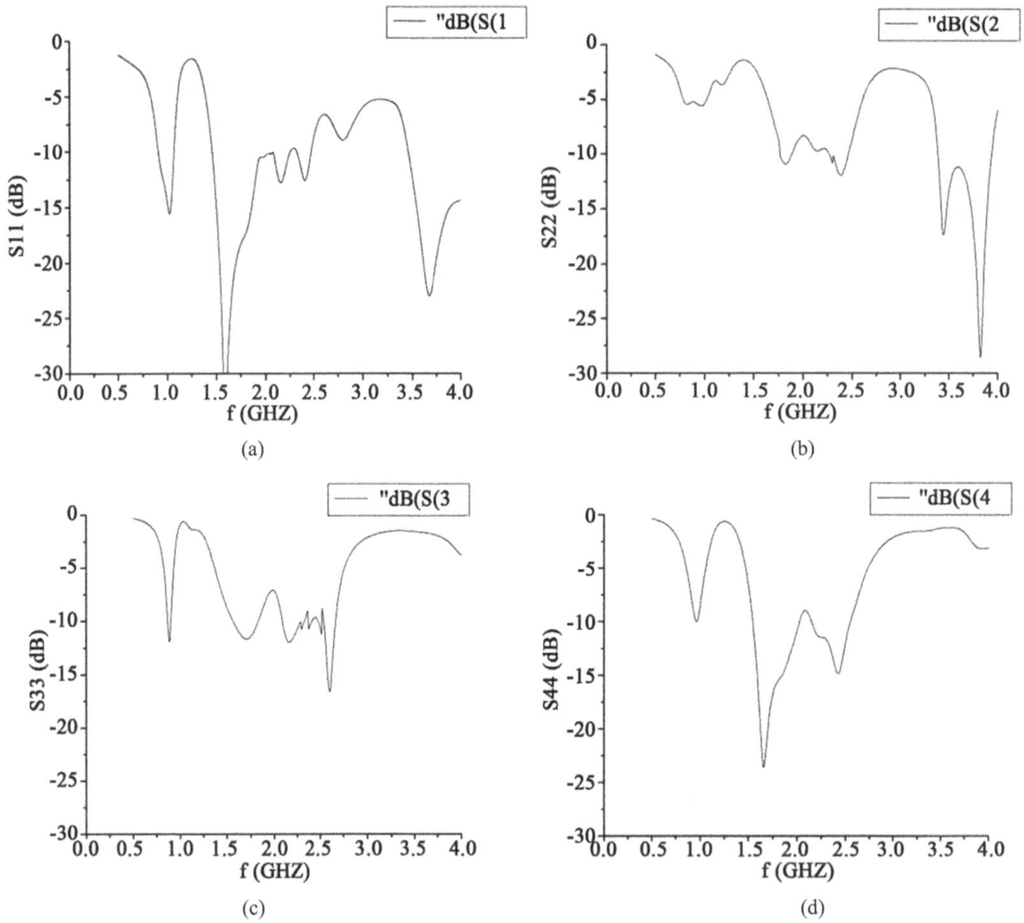

Figure 5. Antenna return loss.

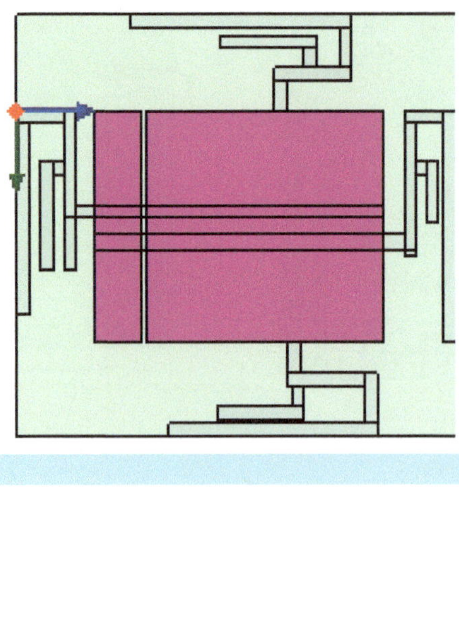

Figure 6. Antenna structure.

5. Conclusion

In this paper, Section 1 mainly introduces the present status of research on MIMO antenna systems at home and abroad. In Section 2, an algorithm calculating envelope correlation coefficient between MIMO antenna elements is proposed. The third chapter and the fourth chapter designed a PIFA-MIMO antennas system and a monopole MIMO antennas system respectively, which were all widely used in the design of antennas system in mobile phone. And based on the formula calculating ρ_e (envelope correlation coefficient) in Section 2, the ρ_e between antenna elements were also calculated in Section 3 and Section 4. The calculation results show that the two antennas systems can meet the performance requirements well.

References

[1] Sun, X.Y. and Xu, P.F. (2015) Analysis of Channel Capacity of Wireless MIMO Systems. *Computer and Telecom*, **7**, 35-38.

[2] Lee, W.-W. and Rhee, B.-H. (2013) Characterization of Performance of a Mobile MIMO Antenna in Free Space. *IEEE Antennas and Wireless Propagation Letters*, **12**, 1153-1156. http://dx.doi.org/10.1109/LAWP.2013.2280465

[3] Tang, T.-C. and Lin, K.-H. (2014) An Ultra-Wideband MIMO Antenna with Dual Band-Notched Function. *IEEE Antennas and Wireless Propagation Letters*, **13**, 1076-1079. http://dx.doi.org/10.1109/LAWP.2014.2329496

[4] Shoaib, S., Shoaib, I., Shoaib, N., Chen, X.D. and Parini, C.G. (2014) Design and Performance Study of a Dual-Element Multiband Printed Monopole Antenna Array for MIMO Terminals. *IEEE Antennas and Wireless Propagation Letters*, **13**, 329-332. http://dx.doi.org/10.1109/LAWP.2014.2305798

[5] Wang, F.Y., Liang, R.X., Gong, S.X., *et al.* (2013) Defected Ground Structure Applied to the Dual-Band MIMO Antennas Systems. *Microwave Journal*, **1**, 25-29.

[6] Prasanna, K.M. and Behera, S.K. (2013) Compact Two-Port UWB MIMO Antenna System with High Isolation Using a Fork-Shaped Structure. *IEEE* 2013 *International Conference on Communications and Signal Processing* (*ICCSP*), Melmaruvathur, 3-5 April 2013, 726-729. http://dx.doi.org/10.1109/iccsp.2013.6577151

[7] Ren, J., Hu, W., Yin, Y.Z. and Fan, R. (2014) Compact Printed MIMO Antenna for UWB Applications. *IEEE Antennas and Wireless Propagation Letters*, **13**, 1517-1520. http://dx.doi.org/10.1109/LAWP.2014.2343454

[8] Kiem, N.K., Phuong, H.N.B., Hieu, Q.N. and Chien, D.N. (2013) A Compact Printed 4 × 4 MIMO-UWB Antenna with WLAN Band Rejection. 2013 *IEEE Antennas and Propagation Society International Symposium* (*APSURSI*), Orlando, 7-13 July 2013, 2245-2246. http://dx.doi.org/10.1109/APS.2013.6711781

[9] Khan, M.S., Capobianco, A.D., Asif, S., Iftikhar, A., Ijaz, B. and Braaten, B.D. (2015) Compact 4 × 4 UWB-MIMO Antenna with WLAN Band Rejected Operation. *Electronics Letters*, **51**, 1048-1050. http://dx.doi.org/10.1049/el.2015.1252

[10] Li, L. (2012) Research on Correlation Matrix and Correlation Coefficient of Receiving Signal in Antenna. http://www.docin.com/p-347946681.html

29

Universal Tracing Interface for Multicore Processors

Janne Vatjus-Anttila, Mika Hoppari, Lance Fono, Kari Kolehmainen, Subayal Khan

Communication Systems, VTT Technical Research Centre of Finland, Oulu, Finland
Email: janne.vatjus-anttila@vtt.fi, mika.hoppari@vtt.fi, lance.fono@vtt.fi, kari.kolehmainen@vtt.fi, subayal.khan@vtt.fi

Abstract

Application developers of today need to produce code which is error-free, and whose performance is optimized for plethora of devices. Performance of application code is studied e.g. by analyzing performance data obtained by executing application with tracing tool. Developers typically have their favorite tools which they prefer to use but unfortunately target devices are based on different computing platforms that have different performance probes which cause difficulties for using same tool with different multicore platforms. Universal Tracing Interface for Multicore Processors (UTIMP) aims to provide an unchangeable tracing interface enabling developers to perform required tracing tasks with the UTIMP, utilizing the favorite tool when possible, for different multicore platforms.

Keywords

Tracing, Performance, Platform, Probe, Toolchain, Code Generation

1. Introduction

Performance tracing is a critical task in nowadays software development. Current and future systems are so complex [1] that it's practically impossible for developers to take all factors into account and write an optimal application at once. Instead, it's typical that developers incrementally optimize their code based on performance testing results. Thus, it's important that developers have good tools for obtaining performance traces from the execution of their application. Luckily, there are a lot of tracing tools that are able to use performance probes like platform hardware timers to obtain performance data without interfering tracing overheads. For example, Linux kernel provides drivers for Performance Monitoring Units (PMU) [2] of many processor families, *i.e.* an interface to the processor hardware performance counters.

Unfortunately, the set of available performance hardware counters and software counters is not same in all multicore processors, meaning that probes differ between platforms and thus it's not possible to use the same tools in different platforms in similar way without a tool update. And when crucial performance counter is missing update potentially can't be based on HW counter support!

Open source tools, e.g. perf [3], give the possibility for making required corrections or adding missing support to tool sources but completing such an updating task correctly may be easier said than done. Altogether the original performance optimization task doesn't really proceed when resources need to be focused on tracing tool updating or learning an alternative tool(s) which naturally extend(s) application time to market and increase(s) developing expenses.

Especially commercial tools [4] can be even more problematic to use on various platforms because source code tends to be unavailable. They either work or don't work on different multicore platforms and acquiring support from the tool vendor may be the only way to get the tool working on the desired platform. Such support is often expensive as is purchasing licences of other tools which work better on the targeted platform. Due to these problems of various performance tracing and profiling tools, there is a serious call for a universal tracing interface that can be used via handy UI/client and exploit the most suitable probe seamlessly.

Ongoing UTIMP (Universal Tracing Interface for Multicore Processors) development aims to provide a solution to the problem by providing a universal interface for performance tools. It abstracts away the underlying probes and tools and provides an unchanging service interface. Thus, all tools that are modified to support the UTIMP interface can be used via UTIMP client(s) and when developer needs to test his application in a new platform he doesn't need to change the performance tracing tool which automatically picks the most suitable tool, and probe from a set of applicable tools to requested tracing task.

The rest of the paper is structured as follows. Related work is described in Section 2. Then in Section 3, a description of the UTIMP prototype and its further development plans are presented. In Section 4, example use-case is observed. Conclusions are briefly discussed in final section.

2. Related Work

There are many existing tracing interfaces and a number of specific profiling toolsets but only few of them seem to provide interface for easy integration. Some of the tools are meant to be used in limited group of computing platforms and operating systems or they support only one or a few tracing types e.g. kernel and/or user events tracing. However, main thing is that performance tracing tools for specific and rather versatile usages are available for multicores and distributed cyber-physical systems which are in theory possible to integrate with UTIMP. Some interesting work done by others is observed in next paragraphs.

StatsD is a network daemon that aggregates and summarizes application metrics. Using StatsD requires developers to instrument their own code with language-specific StatsD libraries. The libraries communicate with a StatsD daemon via UDP datagram. StatsD provides the following very simple protocol definition:

<metricname> : <value> | <type>

The StatsD daemon listens to a predefined network port and it prepares each entry for aggregation. The simplicity of the protocol and the standard network interface that StatsD exposes makes it an easy to use, platform independent solution for performance monitoring. Setup of the daemon however is platform specific and additional libraries may need to be independently installed before it works properly on the target system. Furthermore, the trace data itself needs to be collected through the developer's instrumented code or by using a separate tool. Another drawback is that the daemon does not buffer data but does allow sampling events if there is high throughput and this could lead to potentially heavy traffic with large datasets [5] [6].

CollectD is a daemon that collects performance monitoring statistics over a set period of time and allows the developer to store the collected data in a number of different ways. It is written entirely in the C programming language in order to maximise its performance and portability across different platforms. Furthermore, the CollectD daemon is able to be run standalone without requiring extra scripting languages or cron job setups. There are over 90 different CollectD plugins that can be used to source different kinds of system statistics and metrics, making it a very versatile solution. In order to get visualisations with CollectD however, the developer is required to use an independent tool and while there are several front end options to do this [7], these are not officially supported by CollectD.

The open source data collection engine Logstash claims to be able to unify arbitrary data from disparate sources and normalise such data into the developer's chosen destination. Originally built to collect and analyse log files, Logstash's capabilities have grown to include a number of input, filter and output plugins, ensuring that the developer has a vast selection of monitoring opportunities.

One of Logstash's greatest assets is its ability to handle a very wide range of known tracing formats (web logs, networking logs, firewall logs, syslogs and Windows events logs to name a few) while it can also handle arbitrarily formatted data. Furthermore, it can be used to collect metrics from 3rd party tracing tools (including CollectD) and application platforms over TCP and UDP.

While Logstash can parse, prepare and deploy a wide variety of traces into the developer's desired format, it still must be routed to a 3rd party tool for analysis, monitoring and visualisation. There are however, a variety of both proprietary and open source services that utilise Logstash including analysis tools such as Elasticsearch, MongoDB and Riak as well as monitoring monitoring tools like Nagios, Ganglia, Zabbix, Graphite, Datadog and Cloud Watch [8].

The StatsD, CollectD and Logstash provide interface or plugin which resembles the idea of universal interface of UTIMP. The UTIMP aims to provide as universal interface as possible for tools and probes which can be used with several different multicore platforms and for several tracing purposes like application and kernel tracing and to obtain several sorts of tracing data e.g. CPU usage, memory usage, network usage, function call traces etc.

3. UTIMP

3.1. Overview

According our master plan, UTIMP offers a generic interface for one or more clients, such as performance logging and visualisation tools, to monitor the activities of a multicore system. The UTIMP collects the tracing data of the individual cores, pre-processes it, and makes it accessible for the clients. Status of UTIMP is best described with, work in progress, which means that prototype of UTIMP is workable but not yet finalised tool and therefore description of the universal tracing interface is not included in details. The current UTIMP architecture is presented in **Figure 1**.

UTIMP is planned to reside in the multicore processor on the main controller or on one of the cores. Alternatively it could run on the controlling host of the multicore system as a whole or reside partially in the target and

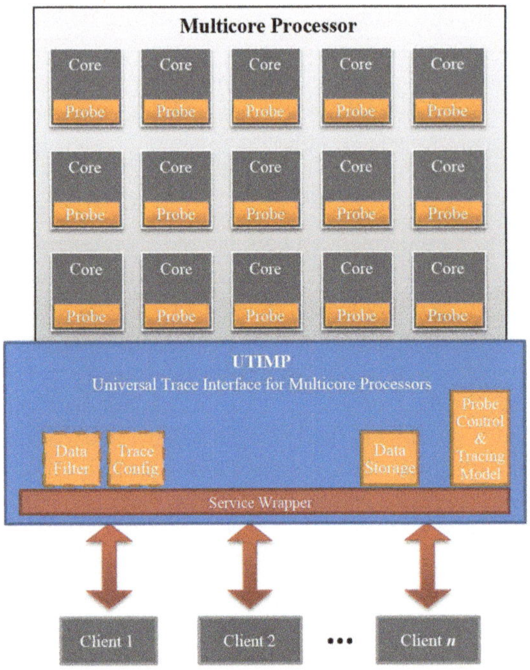

Figure 1. UTIMP architecture.

partially the on host or in a cloud providing the service. Exploiting these services happens via separate clients which communicate via sockets with UTIMP server. The UTIMP test client GUI, which enables basic functionality for trace generation *i.e.* connect to UTIMP, probe selecting, probe configuration, start tracing, stop tracing, show trace and log of recent tracing events is presented in **Figure 2**.

The UTIMP functionality is and will be further divided into submodules and components. In different modules the trace data can be filtered, stored, combined, trigger events, and linked to the application state. In addition to trace data related modules there is need for modules which manage parameters of the core specific probes etc.

Components are planned to avoid any unexpected usage of resources, and any undescribed and varying behaviour. Side-effects like CPU cycles, interrupts or cache effects, or memory usage cannot be avoided but are minimized and defined accurately.

3.2. Modules

The UTIMP consists of several different modules that are each responsible for one core function. Every module is self-monitoring their status, such as memory allocations and changing their behaviour to prevent overhead. At the moment UTIMP contains three kinds of modules: Probe control, Data Storage, and Probe modules. Ongoing development provides more modules enabling new features, for example filtering and configuring functionalities are being sketched and tested. Communication between modules is built on the MPI interface [9]. This enables flexible communication between modules also in heterogeneous multicores and gives the possibility for distributing modules among target, cloud and host nodes.

The Probe Control Module is a generic interface that enables usage of the probes to the rest of the UTIMP components. It allows dynamically configuring parameters of preconfigured individual probes, e.g. sampling frequency or traces to be published. Pre-configuration can be passed via the GUI client or as a text file. In addition, the controller is responsible for dynamically adapting the UTIMP resource usage to minimize the caused overhead. Thus, all other modules report their resource usage to the control module.

The Data Storage Module takes care of UTIMP memory usage. It uses internal logic for storing the data either in memory or disk or for uploading it across network. It also takes care about storing the monitoring data reported by probes. The Probes are planned not to store large amounts of data by themselves but instead by sending data to the Data Storage Module which is aware of permitted and desired behaviour and resource usage. The Data Storage Module is responsible for not overloading the rest of the system.

The Probe Module is present in each core of the multicore processor to capture the traces of interest like memory usage. They are implemented as light-weight as possible to minimise the overhead. Probes expose a generic probe interface for the usage of other UTIMP modules. Probes measure targets and deliver results with associated metadata to other modules for further handling. Probes may only do short-term buffering, sufficient

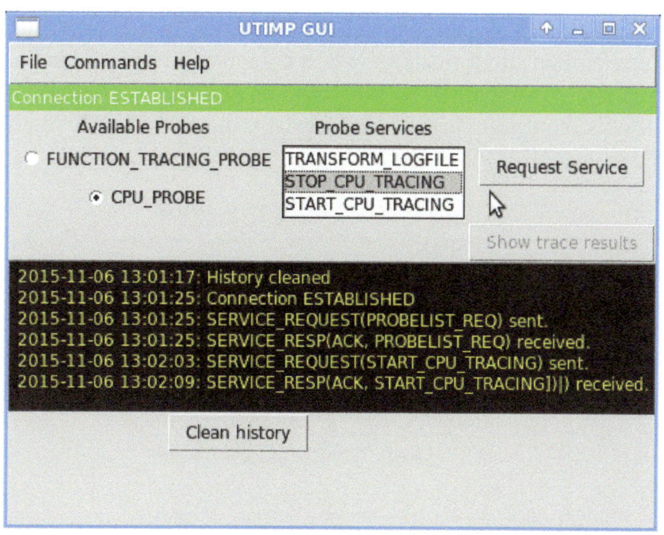

Figure 2. UTIMP client screenshot taken from CPU tracing process.

to guarantee resource sharing and data harvesting un-synchronization and not losing data.

At the time of writing several probes have already been implemented and tested. There are probes for tracing and monitoring CPU load, function calls, and different kinds of user and kernel space events.

The Function Tracing Probe is based on the GNU C/C++ [10] compiler option-finstrument-functions, which makes it useful in any target containing a gcc or gpp compiler. When the flag is used, the compiler takes care of instrumentation and UTIMP records the function trace during the target application execution.

The CPU Probe monitors the load of all CPU cores by using the mpstat tool [11]. Event monitoring probes are monitoring user and kernel space events by using the Performance Application Programming Interface (PAPI) [12] and the Linux Trace Toolkit next generation (LTTng) interface [13]. The PAPI probe tracks hardware events that occur in User Space while LTTNG probe enables also the Linux kernel and libraries tracing.

Probes send trace data to Data Storage Module which stores the traces as text files. In addition to the tracing data itself trace files include also information of the computing node's resources, e.g. CPU properties and memory properties.

Figure 3 shows an example of a traced event; single function excerpt of a function trace. Function tracing data is written for each function call. It contains the function name and all information fields selected in the trace configuration. Fields have self-explanatory names that make them human readable and this is handy during development phase. They provide the required data for various parsing algorithms. For now, light function duration and call count derivative parsing algorithms have been exploited, e.g. max duration, total duration, max call count etc. Possible trace fields missing from the excerpt are Exit_time, File_name and Event_type.

3.3. Universal Interface

Current textual UTIMP trace format defines the other end of the interface. Performance extraction tools tend to have their own output formats for the data they provide which does not suit to UTIMP trace fields automatically. In the course of the UTIMP prototype development, different ways were experimented for producing UTIMP traces from different performance and tracing tools.

Most of the performance tools generate output data to files or standard output in textual format which gives the possibility to parse interesting items and write them directly to UTIMP trace format. Linux command line tools like grep and sed for instance provide handy ways for this sort of interface which can be done for just about any textual output format. This is fast way to test and doesn't even necessarily require the performance tool source code editing. For instance early version of CPU probe was implemented with this method. However, this method is not efficient alternative because the interesting performance data is formatted at least in the performance tool and command line and possibly also in UTIMP if parsed data propagates through it to trace file.

Another considered method was intermediate data format. There are well known file formats like xml, ini, json, csv etc. which could be used to dive the interface in two parts. UTIMP would only have one implementation per file format so it wouldn't have to be tuned each time new tool is integrated. Some of the performance tools already output these familiar file formats so in best case integrating wouldn't require work at all to the tool end nor use of command line parsing tools. So this could be universal for some performance tools but not universal interface for all tools. And intermediate file format requires that data formatting is done in writer end reader ends so it is not the most efficient method.

```
{
 "CoreID": "2",
 "Function_name": "DecoderEngine",
 "Entry_time": "15:33:23.581701",
 "Duration": "163123598ums",
 "ThreadID": "140611729835872",
 "Parent_name": "main",
 "Function_Address": "0x411b40"
},
```

Figure 3. Function trace excerpt which presents single log of Decoder Engine function call.

The universal interface to which we have concluded does not rely on command line tools or intermediate file formats although both methods have good features. Our universal interface is UTIMP API which provides C/C++ function interface through which performance data is passed directly from performance tool to UTIMP without unnecessary formatting that produces the inefficiencies and extra overhead. The downside of the method is that source code of the integrated performance tool is necessary in order to build the most efficient performance probe. Development of the API is in progress and while it is still partly un-mature its functionality is not described further. Early version of it has been tested with the Function Tracing Probe which differs from the basic use of API as source code instrumentation with API functions is not needed. Instead new functions, __cyg_profile_func_enter and __cyg_profile_func_exit, for entry and exit function calls generated by the compiler option are instrumented with UTIMP API function calls.

3.4. Further Development

The current version of UTIMP provides basic functionality of a generic tracing interface and a couple of tools and probes have been integrated into it. However, development still continues and new features will be added to future versions.

As presented in **Figure 4**, a lot of new modules are planned and under design in upcoming versions. Those modules will enable new features and services for clients. New features will enable handy trace filtering, remote access and control, context linking, event management, syncing, and security handling. Further development includes also the addition of probes in order to increase tracing support for a larger set of computing platforms.

In addition, existing modules and probes will get upgrades as thus far they have not been optimised. For example, a probing algorithm will be added to analyse the overheads caused by each tool, to see if there are certain trace services that are inefficient and may need optimizing. E.g. the current compiler based function trace implementation overhead can disturb functionality of timing critical applications. The result of the proposed algorithm is used in the control module to adjust the responsible probe.

In the long term, one way to control side-effects and overheads is through the installation location and flexibility of UTIMP components. For instance, a resource restricted target may prefer a minimal UTIMP installation. Another alternative is to configure UTIMP to minimize certain kinds of overhead. For example if network usage overhead must be minimised, an alternative storing configuration is selected and the generated trace is stored temporarily in the target device and passed to clients after tracing is complete.

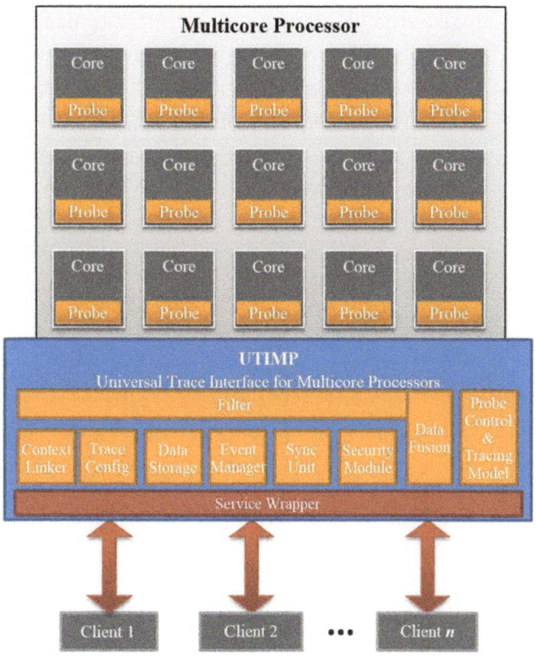

Figure 4. In future a lot of new functionality will be added as new modules.

Further development of remote features is especially in our interests and a cornerstone of our development vision. Client-server socket communication, which is part of implemented features, offers the first step towards remote tracing and comprises a basis for distributed performance monitoring system which is a definitive requirement due present remote tracing needs [14].

Eventually we plan to explore yet unexploited possibilities of distributed performance evaluation by implementing the UTIMP as a distributed software system consisting of a device controller (subsystem) consisting of probes for extracting performance data and a server sub-system (cloud based) for visualization. According to this vision, UTIMP can evolve to serve performance evaluation needs of current local and distributed multicore processor systems as well as boost technologies required in the forthcoming and already appearing IoT and cloud era. **Figure 5** highlights the architecture of the envisioned distributed UTIMP system.

Another tempting framework for new client variants are integrated development environments (IDE) which provide appropriate functionality for productive software development, debugging as well as optimization according visual profiling and tracing results. Integrating the IDE to some wrapper providing socket connection to a remote UTIMP can be implemented as a plugin for IDEs providing the plugin feature. For example, such a plugin feature for Android Studio which is currently under development, allows the IDE to receive trace data from Android devices using modified UTIMP probes [15].

4. Use Case Example

In this use case UTIMP is used to generate a function trace out of an image processing target application, Bit plane encoder/decoder [16], as part of Legacy Code Tool chain (LCTC) that detects possible bottlenecks from a generated trace, creates a graphical model of detected problem function, generates OpenMP parallelised code from a refined model to improve performance of the target application and enables testing, including execution measurement feedback as SOAP [17] messages to the model, of the parallelised application. The work flow of LCTC is visualised in **Figure 6**. In the figure, modelling related phases are highlighted.

Parallelisation is based on OpenMP 3.0 [18] and in particular to the exploitation of OpenMP tasks. An Overall sketch of LCTC prototyped in the PAPP [19] project is depicted in **Figure 7**. LCTC was used in an Ubuntu 10.04 Virtual Box image on a Windows 7 host machine.

LCTC consists of a few tools which include UTIMP, Utimp Parser and M-Net [20] domain specific language (DSL). DSL is developed and applied with the Meta Edit+ [21] domain specific modelling tool *i.e.* Meta Edit+ is part of LCTC also.

The purpose of LCTC is to enable effective target application optimization. To enhance the overall optimization cycle some work phases were automated, e.g. graphical function models are automatically generated and require just minor layout refining from the user. M-Net out of the box functionality did not include such an import generator so M-Net was extended with one among other extensions which were mainly needed for tool

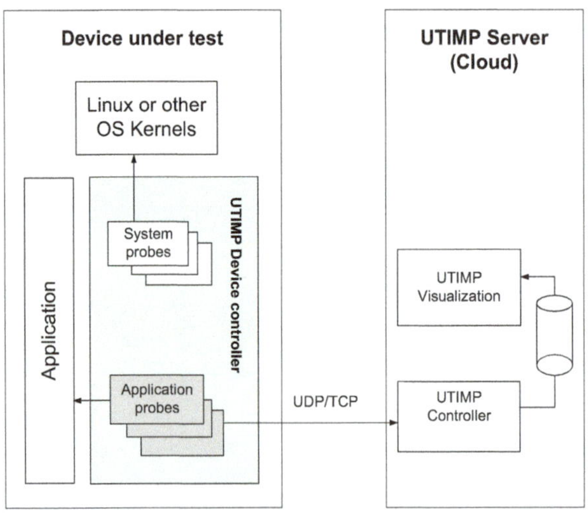

Figure 5. A high-level representation of the main components of envisioned UTIMP architecture.

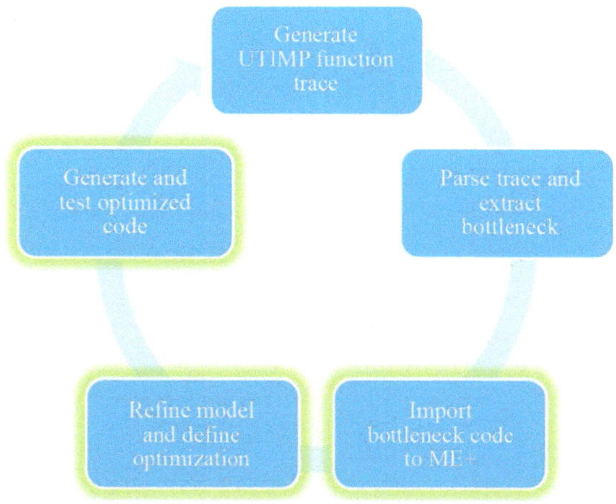

Figure 6. Optimization work flow.

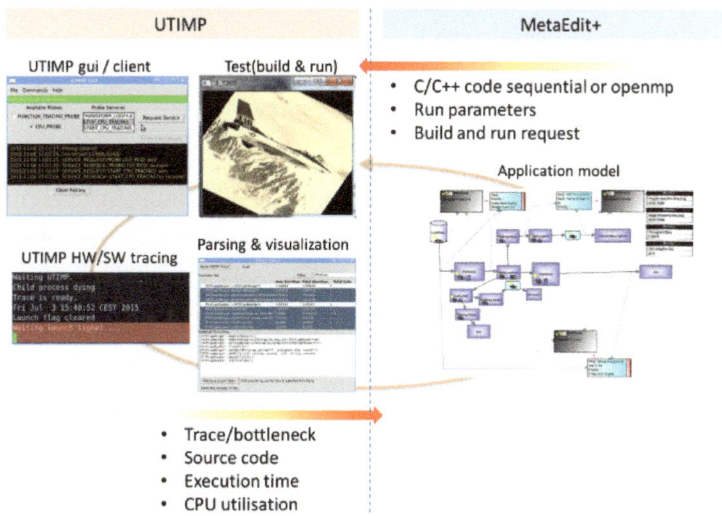

Figure 7. UTIMP integration to legacy code toolchain.

chain integration. In the end, all phases starting from building a tracing version of the target application to the testing of parallelised target application can be done via DSL generators.

Steps performed in the use case were as follows. The UTIMP configuration file was defined in the M-Net and passed to UTIMP and a tracing version of the image processing target application decoding case was started. A UTIMP function trace, resulted from executed decoding case, was then parsed with Utimp Parser and the most time consuming functions were sought. A screen capture from parser tool is presented in **Figure 8**.

From the parsing results we first picked the Decoder Engine function which obviously was a decoding top level function. Our interest was limited to parallelizing loops and the picked function as well as its sub functions included plenty of those but were too complex for the code generator. Interest was refocused to the function from which Decoder Engine function was called and it was imported to Meta Edit+ with M-Net DSL generator.

After automatized import phase, a task needs to be selected for parallelisation. It is done with the Optimizer object by placing it under selected task. The graphical M-Net task model highlighted a "while loop" task as a suitable OpenMP optimization target. The Optimizer object was used to optimize the highlighted task, *i.e.* while loop which included the Decoder Engine function call. **Figure 9** shows a clip of the task graph after placing the Optimiser object and Measurement object used for timing over the "while loop" task.

With the Optimizer object, the level of parallelism and use of OpenMP pragmas and clauses were defined. The Measurement object enabled extracting the "while loop" execution time according to which OpenMP variants were evaluated. The call count and iteration count can also be measured with the Measurement object but they weren't measured in this case.

	Max Duration
DecoderEngine	161112283
ACBpeDecoding	51458091
StagesDeCoding	5568326
DecodingOutput	5166148

Figure 8. UtimpParser parsing result extracted from decoding case function trace file.

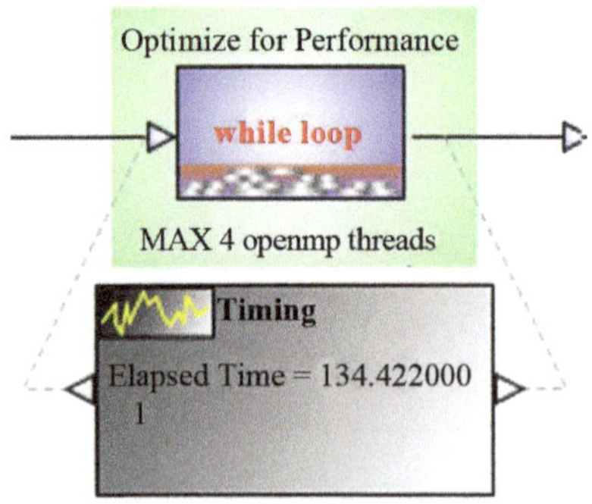

Figure 9. Model refining with optimiser and measurement objects.

When refining of the graphical model was done, a Code generator was utilized. The "while loop" excerpt of the parallelised code generated from the instrumented task graph is presented in **Figure 10**. The rest of the code instrumentation, not included in the excerpt, enables timing of the while loop, feedback *i.e.* passing timing values to the Measurement object and inclusion of the necessary header files.

The parallel code was generated and tested via the M-Net generators which updated the source code with parallelised code and started application build- and run scripts. The code was runnable and different versions of the code were generated and tested. Timing data feedback to the M-Net task model revealed the best code version and level of parallelisation that gave best speedup in comparison to sequential code variant.

The sequential version of the target application, reads in several image files in a while loop, and processes them one by one. Our parallelised version changed this sequential processing to parallel processing. After LCTC optimization, a set of tests with 20 input images were made to collect results for a final comparison. Input images included 8 MB and 1 MB images, 10 of both sizes. Images were first encoded and then decoded back. Images resulting from the decoding were equal with original input images which validated the functionality of each tested OpenMP version. Execution times of the parallelised versions as well as the sequential version were measured for the decode direction. Each measurement was repeated 5 times and averages were calculated to increase reliability of measurement results. The performance gain reached with the use of OpenMP in decoding case can be observed from **Table 1**.

The best speedup in comparison to the sequential case was reached with a two thread variant of the while loop, *i.e.* speedup was 48.09 s - 32.62 s = 15.47 s (~ ×1.5 performance) which equates to about a 32% reduction to the decoding execution time. The generated code provided some speedup but 3 and 4 thread variants could not exploit fully the quad core laptop resources, which in this case was likely to be caused by a problem in the virtual machine configuration. Even so, presented use case gave an example of applying UTIMP to application performance optimization as part of LCTC.

5. Conclusions

In this paper, UTIMP, a universal tracing interface for multicore processors has been presented. It aims to provide a universal interface which allows integration of existing tracing and profiling tools and probes to UTIMP thus providing performance evaluation services that developers can exploit in different platforms via a single, possibly browser based client. This way it makes developers work easier because they do not need to study new tools or add missing support to their favourite existing tools when testing performance of the developed code in different platforms.

```
omp_set_num_threads(4);
#pragma omp parallel
{
#pragma omp single nowait
{
    StructCodingPara *pCod;
    while ((pCod = getImageCodParams(f, fEncoding)) != NULL)
    {
#pragma omp task untied
{
        if (fEncoding)
            EncoderEngine(pCod);
        else
            DecoderEngine(pCod);

        free(pCod->Bits);
        free(pCod->PtrHeader);
free(pCod);
}
    }
}
#pragma omp taskwait
}
```

Figure 10. Parallelized while loop.

Table 1. Application execution times and CPU loads with different levels of parallelism.

Measurement	Code variants				
	Sequential	OMP 1 thread	OMP 2 threads	OMP 3 threads	OMP 4 threads
Decode time	48.09 s	49.71 s	32.62 s	34.19 s	43.46 s
CPU load	98%	98%	182%	226%	230%

In its current state, UTIMP already provides a basic interface for tracing to which a couple of tools have been integrated, e.g. a working function call tracing service can be used to produce informative traces. The use case example showed that the UTIMP prototype can be used to collect traces and parse out time-consuming code which needs optimization. Further development of UTIMP is under work and applying it to largely used linux/unix variants like Android and iOS has already begun.

Acknowledgements

This work is supported by the European Commission, Tekes—the Finnish Funding Agency for Technology and Innovation, and VTT under the grant agreement ARTEMIS-2012-1-295440 PAPP.

References

[1] Increased Complexity in Embedded Software Systems (2014). http://www.ni.com/white-paper/52165/en/

[2] Andrews, J. (2015) System Performance Analysis and the ARM Performance Monitor Unit (PMU). https://community.arm.com/groups/soc-implementation/blog/2015/02/19/system-performance-analysis-and-the-arm-performance-monitor-unit-pmu

[3] Linux Kernel Profiling with Perf (2015). https://perf.wiki.kernel.org/index.php/Tutorial

[4] List of Performance Analysis Tools (2015). https://en.wikipedia.org/wiki/List_of_performance_analysis_tools

[5] StatsD—README.md. (2015). https://github.com/etsy/statsd/blob/master/README.md

[6] Jackson, K. (2012) Open Stack Cloud Computing Cookbook. Packt Publishing Ltd., Olton, Birmingham.

[7] List of Front-Ends (2015). https://collectd.org/wiki/index.php/List_of_front-ends

[8] Logstash Introduction (2015). https://www.elastic.co/guide/en/logstash/current/introduction.html

[9] Squyres, J. (2011) Why MPI? Cisco Blog/High Performance Computing Networking 2011. http://blogs.cisco.com/performance/why-mpi

[10] GNU gcc Manual (2015). https://gcc.gnu.org/onlinedocs/gcc/

[11] MPSTAT Manual Page. http://sebastien.godard.pagesperso-orange.fr/man_mpstat.html

[12] Browne, S., Dongarra, J., Garner, N., Ho, G. and Mucci, P. (2000) A Portable Programming Interface for Performance Evaluation on Modern Processors. *International Journal of High-Performance Computing Applications*, **14**, 189-204. http://dx.doi.org/10.1177/109434200001400303

[13] Desnoyers, M. and Dagenais, M.R. (2006) Low Disturbance Embedded System Tracing with Linux Trace Toolkit Next Generation. *Proceedings of the Embedded Linux Conference*, San Jose, 11-12 April 2006, 1-4.

[14] Iosup, A., Ostermann, S., Yigitbasi, N., Prodan, R., Fahringer, T. and Epema, D. (2010) Performance Analysis of Cloud Computing Services for Many-Tasks Scientific Computing. IEEE TPDS, Many-Task Computing, November 2010.

[15] Creating Plugins for Android Studio. http://yvonne.idescout.com/2015/08/writing-plugins-for-android-studio.html

[16] CCSDS 122.0-B-1 Recommended Standard. http://public.ccsds.org/publications/archive/122x0b1c3.pdf

[17] The gSOAP Toolkit for SOAP and REST Web Services and XML-Based Applications. http://www.cs.fsu.edu/~engelen/soap.html

[18] Version 3.0 Complete Specifications (2008). http://openmp.org/wp/openmp-specifications/

[19] Portable and Predictable Performance on Heterogeneous Embedded Manycores. http://www.papp-project.eu/

[20] Yrjönen, A. and Merilinna, J. (2009) Extending the NFR Framework with Measurable Non-Functional Requirements. *Proceedings of the 2nd International Workshop on Non-Functional System Properties in Domain Specific Modeling Languages*, Denver, 4 October 2009. http://sunsite.informatik.rwth-aachen.de/Publications/CEUR-WS/Vol-553/

[21] White Paper, "Domain-Specific Modeling with Metaedit+: 10 Times Faster Than UML". https://www.metacase.com/papers/

30

Energy-Efficient Content-Based Routing in Internet of Things

Samia Allaoua Chelloug

Networks and Communication Systems Department, College of Computer and Information Sciences, Princess Nourah Bint Abdul Rahman University, Riyadh, KSA

Email: SAChelloug@pnu.edu.sa

Abstract

The convergence of the Internet, sensor networks, and Radio Frequency Identification (RFID) systems has ushered to the concept of Internet of Things (IoT) which is capable of connecting daily things, making them smart through sensing, reasoning, and cooperating with other things. Further, RFID technology enables tracking of an object and assigning it a unique ID. IoT has the potential for a wide range of applications relating to healthcare, environment, transportation, cities... Moreover, the middleware is a basic component in the IoT architecture. It handles heterogeneity issues among IoT devices and provides a common framework for communication. More recently, the interest has focusing on developing publish/subscribe middleware systems for the IoT to allow asynchronous communication between the IoT devices. The scope of our paper is to study routing protocols for publish/subscribe schemes that include content and context-based routing. We propose an Energy-Efficient Content-Based Routing (EECBR) protocol for the IoT that minimizes the energy consumption. The proposed algorithm makes use of a virtual topology that is constructed in a centralized manner and then routes the events from the publishers to the intended interested subscribers in a distributed manner. EECBR has been simulated using Omnet++. The simulation results show that EECBR has a significant performance in term of the energy variance compared to the other schemes.

Keywords

Internet of Things, Publish/Subscribe, Routing, Content-Based Routing, Energy, Virtual Topology, Omnet++

1. Introduction

IoT is typically composed of smart objects that are equipped by sensors and RFID systems [1] such that the connectivity requirements between the IoT smart objects are achieved over advanced communication standards like Wifi, Wimax, LTE, and cloud computing [2] [3]. More precisely, the IoT sensor nodes are provided with one or more integrated sensors with limited computing, memory and power capabilities. They have also the ability to communicate in a short-range distance [4]. On another hand, each RFID includes a reader to uniquely identify tagged objects or people [5]. After their deployment, IoT devices are expected to be intelligent by handling, transmitting, and reasoning about the occurred event. Consequently, the event manager is a core element of the IoT architecture that may successfully integrate a publish/subscribe middleware system which is a software that coordinates the communication between heterogeneous devices [1] [6]. Multiple protocols have emerged to address the routing problem in IoT. Most of them extend IPV6 [7] and rely on IEEE 802.15.4 [8] standard that is suitable for tiny sensor networks allowing their radio components to alter between the active and idle state according the medium's conditions. The literature on routing protocols for IoT systems depicts that a large number of protocols implement clustering techniques to route efficiently data across the IoT devices. A more frequently issue in IoT concerns events' routing that should be delivered at real-time to accomplish the requirements of some applications. We mention that event-based routing protocols have already been applied to distributed systems. The idea of event-based routing is to allow a set of devices to specify interests in certain event conditions. If an event has been trigged, the subscribers are interested to that event will be notified. The taxonomy of event-based routing protocols includes content and context based routing [9]. Our idea in this paper is to assume an IoT system that is provided by a publish/subscribe middleware and to propose a protocol for routing events in such environment. Our protocol is simple and uses the concept of content-based routing which is carried in a distributed manner via two simple rules. The fundamental characteristic of energy-efficiency of our protocol has been achieved through the construction of a virtual topology [10] that balances the energy load between IoT devices due to the self-organization property of the obtained hierarchy. This paper is organized as follows: Section 2 presents basic concepts behind IoT. Section 3 surveys IoT routing protocols. Section 4 presents potential event-based routing protocols that are addressed in distributed systems. Section 5 is concerned with the assumptions of EECBR. It details also the steps that are required for constructing the virtual topology and the rules that are used for distributed content-based routing. Then, Section 6 illustrates our simulation results. Finally, Section 7 concludes the paper.

2. IoT Concepts

IoT has been attracting much interest of researchers in the last few years. This is so as it enables a set of things/objects to be:

- Pervasive: IoT devices are able to interact with the environment by sensing, and reasoning about data that have been produced. They may be connected through wired or wireless networks to exchange information and interact with the environment for taking the best decision at real-time.
- Identified via a unique address. This desirable characteristic is obtained by IPV6 which grants an expanded addressing space.
- Cooperative with other things by granting them an access to the local information to create new applications or services [11].

In this sense, IoT creates a worldwide network of interconnected objects that should be uniquely addressable. Computers, smartphones, vehicles, homes appliances, cameras are examples of such things/objects. Moreover, IoT interconnects generic objects [1]. For example, smart refrigerators trace and report the availability and expiry dates of food items. They also rely on IoT network to place an order to grocery shops once a certain limit of the supply of food items is reached [12]. The integration of many technologies is the power behind IoT. Some of these technologies help to acquire and process contextual information while others improve security and privacy. Namely, sensor networks and RFID technologies play a major role for IoT systems. More specifically, RFID includes a tag that is equipped with an antenna for object identification. A sensor network based on RFID provides not only the possibility to identify objects but also to track their behavior or measure some parameters that are related to the environment [1] [5]. In addition, cloud computing is also exploited in some IoT systems to create contents and applications for the users. Cloud computing is an emerging model that has been used by industrial professionals and academics to provide a set of services over a public or a private cloud. Such services may refer to a hardware, a software or a system. A cloud computing model may contain a collection of physical data centers and virtualized computers to handle data and computation in a distributed and parallel manner. The idea of virtualization was introduced to provide high-speed computation by dividing a single job into many tasks that may be executed by the virtual machines. Moreover, cloud computing services may be categorized into: Saas, Paas, and Laas. The first service concerns web-based access and storage; while, the second one allows the developers to host web applications. Finally, Saas is a set of applications that are accessible from various client devices [13] [14]. The motivation behind integrating IoT and cloud computing is to enhance an IoT environment by taking benefit from the storage and processing capacities of cloud computing. This helps a lot to avoid sensors' constraints. The cloud computing may also benefit from IoT by providing services to smart objects [15] [16].

IoT applications are numerous. Broadly speaking, IoT would be adopted in [1] [5] [12]:

- Transportation: where advanced cars, trains, roads, and trails would be equipped by sensors and tags and communicate with traffic control sites.
- Smart homes and offices: in this case IoT enables controlling the room heating to our preferences. It can also be used to change the room lightning according to the time and day.
- Cities: the control of a parking and the monitoring of material conditions in buildings and bridges are two examples of IoT exciting applications.
- Environment: the sensors that are integrated to the things may help hugely to detect combustion gases, fire conditions, and soil moisture.
- Healthcare: IoT supports healthcare applications by sensing medical data of the patients and allowing a remote diagnosis. It makes also the life of elder and disabled people easier. IoT may also be needed for controlling medical fridges.
- Energy: the monitoring and management of energy consumption is a practical application of IoT that yields to an intelligent grid which is based on renewable energy resources. It reacts to power fluctuations to execute efficiently the requests and the actions of the end users and devices [11].

A simplified architecture of IoT has two main layers consisting of two sub-layers each. The perception layer includes all technologies that allow perceiving and collecting data. The network layer takes care of transporting data in a transparent manner using the suitable communication standards including Wifi, Wimax, GPRS, and WSN. Rather than using the data management sub-layer of the service layer which treats complex and uncertain data structures, the application service sub-layer handles transforming data into content and providing an interface to user applications [5]. The data management sub-layer is also called the middleware layer and it represents the most critical layer of the architecture.

IoT has a profound effect on standardisation efforts because it is intended to support many diverse applications, protocols, and technologies as mentioned above. Three European bodies namely CEN, CENELEC, and ETSI are playing a vital role in developing IoT standards but the final objective of standardisation is not achieved yet. The standardisation is a bottleneck in the area of IoT for the reason that it requires a formal specification for testing any standard before providing it. We note that the test specification is resource consuming. In fact, the standardisation bodies are focusing on elaborating an interoperable protocol stack that expands HTTP, TCP, IP stack [17] to integrate sensor data within RFID and support new application in collaboration with some working groups of experts. The working groups are responsible for creating large documents, technical specifications and technical reports for any proposed standard.

3. IoT Routing Protocols

Routing is a topic that has attracting the research community in last years and intense works have been devoted to this field. Traditional routing protocols for Ad hoc networks may fall into three categories: reactive, proactive, and hybrid protocols [18] [19]. The first category establishes a route on demand; while, the second one maintains the topology and updates the routing tables periodically. The hybrid protocols perform both reactive and proactive routing to reduce the delay. Later, new protocols for sensor networks have been proposed. Two features are handled in such protocols: data exchange is performed hop by hop and the lifetime of the network is maximized [20]. Routing in IoT systems is highly related to routing in Ad hoc and sensor networks. Energy consumption of sensors, mobility of things, and the type of the IoT's middleware are three primary concerns that may affect routing in IoT. 6LoWPAN (IP6 over power personal area networks) [21] is a major routing protocol for IoT systems. It has been defined by the Engineering Task Force (IETF) to route data through the Internet among non IP sensors. It relies on IEEE 802.15.4 [8] that is suitable for low power sensors. The network topology of 6LoWPAN includes a set of reduced function sensors that are connected to full function sensors. Moreover, a gateway is placed between two different network domains. By another hand, the network stack of 6LoWPAN is composed of seven layers: the IEEE 802.15.4 physical and Mac layers, the adaptation layer, IPV6 [22], the transport, and the application layers. More specifically, the physical layer of IEEE 802.15.4 provides 27 channels that operate in different frequency bands with varying data rates (20 - 250 Kbps). The Mac layer manages the medium via the CSMA/CA protocol [23] and ensures also the device's association, disassociation, and synchronization. The function of the adaptation layer is to adapt IPV6 packets and fit them into the IEEE 802.15.4 format. It also ensures the fragmentation of IPV6 packets into Mac frames. The transport layer of the network stack adopts a simple transmission protocol which is the User Datagram Protocol (UDP) [24].

The routing process in 6LoWPAN starts when a reduced function sensor has to send a packet to another IP sensor. In this case, the higher full function sensor that is connected to the reduced function sensor will be responsible to send the packet hop by hop until it reaches the gateway. Actually, the gateway uses the IP address to locate the domain of the remote IP sensor. Further, 6LoWPAN ad hoc on Demand Distance Vector Routing (Load) has been adopted for 6LoWPAN routing. Like AODV [25], Load uses route request and route reply messages but it does not use the sequence number. It also relies on the Link Layer Notification messages that approve the reception of Mac messages. It creates a mesh topology and runs on full function devices. A route is selected by Load if it includes a certain number of links such that their accumulated route cost (LQI) [26] is worse than a certain threshold. The route should also include less hops between the source and the destination. Hierarchical routing (HiLow) is another protocol that is used in 6LoWPAN to minimize the delay. The idea is to build a hierarchy and then 6LoWPAN device will either join an existing parent or become a parent. According to [27] 6LoWPAN is used for networks with high processing capabilities. For this end, Protocol for Low Power and Lossy Networks (RPL) [28] has been designed for constraint devices in power, computation, and memory capabilities. RPL is a distance vector routing protocol that is based on IPV6. It builds a Destination Oriented Directed Acyclic Graph (DODAG). Many metrics may be used to construct a DODAG: the Expected Number of Transmissions (ETX) [29], the remaining energy of the devices… Energy-Efficient Probabilistic Routing (EEPR) [30] is an alternative solution for routing in an IoT environment. It is based on the same idea of AODV but the transmission of a RREQ packet follows a certain forwarding probability that depends on the residual energy and the ETX metric The proposed mathematical equation shows that the forwarding probability will be high if the ETX metric is low and the residual energy is high. The simulation results show that the variance of the residual energy of EEPR is smaller than the variance of the residual energy of AODV. The work in [31] proposed the Context Awareness in Sea Computing Routing Protocol (CASCR). It exploits the idea of sea computing that was first announced at the Chinese Academy of Science High Technology Planning Seminar in 2010. The sea computing refers to the embodiment of autonomous devices into various things. It produces effective intelligent decisions through local interactions of the devices. CASCR associates a state and some operations to each device. The states include: full-working, serving, single working, sleeping, and hibernating. The set of operations contains: gather information, transmit information, apply information fusion, and generate a control operation. In particular, CASCR predicts the new state of each device using Markov chains [32] that defines the new state as a function of the device's history. The device holding a routing request should send it to its first hop neighbors that are provided with a context data table which specifies the network topology in term of: neighbors, subordinators, superior, colleagues and disabled devices. The context data table is used to find the next hop. The simulation results show that CASCR has low energy consumption. The authors of [33] involved in developing a cross-layer protocol to meet the performance parameters of IoT applications: minimum date delivery rate, and maximum packet delay. The proposed protocol runs on the top of IEEE 802.15.4. It combines the network topology information, the performance requirements and the link performance to decide on the next hop for routing data. It extends the ETX metric by considering the packet loss which may be generated by the MAC contention. The authors of [33] conducted a simulation and demonstrated that the reliability metric is better than the ETX metric. The author of [34] addressed traditional routing protocols for Adhoc networks such as AODV, DSR, and OLSR [35] and discussed their performance in an IoT environment. Three metrics were used:

- Routing overhead: which is the result of the number of route packets by the number of data packets.
- Average end to end delay: concerns the total delay that is generated by route discovery, queuing…
- Throughput: it is the result of the number of packets received by the destination by the number of packets sent by the source.

The paper of [34] contains simulation results that are based on the random waypoint mobility model. It illustrates that the overhead of the three protocols depends on the number of nodes. The percentage of mobile nodes is another parameter of the simulation. However, we mention that the authors did not define their network topology nor the IoT architecture.

4. Event-Based Routing

Publish/subscribe is an efficient way for asynchronous communications. It allows to specify interests in a certain event conditions. If an event has been trigged, the subscribers that are interested to that event will be notified [36]. There are two basic types of publish/subscribe systems: topic and content-based publish/subscribe. In the former scheme, the system's device subscribes to topics, which are identified by keywords; whereas in the latter scheme, the subscription concerns the actual content of the considered event. The drawback of topic-based publish/subscribe concerns the accuracy. Some devices may be notified about events they are not interested in. In return, content-based publish/subscribe has been widely used but it lacks efficiency due to the mapping between subscriptions and events [37]. Our focus in this section is to study the problem of event delivery for publish/subscribe system. Our motivation stems from the architecture of IoT systems that requires a middleware to coordinating between different components of an IoT system and providing abstractions in term of syntax and semantic for a wide range of diverse devices [1] [6]. In reviewing the literature on middleware for IoT systems, we considered that many middleware systems are based on publish/subscribe communications for IoT devices to ensure a real-time tracking of the events [38]-[41]... So, we will present existing event routing protocols for publish/subscribe systems which are classified into namely, filter and multicast-based routing as shown in **Figure 1**. The idea of filter-based routing is to impose a certain forwarding scheme: the sub-server that is aware about an event should only forward it to its sub-servers leading to the subscribers that are interested to a specific event. The concept of multicast-based routing is different because the routing decision is made only once at the publisher. Multicast groups are constructed and the events are broadcasted to the appropriate groups. The author of [42] distinguished three categories for event routing: flooding, selective-based routing that include rendez-vous and filter-based routing, and gossiping based routing that is probabilistic and suitable for dynamic environments.

An interesting recent technique for content-based routing is called Kyra of [42]. It generates a two level topology through the Hierarchical Agglomerative Clustering (HAC) algorithm [43] that groups the servers into

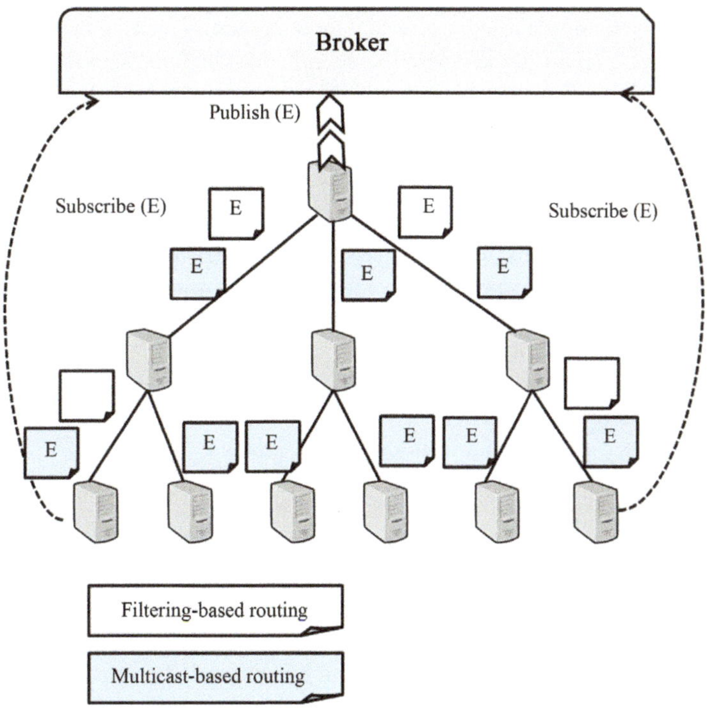

Figure 1. Filter and multicast-based routing.

clusters. Then, Kyra builds a spanning tree [44] that includes clique servers which know about each other. Typically, Kyra filters the received events and forwards them along the tree. The key feature of Kyra is its ability to balance the routing load over the sub-servers. The authors of [45] proposed the Context and Content-Based Routing Protocol (CCBR) for multi-sink mobile wireless sensor networks. It capitalizes on the broadcast feature of wireless networks. Each network node maintains a distance and a content table. The first table keeps track of the estimation of the distance between the considered node and the network's sinks. The content table serves for storing the interests of sinks that are relevant for the considered node. According to CCBR, a node that receives a packet should compare the distance included in that packet's header to the distance that separates itself from the sink. The packet will be dropped if its distance is greater than the packet distance. Otherwise, it will update the distance in the header and schedule the packet for transmission. The results in [45] show that CCBR's delivery rate is good and the routing cost is low. Another technique for event routing has been proposed in [46]. The Topic-based Event Routing (Tera) is a powerful technique for large scale dynamic peer to peer systems where the network nodes can subscribe/unsubscribe at any time. It associates a tag to each event. The tag represents a topic and it should be used to deliver the events to all subscribers that have expressed their interests in the corresponding topic. Tera is based on a distributed algorithm Overlay Maintenance protocol (OMP) [47] and adopts inner and outer cluster routing such that the inner routing is just a broadcasting process. The scalability of Tera was tested in [46] and it demonstrates a good scalability. More importantly, [48] provided a formal description of a content-based routing scheme that includes distributed brokers which are connected to an acyclic topology and each broker handles a private routing table. The suggested protocol in [48] enables a broker to send a notification to its local clients upon receiving a publish message. The authors of [48] described also formally a simple filter-based routing protocol that floods new canceled subscriptions to all other brokers.

5. EECBR

Our proposed protocol aims at providing an efficient routing service for IoT sensors which are responsible for collecting and routing data. Different from routing protocols for wireless sensor networks, EECBR takes advantage of the publish/subscribe middleware to achieve two functions without any prior knowledge of the location of the underlying sensors:

1) Organizing IoT' sensors via a virtual topology that is created according to the set of sensors that are interested to a specific event. It also considers the energy of the sensors.
2) Routing the events hop by hop through the virtual topology until reaching the interested sensors.

Assume the scenario of a smart home where the equipment integrates sensors like: door control sensor, motion sensor, light sensor, humidity sensor, window control sensor… and RFID systems. The scope of such system concerns ensuring the safety and the security of the house. The integrated sensors are connected to the Internet to allow a remote control of the house. In such scenario, the activity of some sensors is related to other ones. For example, the light sensor depends on the motion sensor. Further, some sensors may adjust their power of transmission according on the events provided by the light and the humidity sensors. So, our idea to use publish/subscribe communications stems from the possibility of registering dependent sensors to the interested events. Upon the occurrence of an event, the dependent sensors are notified and the virtual topology is used to relay the events form the publishers to the interested sensors.

Our protocol is based on the following assumptions:

- Our IoT system includes sensors that are integrated to the things.
- Each sensor is aware about its neighbors.
- Each sensor has a battery with an initial random energy amount.
- The energy is decreased for each transmission.
- The topology of the things is ad hoc.
- The things are static.
- The links between the things are reliable with no data loss.
- The IoT system is provided by a publish/subscribe middleware and the role of the network layer is to deliver the published events to their adequate destinations.

EECBR is a simple protocol that tackles the problem of routing IoT events in an energy-efficient manner by balancing the energy consumption of the IoT devices. It first creates a virtual topology. The virtual topology may be a tree or a set of clusters. It depends on the network's topology. Starting from a set of sensors which are interested to a specific event, EECBR marks a sensor as a leader if it covers the maximum set of interested sensors that are located in its communication range. EECBR may add more levels depending on the topology of the network.

Different from many protocols that establish a scheduling and decide on awake/sleep sensors by electing some leaders for monitoring a certain area, our contribution builds a virtual topology to organize sensors and reduce the communication cost. The leaders may ensure two functions: collecting data that is gathered by sensors that are located in the range of the leader, relay events to the sensors that are interested to them. The context of our proposed protocol concerns the second function. EECBR calculates the contribution of each sensor for deciding on the leaders. The contribution of a sensor s depends on: the number of uncovered sensors that are interested to a specific event/uncovered leaders of level (i-1) that are located within the range of s, and the energy of s. a sensor becomes a leader if its contribution if it has the largest contribution. This enables to broadcast an event to the intended receivers. The algorithm for EECBR is illustrated in **Figure 2**.

Begin

S_i $i = 1: \dots n$: Set of uncovered interested subscribers.

C_i : Contribution of the i[th]device .

e_i: energy of the i[th] device.

T_i: State of the i[th] device.

$C_i = {|S_{ik}|}/{e_i}$

S_{ik} : The set of uncovered interested subscribers that are neighbors to the i[th] device.

$Li = 1: \dots m$: Set of uncovered leaders.

C_{li} : L-Contribution of the i[th]device .

$C_{li} = {|L_{in}|}/{e_i}$

L_{in} : The set of uncovered leaders that are neighbors to the i[th] leader.

// **Construction of the first level of the virtual topology//**

While$|S_i| \neq 0$

For each device i which is not a leader

compute C_i ;

End For

Select the device j with the highest contribution;

$Tj = leader$;

Update the list of uncovered subscribers;

End While

//Construction of the eventual other levels of the virtual topology//

While more levels can be included

// The number of levels depends on the network's topology.

For each device i which is not a leader

compute C_{li} ;

End For

Select the device j with the highest L-contribution;

$Tj = leader$;

Update the list of uncovered leaders;

End While

End

Figure 2. EECBR.

The probability for constructing a tree is very high for fully and partially connected networks. Our virtual topology is bottom-up. It starts with the interested subscribers. Our idea is closely related to [49] which maximizes the network lifetime by finding the covering nodes for a set of targets. So, we proceed to define the contribution of each device as the set of uncovered interested subscribers by the device's energy and the device with the highest contribution is selected to be a leader.

Upon constructing the first level of the virtual topology more levels may be included. The routing process is performed in a distributed way using the following rules:

1) Each publisher sends the generated event (E) to its leader.
2) Each leader sends the received event (E) to its connected interested subscribers and any other leader that is connected to it.

6. Simulation Results

We have conducted our simulation runs through the Omnet++ simulator. OMNeT++ standing for Objective Modular Network Test-bed in C++ is an object-oriented modular discrete event network simulation framework that has a generic architecture. OMNeT++ itself is not a simulator of anything in particular, but rather provides infrastructure and tools for writing simulations. OMNeT++ is currently gaining widespread popularity as a network simulation platform in the scientific community as well as in industrial settings. It provides a discrete event simulation environment. OMNeT++ provides a component architecture for models. The Modules are programmed in C++ and assembled into components using NED. The main characteristic of OMNET++ is that the reusability of models comes for free. Omnet++ is successfully used for complex and queuing systems. Ad hoc and sensor networks [50]… We simulated five scenarios that are summarized in **Table 1**. The simulation of the scenarios that require the positions of the devices has been performed through the integration of Mixim [51]. The networks topology includes 20 devices.

We have calculated the variance of the energy per device as a metric to compare the different scenarios. By definition, the variance measures how far a set of numbers is spread out. A variance of zero indicates that all the values are identical. A small variance indicates that the data points tend to be very close to the mean and hence to each other, while a high variance indicates that the data points are very spread out around the mean and from each other. Equation (1) states that the variance depends on the mean. In case of many variables, we need to include their numbers in the variance equation.

$$\sigma^2 = (x - \mu)^2 \quad (1)$$

We considered the variance between the initial and the remaining energy for each device. **Figure 3** shows that EECBR has the lowest energy variance. The performance of the distance-based routing (**Figure 4**) in term of the energy variance is better than the performance of the progress, greedy and distance-energy schemes (**Figure 5**, **Figure 6** and **Figure 7**) respectively. **Figure 6** and **Figure 7** illustrate that the performance of the greedy and the distance-energy routing schemes are approximately similar. This means that the energy is more important than the distance for the routing decision.

Table 1. Simulation scenarios.

Scenario	Description
EECBR	In this scenario the events are routed using our virtual topology and our EECBR rules.
Progress-based routing	It this case, geographic routing is performed to route the events from the publishers to the interested subscribers. The progress-based routing is a type of greedy routing where the events are routed to the best neighbors. The process is repeated until reaching the destination. The progress is defined as the distance between a node S and the projection A' of a neighbor node A onto the line connecting S and final destination D [52].
Distance-based routing	It this case, geographic routing is performed to route the events from the publishers to the interested subscribers. The neighbor which is closest to the destination is selected [52].
Greedy routing	In this case, each publisher sends its events to the neighbor with the highest energy [52].
Distance-energy routing	In this case, a metric that combines the distance and the energy has been defined as a * distance + b * energy where a and b are two random variables between 0 and 1. The neighbor having the best distance-energy metric is selected [53].

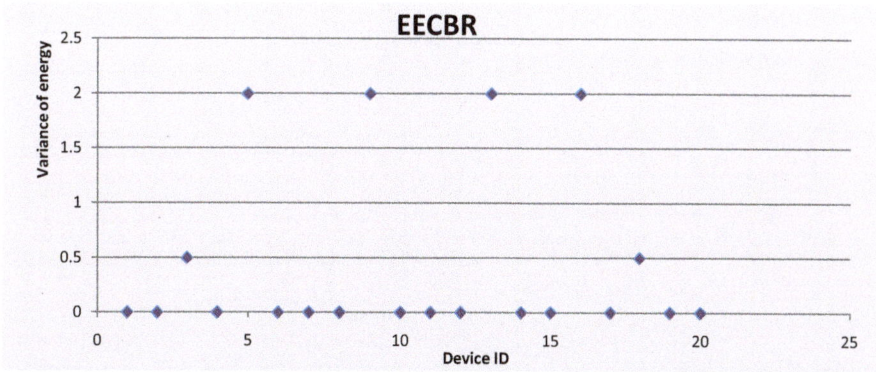

Figure 3. Energy variance of EECBR.

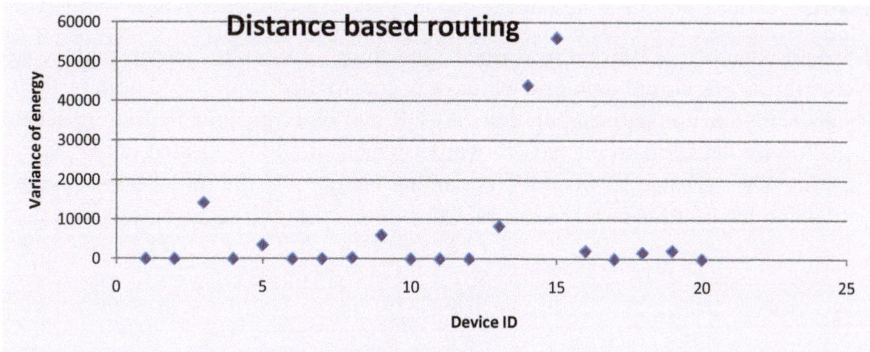

Figure 4. Energy variance of distance-based routing.

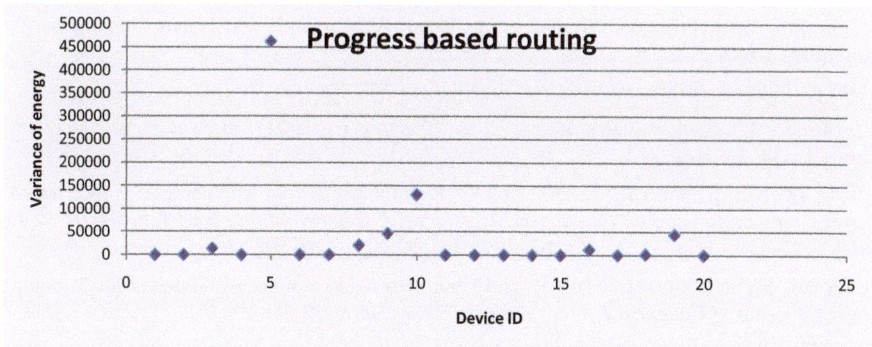

Figure 5. Energy variance of progress-based routing.

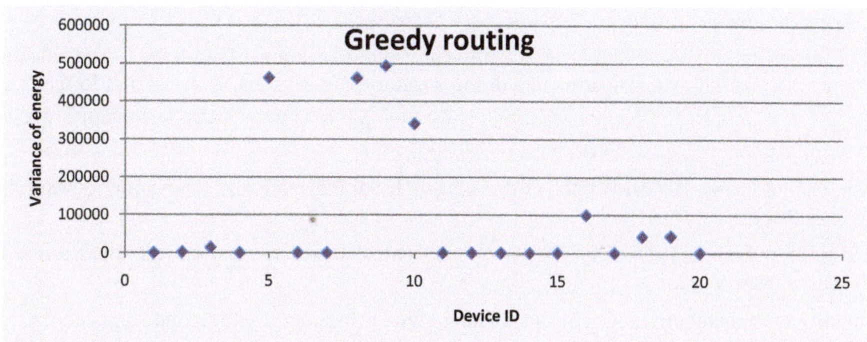

Figure 6. Energy variance of greedy routing.

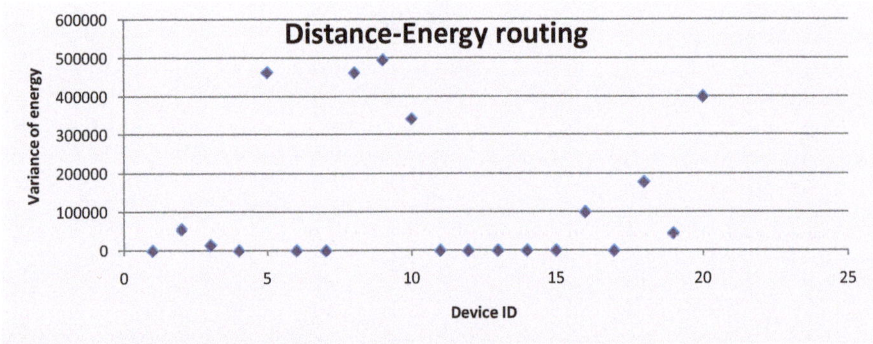

Figure 7. Energy variance of distance-energy routing.

7. Conclusion

In this paper, we introduced EECBR that performed a matching between IoT events and the subscribers and then routed the events through a virtual topology which was constructed from the bottom to the top. Our Omnet++ simulation results show the comparison between EECBR and other geographic routing schemes that use to the position information for deciding on the next forwarder which should be located in the way between the source and the destination. The variance between the remaining energy and the initial ones is very small in EECBR. This allows extending the lifetime of IOT sensors. Our future work will focus on solving the cross-layer issue in IoT routing.

References

[1] Atzori, L., Iera, A. and Morabito, G. (2010) The Internet of Things : A Survey. *Computer Networks*, **54**, 2787-2805. http://dx.doi.org/10.1016/j.comnet.2010.05.010

[2] Souza, A.M.C. and Amazonas, J.R.A. (2015) A New Internet of Things Architecture with Cross-Layer Communication. *Proceedings of the 7th International Conference on Emerging Networks and Systems Intelligence Emerging* 2015, Nice, 19-24 July 2015, 1-6

[3] Michael, M.P. (2007) Architectural Solutions for Mobile RFID Services on Internet of Things.

[4] Akyildiz, I.F., Su, W., Sankarasubramaniam, Y. and Cayirci, E. (2002) Wireless Sensor Networks : A Survey. *Computer Networks*, **38**, 393-422.

[5] Jia, X., Feng, Q., Fan, T., and Lei, Q. (2012) RFID Technology and Its Applications in Internet of Things (IoT). 2012 *2nd International Conference on Consumer Electronics, Communications and Networks* (*CECNet*), Yichang, 21-23 April 2012, 1282-1285. http://dx.doi.org/10.1109/CECNet.2012.6201508

[6] Bandyopadhyay, S., Sengupta, M., Maiti S., and Dutta, S. (2011) Role of Middleware for Internet of Things: A Study. *International Journal of Computer Science & Engineering Survey*, **2**, 94-105. http://dx.doi.org/10.5121/ijcses.2011.2307

[7] Deering, S. and Hinden, R. (1998) Internet Protocol Version 6 Specification, RFC 2460.

[8] Engineers, E. (2003) IEEE Standards. "Part 15.4: Wireless Medium Access Control (MAC) and Physical Layer (PHY) Specifications for Low-Rate Wireless Personal Area Networks (LR-WPANs)", IEEE-SA Standards Board.

[9] Baldoni, R. and Virgillito, A. (2005) Distributed Event Routing in Publish/Subscribe Communication Systems: A Survey. Technical Report TR-1/06. Dipartimento di Informatica e Sistemistica, Universita di Roma "La Sapienza"

[10] Kamal, J.N.A.A.E. (2008) Efficient Virtual-Backbone Routing in Mobile Ad Hoc Networks. *Computer Networks*, **52**, 327-350. http://dx.doi.org/10.1016/j.comnet.2007.09.007

[11] Vermesan, O. and Friess, P. (2014) Internet of Things—From Research and Innovation to Market Deployment. River Publishers, Aalborg.

[12] Kopetz, H. (1997) Real-Time Systems: Design Principles for Distributed Embedded Applications. Real-Time Systems Series, Springer, New York.

[13] Antonopoulos, N. and Gillam, L. (2012) Cloud Computing: Principles, Systems and Applications. Springer, London.

[14] Buyya, R., Yeo, C.S., Venugopal, S., Broberg, J. and Brandic, I. (2009) Cloud Computing and Emerging IT Platforms: Vision, Hype, and Reality for Delivering Computing as the 5th Utility. *Future Generation Computer Systems*, **25**, 599-

616. http://dx.doi.org/10.1016/j.future.2008.12.001

[15] Suciu, G., Vulpe, A., Todoran, G., Cropotova, J. and Suciu, V. (2011) Cloud Computing and Internet of Things for Smart City. 1409-1416.

[16] Botta, A., De Donato, W., Persico, V. and Pescap, A. (2014) On the Integration of Cloud Computing and Internet of Things. *Proceedings of the* 2014 *International Conference on Future Internet of Things and Cloud* (*FiCloud*), Barcelona, 27-29 August 2014, 23-30. http://dx.doi.org/10.1109/FiCloud.2014.14

[17] Buchanan, B. (1999) Handbook of Data Communication and Networks. Springer, New York. http://dx.doi.org/10.1007/978-1-4757-0905-6

[18] Mbarushimana, C., Shahrabi, A. and Glasgow, G. (2007) Comparative Study of Reactive and Proactive Routing Protocols Performance in Mobile Ad Hoc Networks. *Proceedings of the* 21*st International Conference on Advanced Information Networking and Applications Workshops*, **2**, 679-684. http://dx.doi.org/10.1109/AINAW.2007.123

[19] Abolhasan, M., Wysocki, T. and Dutkiewicz, E. (2004) A Review of Routing Protocols for Mobile. *Ad Hoc Networks*, **2**, 1-22. http://dx.doi.org/10.1016/S1570-8705(03)00043-X

[20] Al-Karaki, J. and Kamal, A. (2004) Routing Techniques a Wireless Sensor Network: A Survey. *IEEE Wireless Communications*, **11**, 6-28. http://dx.doi.org/10.1109/MWC.2004.1368893

[21] Shelby, Z. and Bormann, C. (2009) 6LoWPAN : The Wireless Embedded Internet. John Wiley & Sons, Chichester. http://dx.doi.org/10.1002/9780470686218

[22] Ee, G.K., Ng, C.K., Noordin, N.K. and Ali, B.M. (2010) A Review of 6LoWPAN Routing Protocols. *Proceedings of the Asia Pacific Advanced Network*, **30**, 71-81. http://dx.doi.org/10.7125/APAN.30.11

[23] Georgiadis, L. (2002) Carrier-Sence Multiprle Access (CSMA) Protocols, 1-18.

[24] Postel, J. (1980) User Datagram Protocol, RFC 768.

[25] Perkins, C.E., Park, M. and Royer, E.M. (1999) Ad-Hoc On-Demand Distance Vector Routing. *Proceedings of the* 2*nd IEEE Workshop on Mobile Computing Systems and Applications*, New Orleans, 25-26 February 1999, 90-100. http://dx.doi.org/10.1109/mcsa.1999.749281

[26] Yang, T., Yang, Q. and Cheng, L. (2015) Experimental Study: A LQI-Based Ranging Technique in ZigBee Sensor Networks. *International Journal of Sensor Networks*, **19**, 130-138.

[27] Tran, T. and Thuy, T. (2011) Routing Protocols in Internet of Things. Report, ESG Seminar.

[28] Klein, B.A. (2011) RPL: IPv6 Routing Protocol for Low Power and Lossy Networks. 59-66.

[29] Das, S.M., Pucha, H., Papagiannaki, K. and Hu, Y.C. (2007) Studying Wireless Routing Link Metric Dynamics. *Proceedings of the* 7*th ACM SIGCOMM Conference on Internet Measurement*, San Diego, 23-26 October 2007, 327-332. http://dx.doi.org/10.1145/1298306.1298352

[30] Park, S., Cho, S. and Lee, J. (2014) Energy-Efficient Probabilistic Routing Algorithm for Internet of Things. *Journal of Applied Mathematics*, **2014**, Article ID: 213106. http://dx.doi.org/10.1155/2014/213106

[31] Chen, Z., Wang, H., Liu, Y., Bu, F. and Wei, Z. (2012) A Context-Aware Routing Protocol on Internet of Things Based on Sea Computing Model. *Journal of Computers*, **7**, 96-105. http://dx.doi.org/10.4304/jcp.7.1.96-105

[32] Wilmer, E.L. (2009) Markov Chains and Mixing Times. AMS Publisher, Providence.

[33] Di Marco, P., Athanasiou, G., Mekikis, P. and Fischione, C. (2015) MAC-Aware Routing Metrics for the Internet of Things. Elsevier, Amsterdam, 1-31. http://dx.doi.org/10.1016/j.comcom.2015.05.010

[34] Xin, H. (2015) Routing Protocols Analysis for Internet of Things. 2-5.

[35] Mohapatra, S. and Kanungo, P. (2012) Procedia Engineering Performance Analysis of AODV, DSR, OLSR and DSDV Routing Protocols Using NS2 Simulator. *Procedia Engineering*, **30**, 69-76. http://dx.doi.org/10.1016/j.proeng.2012.01.835

[36] Eugster, P.T.H., Felber, P.A. and Kermarrec, A. (2003) The Many Faces of Publish/Subscribe. *ACM Computing Surveys*, **35**, 114-131. http://dx.doi.org/10.1145/857076.857078

[37] Shen, H. (2010) Content-Based Publish/Subscribe Systems, Handbook of Peer-to-Peer Networking. Springer Science and Business Media, LLC, New York, 1333-1366. http://dx.doi.org/10.1007/978-0-387-09751-0_49

[38] Souto, E., Guimara, G., Vasconcelos, G., Vieira, M., Rosa, N., Ferraz, C. and Kelner, J. (2006) Mires: A Publish/Subscribe Middleware for Sensor Networks. *Personal and Ubiquitous Computing*, **10**, 37-44. http://dx.doi.org/10.1007/s00779-005-0038-3

[39] Zhang, W. and Hansen, K.M. (2008) An {OWL/SWRL} Based Diagnosis Approach in a Pervasive Middleware. *Proceedings of the* 20*th International Conference on Software Engineering and Knowledge Engineering*, San Francisco, 1-3 July 2008, 893-898.

[40] Moreira Sa de Souza, L., Spiess, P., Koehler, M., Guinard, D., Karnouskos, S. and Savio, D. (2008) SOCRADES: A Web Service Based Shop Floor Integration Infrastructure. *Lecture Notes in Computer Science*, **4952**, 50-67. http://dx.doi.org/10.1007/978-3-540-78731-0_4

[41] Žarko, I.P., Pripužić, K., Serrano, M. and Hauswirth, M. (2013) IoT Data Management Methods and Optimisation Algorithms for Mobile Publish/Subscribe Services in Cloud Environments. *Proceedings of the* 2014 *European Conference on Networks and Communications* (*EuCNC*), Bologna, 23-26 June 2014, 1-5.

[42] Cao, F. and Singh, J.P. (2004) Efficient Event Routing in Content-based Publish-Subscribe Service Networks. *Proceedings of the Twenty-Third Annual Joint Conference of the IEEE Computer and Communications Societies*, Hong Kong, 7-11 March 2004, 929-940.

[43] Lung, C., Zhou, C. and Yang, Y. (2007) Applying Hierarchical Agglomerative Clustering to Wireless Sensor Networks. *Proceedings of the International Workshop on Theoretical and Algorithmic Aspects of Sensor and Ad-Hoc Networks*, Miami, 28-29 June 2007, 97-105.

[44] Pettie, S. and Ramachandran, V. (2002) An Optimal Minimum Spanning Tree Algorithm. *Journal of the ACM*, **49**, 16-34. http://dx.doi.org/10.1145/505241.505243

[45] Cugola, G. and Migliavacca, M. (2009) A Context and Content-Based Routing Protocol for Mobile Sensor Networks. *Proceedings of the Sixth European Conference on Wireless Sensor Networks*, Cork, 11-13 February 2009, 69-85. http://dx.doi.org/10.1007/978-3-642-00224-3_5

[46] Baldoni, R., Beraldi, R., Quema, V., Querzoni, L. and Tucci-Piergiovanni, S. (2007) TERA: Topic-Based Event Routing for Peer-to-Peer Architectures. *Proceedings of the* 2007 *Inaugural International Conference on Distributed Event-Based Systems*, Toronto, 20-22 June 2007, 2-13. http://dx.doi.org/10.1145/1266894.1266898

[47] Calomme, S. and Leduc, G. (2008) An Overlay Maintenance Protocol for Overlay Routing on Top of Ad Hoc Networks. *Lecture Notes in Computer Science*, **4982**, 48-59. http://dx.doi.org/10.1007/978-3-540-79549-0_5

[48] Von, V. (2002) Large-Scale Content-Based Publish/Subscribe Systems. Dissertation, Darmstadt University, Darmstadt.

[49] Cardei, M., Wu, J., Lu, M. and Pervaiz, M.O. (2005) Maximum Network Lifetime in Wireless Sensor Networks with Adjustable Sensing Ranges. *Proceedings of the IEEE International Conference on Wireless and Mobile Computing, Networking and Communications*, Montreal, 22-24 August 2005, 1-8. http://dx.doi.org/10.1109/wimob.2005.1512935

[50] Varga, A. OMNeT^{++}. http://www.omnetpp.org

[51] Wessel, K., Swigulski, M., Köpke, A. and Willkomm, D. (2009) MiXiM—The Physical Layer: An Architecture Overview. *Proceedings of the 2nd International Workshop on OMNeT*$^{++}$, Rome, March 2009, 1-8.

[52] Stefan, R. (2009) Theory and Practice of Geographic Routing. In: Liu, H., Chu, X.W. and Leung, Y.W., Eds., *Ad Hoc and Sensor Wireless Networks*: *Architectures*, *Algorithms and Protocols*, Bentham Science, Sharjah, 69-88.

[53] Champ, J. and Clement, S. (2008) An Energy-Efficient Geographic Routing with Location Errors in Wireless Sensor Networks. *Proceedings of the International Symposium on Parallel Architectures*, *Algorithms*, *and Networks*, Sydney, 7-9 May 2008, 105-110. http://dx.doi.org/10.1109/i-span.2008.49

31

A Secure DHCP Protocol to Mitigate LAN Attacks

Osama S. Younes[1,2]

[1]Faculty of Computers and Information, Technology Tabuk University, Tabuk, KSA
[2]Faculty of Computers and Information, Menoufia University, Al Minufya, Egypt
Email: usama.younas@ci.menofia.edu.eg

Abstract

Network security has become more of a concern with the rapid growth and expansion of the Internet. While there are several ways to provide security in the application, transport, or network layers of a network, the data link layer (Layer 2) security has not yet been adequately addressed. Data link layer protocols used in local area networks (LANs) are not designed with security features. Dynamic host configuration protocol (DHCP) is one of the most used network protocols for host configuration that works in data link layer. DHCP is vulnerable to a number of attacks, such as the DHCP rouge server attack, DHCP starvation attack, and malicious DHCP client attack. This work introduces a new scheme called Secure DHCP (S-DHCP) to secure DHCP protocol. The proposed solution consists of two techniques. The first is the authentication and key management technique that is used for entities authentication and management of security key. It is based on using Diffie-Hellman key exchange algorithm supported by the difficulty of Elliptic Curve Discrete Logarithm Problem (ECDLP) and a strong cryptographic one-way hash function. The second technique is the message authentication technique, which uses the digital signature to authenticate the DHCP messages exchanged between the clients and server.

Keywords

DHCP, Authentication, Data Link Layer Attacks, ECDLP

1. Introduction

Evolving of computer networks, and the variety of its services and applications, has increased the users need for local area networks (LANs) [1]. LAN technologies, such as Ethernet, are the infrastructure for the Internet that everybody uses without further thought.

Ethernet stands as the dominant networking technology in Local Area Network (LAN), which has been widely used in campus networks, enterprise networks and data centre networks due to its simplicity and auto-configuration characteristics. Its ease of use and low cost rely on broadcast-based service or resource discovery protocols, such as Dynamic Host Configuration Protocol (DHCP), Address Resolution Protocol, Network Time Protocol, and Network Basic Input/Output System.

Network security has become more of a concern with the rapid growth and expansion of the Internet. While there are several ways to provide security in the application, transport, or network layers of a network, the data link layer (Layer 2) security has not yet been adequately addressed [2]. In local networks, security weaknesses in the data link layer enable internal attacks. Although switches and routers have some built-in security features, they are not enough to fully ensure the security of LANs. Moreover, these features require network administrators' involvement and are prone to misconfiguration. In addition, data link layer protocols used in local area networks (LANs) are not designed with security features [2].

DHCP simplifies the access to a network. When a host connects to the network, DHCP [3] automates the assignment of TCP/IP stack configuration parameters such as IP addresses, subnet masks, and default gateway. It is an internet protocol that lets network administrators centrally manage the network. Without using DHCP, the IP address must be manually assigned for each host in a network and if the host moved to a new location in the network, the IP address must be manually configured.

DHCP is one of the most used network protocols for host configuration. It was designed since a long time ago [4]-[6] and it had not major changes, although it is vulnerable to a number of attacks, such as the DHCP rouge server attack, DHCP starvation attack, and malicious DHCP client attack. The main source of these attacks is given by the fact that DHCP does not use any authentication scheme for clients, servers, or exchanged messages.

Much research [7]-[17] has been done to make DHCP protocol more secure. Most of this research is based on the work introduced in [7], which proposed a standard for DHCP message authentication, using the option field in the DHCP message, based on a shared secret key between client and server. The main drawback of the related work introduced in literature is that managing the secrete keys between clients and server did not explain. In addition, some of this research uses a digital certificate to sign the exchanged message between the client and server; however, the size of the digital certificate exceeds the DHCP message and cannot be loaded into it. Few schemes [15] [16] were introduced for key management between clients and server. However, these schemes do not scale up with number of clients and make a central point of failure.

In this work, a new scheme called S-DHCP (Secure DHCP) is proposed to secure DHCP protocol. S-DHCP is an authentication technique, based on an extension of the DHCP protocol, which is used to mutually authenticate the DHCP client and DHCP server as well as to protect the integrity of the DHCP messages. The proposed solution is based on two techniques: entity authentication and key management (EAKM) technique, and message authentication (MA) technique. EAKM technique is based on using Diffie-Hellman key exchange algorithm supported by the difficulty of Elliptic Curve Discrete Logarithm Problem (ECDLP) and a strong cryptographic one-way hash function. It is used to mutually authenticate the client and the DHCP server, exchange the session key, and generate a digital signature for each client and the DHCP server. The MA technique uses the digital signatures generated using the EAKM technique to authenticate the DHCP messages exchanged between the client and server.

The paper is organized as follows. Section 2 describes the functions of the DHCP protocol. DHCP security issues are discussed in Section 3. Related work is discussed in Section 4. The entity authentication technique is described in Section 5. Then, Section 6 shows how message authentication technique uses a digital signature to authenticate the DHCP messages. After that, Section 7 analyses the security problems of DHCP according to the proposed solution. Finally, section analyses the performance of the proposed techniques.

2. Dynamic Host Configuration Protocol

DHCP [3] Stands for Dynamic Host Configuration Protocol. It allows a host to obtain an IP address dynamically when it connects to the network. This service automates the assignment of IP addresses, subnet masks, gateway and other IP networking parameters. DHCP distributed addresses are not permanently assigned to hosts but are only leased for a period of time. If the host is powered down or taken off the network, the address is returned to the range for reuse. This is especially helpful with mobile users that come and go on a network.

The operation of DHCP protocol is shown in **Figure 1**. When a DHCP-configured device boots up or connects to the network, the client broadcasts a DHCPDISCOVER packet to identify any available DHCP servers

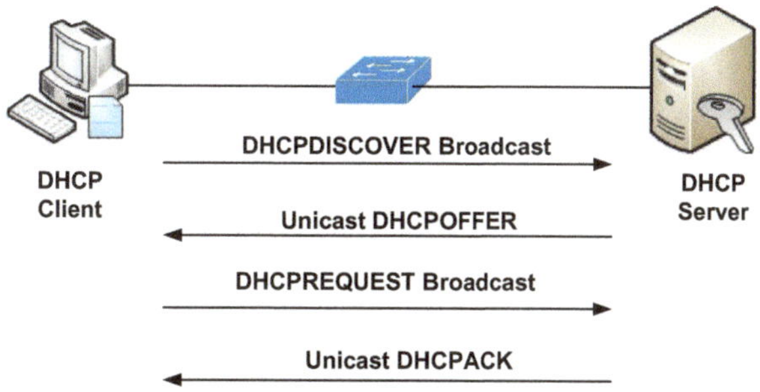

Figure 1. DHCP protocol operations.

on the network. A DHCP server replies with a DHCPOFFER, which is a lease offer message with an assigned IP address, subnet mask, DNS server, and default gateway information as well as the duration of the lease.

The client may receive multiple DHCPOFFER packets if there is more than one DHCP server on the local network, so it must choose between them, and broadcast a DHCPREQUEST packet that identifies the explicit server and lease offer that the client is accepting. A client may choose to request an address that it had previously been allocated by the server. Assuming that the IP address requested by the client or offered by the server is still valid, the server would return a DHCPACK message that acknowledges to the client the lease finalization.

3. DHCP Security Problems

3.1. DHCP Exhaustion Attack

The DHCP server has a pool of IP addresses that are being leased to hosts; but the IP address pool has always limited number of IP addresses. In DHCP exhaustion attack [2], the attacker exhausts the IP addresses in the DHCP address pool. The DHCP server happily hands out the entire set of addresses available to the client's network, because it has no way to differentiate between genuine host and spoofed one. If a legitimate client tries to obtain an IP address, it will have no IP connectivity because the entire of addresses have been allocated to the spoofed clients.

3.2. DHCP Rogue Server Attack

The rogue server is a DHCP server on a network which is not under the administrative control of the network staff. The DHCP rogue server attack [2] is a famous LAN attack in which a malicious user disguises itself as a DHCP server and responds to DHCP requests with a bogus IP address. When clients connect to the network, both the rogue and legal DHCP server receive the DHCP DISCOVER message; then both servers will offer them IP addresses as well as default gateway. The DHCP rogue server responds to DHCP requests with wrong configuration. The wrong information may be wrong default gateway, wrong DNS server, or wrong IP address. When the attacker host (a Rogue DHCP server) make itself a default gateway, it can receive the whole traffic of the network. So, he can analyse and modify all packets sent from the attacked machine and he can steal passwords and privacy information.

3.3. Malicious DHCP Client

A malicious client can gain unauthorized access in a network and then it can use the network services without being allowed to. More, it can launch a DHCP starvation attack, exhausting the DHCP server available IP addresses. An unauthorized client can also install and configure a rogue DHCP server and then realize the attacks specific to an illegitimate DHCP server.

4. Related Work

In RFC 3118 [7], two standard techniques used the option field for authentication of DHCP messages; configuration token and delayed authentication. The configuration token scheme is based on sharing a secret token between the client and server. This scheme can only protect DHCP server that has inadvertently been instantiated.

It does not support DHCP message authentication. The delayed authentication scheme uses the HMAC technique for DHCP message authentication. It uses a pre-shared secret key with MD5 message-digest algorithm to generate the MAC (Message Authentication Code) for DHCP messages. Although this scheme tried to solve the problem of DHCP entity authentication and message authentication using a shared secrete key, it did not show how the secrete key is managed, especially for a large number of clients. In addition, the security of the MD5 hash function is severely compromised. A collision attack exists that can find collisions in seconds using a desktop computer [8].

Using the option field for DHCP message authentication is limited because its maximum size is 255 bytes (the length field is one byte). To encode options larger than 255 bytes, a technique was introduced in [9]. The authors used *sname* and *file* fields to store long options using a technique called aggregate buffer.

In [10], the method introduced in [7] has been developed for securing DHCP using a digital certificate. The authors used a trusted server to distribute the digital certificates between the clients and server. Because of its size, the digital certificate cannot fit into one DHCP message. Therefore, the authors introduced a scheme depends on fragmentation of the DHCP message into many messages.

Based on authentication option specification introduced in [7], a method based on using X.509 digital certificate was introduced in [11]. The digital certificate is used to sign the DHCP messages transmitted between clients and server. A common trust server or authority is used to distribute the digital certificate. The authors showed that the digital certificate cannot be loaded in one DHCP message because of its large size. In addition, they indicated that the digital certificates revocation policies are hard to set up. Also, the authors recommended that using certificate based authentication with delay authentication.

Two methods were proposed in [12] to make DHCP more secure; Secure DHCP with Digital Certificates (SDDC) and Secure DHCP with Shared Secrets (SDSS). The SDDC method depends on using digital certificates to authenticate the DHCP messages and entities. However, this method did not take into account the size of digital certificate may exceed the size of the DHCP message. The SDSS method uses a secrete key shared between entities to authenticate the DHCP messages.

Another mechanism called Challenge Handshake Authentication Protocol (CHAP) was introduced in [13] for authenticating the entities using DHCP. This mechanism lets the server generates an encrypted challenge response and attach it to the DHCPOFFER message when it receives the DHCPDISCOVER message. The challenge response is computed using a hash function and a secret key that must be shared between the DHCP clients and server. Also, Other DHCP messages; DHCPREQUEST and DHCPACK, are attached with the challenge response to authenticate the messages and entities. Although, this work did not change the state machine of the client, it introduced a new server; called authentication server, with additional communication between it and the DHCP server. In addition, the authors did not show how the secret key is managed between the servers and clients.

In [14], a method for authentication of clients and DHCP message was proposed based on Kerberos V [15]. The proposed method used the authentication option specification defined in [7]. Kerberos V protocol uses an authentication server and key distribution centre, which issues tickets for the clients and server that are encrypted using secret keys shared between the clients and server. To authenticate the DHCP request and response, the ticket of the client or server is included in the DHCP message. The main disadvantages of this method are that the Kerberos server represents a single point of failure. In addition, the tickets use the timestamp to prevent replay attack, which means that the clock of clients and servers should be synchronized.

To provide mutual authentication for DHCP clients and server and for DHCP message authentication, a scheme was proposed in [16]. This scheme is similar to schemes introduced in [7] [9] [10]. However, this scheme presented a methodology for sharing secrete keys between entities, which is used to compute MAC (Message Authentication Code) for DHCP messages. The scheme supposes that there is a shared secret key between entities, which is used to generate a secret key for each DHCP message using a generated random number. Unfortunately, the authors did not show how the random number is generated and how the new secrete key is generated from the old secret key and the random number.

The authors in [17] proposed a security system that consists of eight units that make only authorized clients to obtain IP addresses. Before connecting to the system, the client must get an ID and a password from the authentication server. This system uses additional message and is not compatible with the standard. In addition, the proposed scheme for managing the shared secret key between clients and the authentication server does not scale up with number of clients and represents central point of failure.

5. Entity Authentication and Key Management Technique (EAKM)

This section explains the proposed authentication and key management scheme. It is based on Diffie-Hellman key exchange algorithm supported by the difficulty of Elliptic Curve Discrete Logarithm Problem (ECDLP) and based on using a strong cryptographic one-way hash function. The significant advantage of ECDLP is the faster calculations compared to other public key cryptosystems with the same security levels. The EAKM technique is inspired by the work introduced in [18]. **Table 1** shows all notations used for the proposed scheme. The overall scheme is illustrated in **Figure 2** and is explained as follows. The proposed scheme consists of three phases: the setup phase, the registration phase, and the authentication phase.

5.1. Setup Phase

In this phase, the clients and server agree on the used elliptic curve parameters. The sever chooses a secrete key

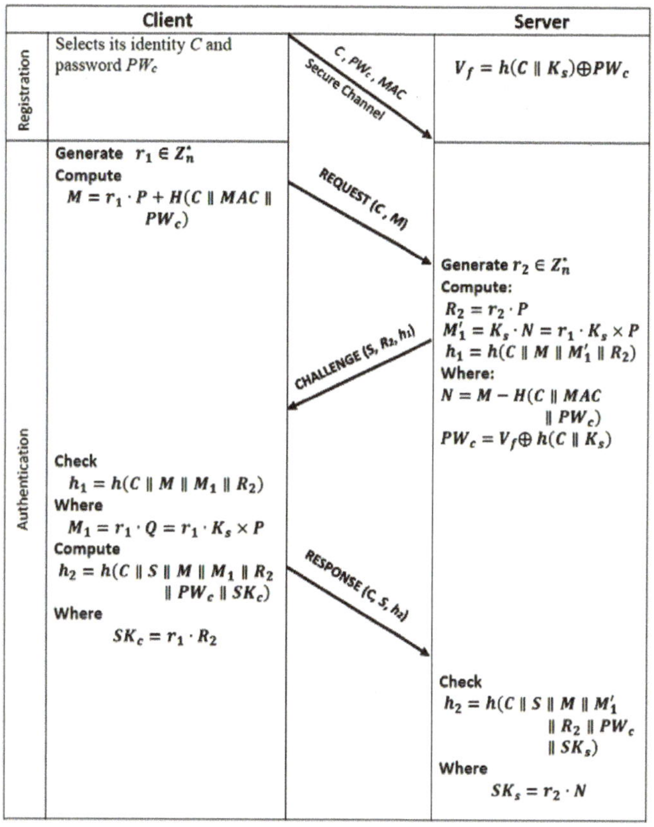

Figure 2. Flow diagram for EAKM technique.

Table 1. The notations used for proposed authentication scheme.

C and S	The ID of the client and server
PW_c	The password of the client
$h(\cdot)$	Strong cryptographic one-way hash function
$H(\cdot)$	Function which makes a point map to another point on elliptic curve
$\|$	Concatenation operation
$\oplus$	Exclusive-or operation
K_s	Server password
SK	Session key
Q	Server public key

Ks, then it selects a point *P* in the elliptic curve to compute the public key $Q = Ks \cdot P$. The server publishes all parameters; *P*, *a*, *b*, *n*, *h*, and *Q*, except the secret key *Ks*.

5.2. Registration Phase

As shown in **Figure 2**, the registration phase consists of two steps:

1) The user or the network administrator chooses the identity *C* and password PW_c for the client. Then, *C*, PW_c, and MAC address of the client are sent to the server over a secure channel, such as a VPN (Virtual Private Network) or SSL (Secure Sockets Layer), or the network administrator insert them manually to the server database.

2) The server computes the password verification code $V_f = h(C \| K_s) \oplus PW_c$ using a strong cryptographic hash function *h* (.) and the server password *Ks*; then it stores the binding (C, V_f, MAC) in its database.

5.3. Authentication Phase

When any client *C* tries to connect to the local area network managed by the authentication server *S*, the authentication procedure between *C* and *S* proceeds as shown in **Figure 2**, which is illustrated as follows:

Step 1: The client generates the random number r_1, and then it computes the parameter $M = r_1 \cdot P + H(C \| PW_c)$ using the parameters *P* and *H* published by the server. The parameter *M*, the client ID (*C*), and the client MAC address are sent to the server in the request message REQUEST (*C*, *MAC*, *M*).

Step 2: If the server received the request message, it checks that the client identifier *C* and the client *MAC* address are stored in its database. If the server did not find *C* and *MAC* address the database, it ignores the request message. Otherwise, the server generates a random number r_2 and use the password verification code V_f, bended with *C* and stored in the database, and the sent parameter *M* to compute the client password PW_c and parameter *N*, M_1' and R_2, which are used to compute the parameter h_1, as shown in **Figure 2**. After calculating the parameter R_2 and h_1, the serve sends them with its identity *S* in a challenge message CHALLENGE (*S*, R_2, h_1).

Step 3: After receiving the challenge message, the client compute the value of $M_1 = r_1 \cdot Q = r_1 \cdot K_s \times P$, which is used to check the value of $h_1 = h(C \| M \| M_1 \| R_2)$. If the computed value for h_1 does not match with the received value, the session between the client and server is terminated. Otherwise, the client computes SK_c, which is used to compute h_2, as shown in **Figure 2**. At the end, the client sends the response message RESPONSE (*C*, *S*, h_2) to the server.

Step 4: In the same way, once the server receives the response message, it verifies the value h_2. First, it computes SK_s. Then it computes $h_2 = h(C \| S \| M \| M_1' \| R_2 \| PW_c \| SK_s)$. If the calculated value for h_2 is equal to the received one in the response message, the server authenticate the client *C* with the common and unique session key $SK = SK_c = SK_s = r_1 \cdot r_2 \cdot P = r_1 \cdot R_2 = r_2 \cdot N$.

As clear from **Figure 2**, a one way hash function with XOR operation is used for encrypting or decrypting all communication messages. Therefore, the computation cost (or computational overhead) is quite low, as explained in Section 8. At the end of the EAKM technique, the digital signature of each client and the server are generated and stored in each of them to be used by the S-DHCP technique, which is explained in the next section.

6. Message Authentication Technique

S-DHCP uses the message authentication technique, which extends DHCP with an integrity/authentication technique for DHCP packets, to mitigate different DHCP attacks. Because S-DHCP is built on top of DHCP, its specification (as for message exchange, timeout, cache) follows the original one for DHCP [4]. Therefore, network clients that do not use S-DHCP module can process the S-DHCP messages. However, in secure LAN, all clients should use S-DHCP. This section describes the message authentication technique.

6.1. Protocol Design Constraints

In order to make the proposed MA technique integrated and compatible with the current DHCP implementation, the following design constraints are adopted:

- The proposed scheme must follow the authentication option format introduced in [7];
- The DHCP client and server state machine must not be changed, *i.e.* any new state must not be introduced;

- Introducing any new DHCP message is not allowed;
- The clients that do not use proposed protocol must be able to obtain configuration information for the server.

In addition to the above, one of the main constrains is that DHCP specification does not permit the fragmentation of DHCP messages. This means that the maximum size of DHCP packet is 1500 bytes; the MTU for Ethernet. By subtracting the length of IP header, UDP header, and DHCP header, the maximum size of option filed in DHCP message is 1236 bytes, which is too short. It cannot allow the transmission of a long encryption key or a digital certificate.

6.2. Protocol Description

Figure 3 shows the DHCP message exchange when S-DHCP is enabled in clients and server. Each DHCP message sent by client or server are processed by the MA scheme before being sent through the network. In the initial state, the DHCP client broadcasts a request, which is DISCOVERY message, searching for DHCP server. The client indicates the use of S-DHCP in option filed in the message. To use the option field in DHCP messages, the format defined in RFC 3118 is followed, which was proposed for entity and message authentication for DHCP protocol. The option field is used as follows:

- For the Protocol field the new value 4 is used which indicates that the authentication information is attached to message;
- For the Algorithm field a new value 5 is used to mark the using of S-DHCP;
- For RDM (Replay Detection Method) field the same value is used as in the specification in RFC 3118;
- For the Replay Detection (RD) field, it contains a monotone strictly increasing counter containing the current time and date which can reduce the danger of replay attacks.

If the DHCP server receives a DHCPDISCOVER message from a client, it passes the message to the DHCP engine if the message's option filed indicates that the protocol S-DHCP is not enabled. Otherwise, the MA scheme processes the message before passing it to DHCP engine. It verifies the value of Replay Detection field. If the verification process fails, the message is dropped. Otherwise, the server starts to prepare the DHCPOFFER message. The server reserves an IP address for the client and makes a lease offer by sending a DHCPOFFER message to the client. This message contains the following:

- The MAC address of the client;
- The offered IP address to the client (*IPc*);
- The IP address of the DHCP server (*IPs*);
- Lease duration;
- The gate way (GW).

In addition, to authenticate and authorize the DHCPOFFER message, the server signature is attached to DHCPOFFER message in the option filed. As shown in **Figure 3**, the server signature attached to DHCPDISCOVER message includes the secrete key shared with the client, the MAC address of the client, *IPc*, *IPs*, GW, and RD. Also, to mitigate the replay attack, the Replay Detection filed is updated with the correct value.

When the client receives the DHCPOFFER message, the MA scheme passes it to the DHCP engine or process it, depending on the authentication option filed is set or not. If this option is set, the MA scheme checks the RD

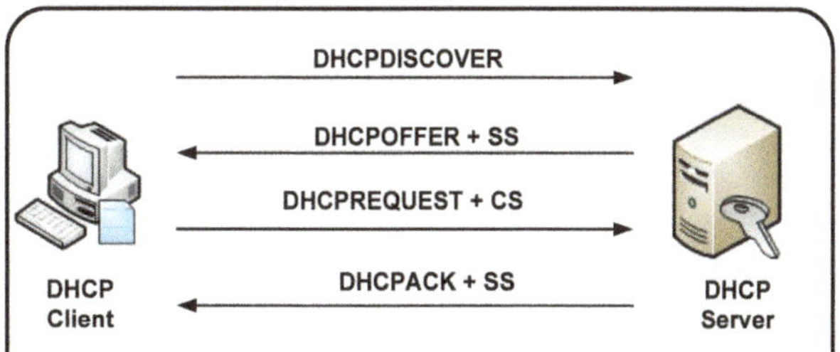

$$\text{CS} = \text{h}\ (C \| R_1 \| R_2 \| SK_s \| PW_i \| \text{MAC} \| IP_c \| IP_s \| \text{GW} \| RD_c)$$

$$\text{SS} = \text{h}\ (S \| R_1 \| R_2 \| SK_s \| PW_i \| \text{MAC} \| IP_c \| IP_s \| \text{GW} \| RD_c)$$

Figure 3. The DHCP message exchange suing MA technique.

filed. If the value of RD is strictly higher than the old value, the processing continues. Otherwise, the message is discarded.

After checking the RD field, the MA scheme verifies the signature of the server and the integrity of the message. First, it extracts the fields from the message required for signature verification, which are MAC, *IPs*, *IPc*, GW, and RD. Then, it calculates *SS*, as shown in **Figure 3**. If the calculated *SS* equals to the sever signature attached to the DHCPOFFER message, and MAC and *IPc* equal to that of the client, it starts to construct the DHCPREQUEST. Otherwise, the message is discarded.

The DHCPREQUEST message is a response to the DHCPOFFER message and it is used to request the offered IP address. Most fields of DHCPREQUEST messages are the same as that of the DHCPOFFER message. The client chooses a new value for Replay Detection filed and update it in the message. To authenticate and authorize the DHCPREQUEST message, the signature of the client is attached to this message, which is shown in **Figure 3**. The CS is generated using the new value for the RD field.

After receiving the DHCPREQUEST message, the DHCP server verifies the RD field. Then it validates the client signature. It extracts the required fields for validation; RD, *IPc*, *IPs*, *GW*, and MAC. Then it generate the signature CS and compare it with the one attached to the DHCPREQUEST message. If the RD field and signature of the client passes the validation procedure, the server starts to prepare the DHCPACK message in the same way as DHCPOFFER message explained above. The DHCPACK message includes all configuration information that the client have requested. When the client receives the DHCPACK message, it must validate it in the same way as explained for the DHCPOFFER message. After validation of the DHCPACK message, the DHCP client starts to configure its network interface with the negotiated parameters.

7. Analysis of Security Threats

This section discusses how the proposed authentication and key management scheme, and MA scheme mitigate main security vulnerabilities that a DHCP server or client can encounter in wireless or wired LAN.

7.1. DHCP Rouge Server Attack

The rogue DHCP server is a DHCP server, set up on a network by an attacker, which is not under the control of network administrators. By placing a rogue DHCP server on the network, the attacker can supply its own system as the default gateway and DNS server resulting in a man-in-the-middle attack.

The proposed S-DHCP scheme mitigates the DHCP rouge server attack because the DHCP server authenticate itself to the client during the authentication process and during DHCP message exchange. During the authentication process, the server is authenticated using the parameter h_1 that depends on v_f and M_1'. To generate M_1', the attacker must know the client password PW_c, Ks, and r_1. Even if the attacker compromised the password of the client, he cannot generate V_f without compromising the secret K_s (which is a high entropy number) and guessing r_1 which are difficult especially with maintaining the secret K_s periodically. On the other hand, if the attacker steals the secret K_s, he needs the password verifier to access the client password.

After the authentication process and during exchanging DHCP messages, if an authenticated client tries to impersonate the DHCP server, it must have the signature of the legitimate DHCP server. To do that, the attacker must compromise PW_c and SK_s. Also, he must guesses r_1 and r_2 assuming that he has R_1 and R_2. To compute SK_s, given the public key Q, the attacker must break the Elliptic Curve Discrete Logarithm Problem (ECDLP), which is very difficult problem. Also, given R_1, R_2 and P, guessing r_1 and r_2 is difficult and cannot be done in polynomial time. In addition, online password guessing cannot succeed, since Ks is a high entropy number that cannot be guessed and the password verifier V_f is not available to attacker.

7.2. DHCP Exhaustion Attack

DHCP starvation attack is designed to deplete all of the addresses within the address space allocated by the DHCP server. The attacker floods the DHCP server with DHCPREQUEST packets of spoofed MAC addresses. Therefore, clients of the victim network are denied an IP address requested via DHCP and thus is not able to access the network. Hence, DHCP starvation can be classified as a denial of service attack. The network attacker can then set up a rogue DHCP server on the network and perform the man in the middle attack, or simply set their machine as the default gateway and can sniff packets.

The proposed scheme can mitigate the DHCP starvation attack. For any received DHCPREQUEST message from any client, the DHCP server verifies the client signature attached to the message. If the DHCPREQUEST message does not contains the client signature, the server that uses S-DHCP discards the request. Otherwise, the server extract the parameters *C*, MAC, *IPc*, *IPs*, and *RDc* form the message header. Then, it uses the parameters *C*, R_1, R_2, *SKs*, and *PWc* bound with *C* and MAC address in its database to generate the signature of the client. After that, it compares between the attached client signature and the generated signature. If there any difference between them, the request message is discarded. Otherwise, the request is accepted.

If the attacker changed the MAC address in the header of the DHCPREQUEST message to initiate the DHCP starvation attack and attached any client signature to the message, the DHCP server will detect the difference between the attached client signature and generated signature. The attacker can do the DHCP starvation attack if and only if he has the signatures of many clients. This means he has to guess the parameters R_1, R_2, *SKs*, and *PWc*, supposing that he knows the parameters *C* and MAC. As explained above, guessing of all these parameters for one client is extremely difficulty. However, to do that for a large number of clients is nearly impossible.

7.3. Malicious User Attack

In the malicious user attack, also called spoofing or masquerading attack, the attacker impersonate the identity of a legitimate user in an illegal manner. The proposed authentication scheme resists against the malicious user attack. To impersonate the user *C*, the attacker must know the parameters *PWc*, *C*, MAC, and *M* of the client. Even if the attacker compromised the password of the user *PWc*, he cannot compute the correct value of *M* because he cannot guess the random number r_1.

7.4. Stolen-Verifier Attack

In the stolen-verifier attack, an attacker may attack the server and steal the database that contains clients' information such as the ID and password verification code. Then the attacker tries to use this information to impersonate the legitimate user. The proposed authentication scheme mitigates this attack, because the attacker cannot compute or guess the client's password even if he has the client ID and password verification code. This because he cannot get the server secret *Ks* which is not stored in the database.

7.5. Brute-Force Password-Cracking Attack

In the brute-force password-cracking attack, an opponent tries repeatedly to guess for the secret parameters such as the password of the client or the server and check them against an available cryptographic hash of the secret parameters. The proposed scheme mitigates this attack because, even if the opponent intercept all exchanged messages by a passive attack and correctly guesses the password, he cannot guess other secret parameters (M, R_2, M_1, M_1'). This is because these parameters are function mapped points to an EC point, which is a very difficult problem.

8. Performance Evaluation

This section shows the results of testing the proposed client and message authentication schemes. For performance evaluation of the proposed sachems, they have been implemented using C language with the open source cryptographic library (Open SSL) [19]. All performance tests are conducted using Ubuntu virtual machines with 3.6 GHz processor and 2 GB of RAM.

The client authentication time (CAT), DHCP message authentication time (MAT), and completion time (CT) are the metrics used for performance evaluation of the proposed schemes. CAT is the time taken to authenticate a client, which is the time needed for processing the request, challenges and response messages. The MAT time is the time needed for processing and exchanging the DHCPDISCOVER, DHCPOFFER, DHCPREQUEST, and DHCPACK messages between the client and server. The time needed for receiving the network configuration parameter from DHCP server is called the completion time. CT equals to the sum of CAT and MAT.

In all experiments, to measure MAT, CAT, and CT; the experiment is repeated many times until 95% confidence interval with 2% error is reached. For repeated experiments and for each metric, the average of the measured values is computed to be used in the analysis and comparisons.

In the first experiment, the number of clients (*N*) that try to access the local area network at the same instant is increased from 5 to 40 clients. In addition, it is supposed that the network is under attacks and the number of attackers is t percent of all nodes, where t is set to 0%, 20%, 40%, and 60%. The attacker impersonates the identity of the legitimate user and tries to deceive the server to get an offer. In this experiment, the metrics CAT, MAT, and CT are measured and the results are shown in **Figures 4-6**, respectively.

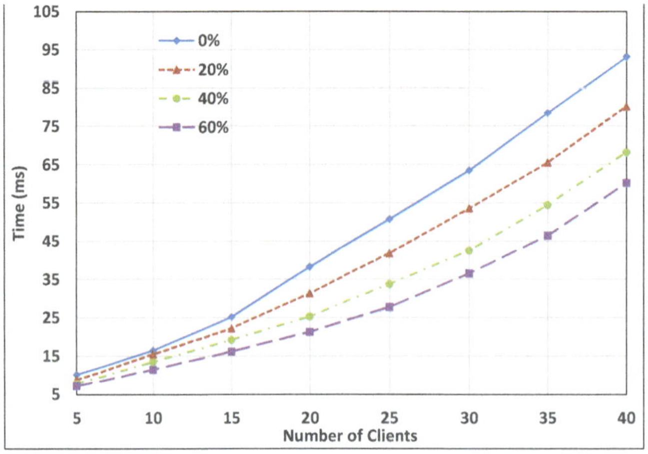

Figure 4. CAT versus N with increasing number of attackers.

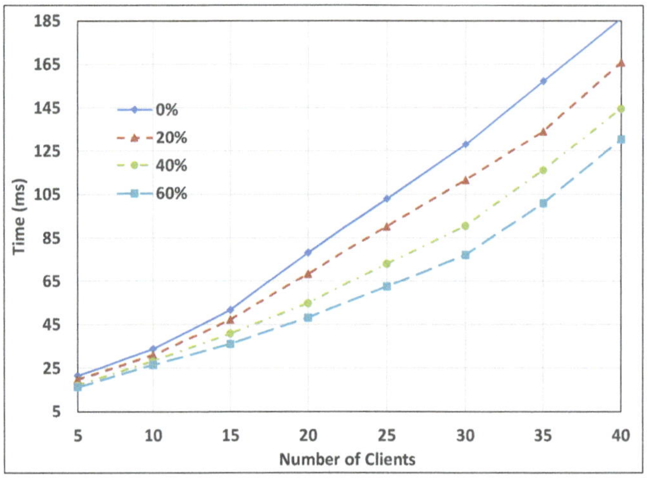

Figure 5. MAT versus N with increasing number of attackers.

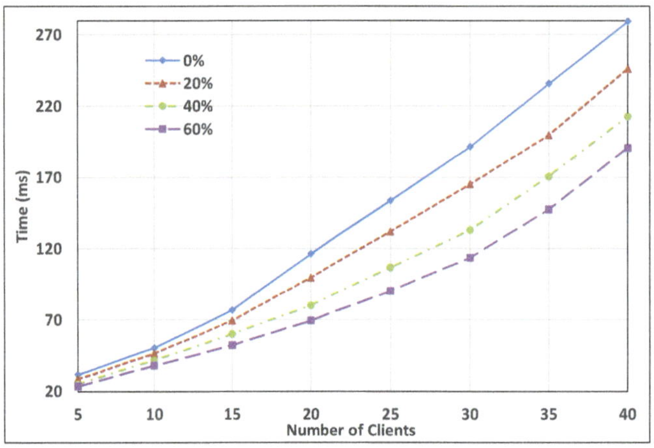

Figure 6. CT versus N with increasing number of attackers.

As shown in **Figures 4-6**, the higher the number of clients the greater the measured value for CAT, MAT, and CT. This is because increasing the number of client increases the time needed for processing messages in the server for either client authentication or message authentication. In addition, for the same number of authenticated clients, it is clear that increasing *t* decreases the value of CAT, MAT, and CT. This because increasing *t* increases the number of attackers and decreases the number of legitimate users. In addition, for attackers, the sever does not process the DHCPREQUEST and DHCPOFFER messages, as explained in the last section. This indicates the effectiveness of the proposed scheme because it prevents the attacker to complete the attack.

To measure the overhead of using the proposed scheme (S-DHCP), MAT is measured in the case of using S-DHCP and standard protocol (DHCP). As the first experiment, *N* is varied between 5 and 40, where there are no attackers in the network. **Figure 7** shows the results of this experiment. As shown in the figure, the overhead increased from 7.8 ms to 66.2 ms with increasing *N* from 5 to 40. Although the proposed scheme makes delay in the process of accessing the network by this overhead, it makes the network more secure by mitigating many attacks, as explained in Section 7. In addition, it is clear that overhead is very small.

Figure 8 compares between the proposed work and the work introduced in [12], which uses the digital certificates to authenticate the DHCP messages. For different number of nodes, the authentication time is measured, where the length of the key used to encrypt the digital certificate is 2048 and 3072 bits. As shown in **Figure 8**, the authentication time in the case of using digital certificate is much higher than that in the case of using the proposed scheme especially with a large number of clients.

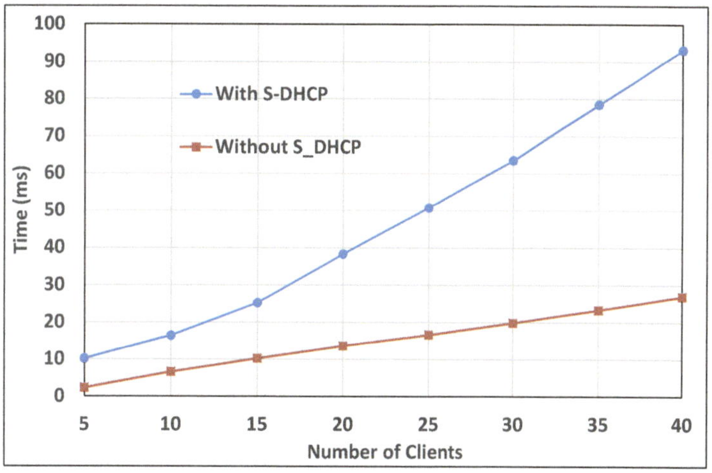

Figure 7. MAT versus *N* with in the case of using DHCP and S-DHCP.

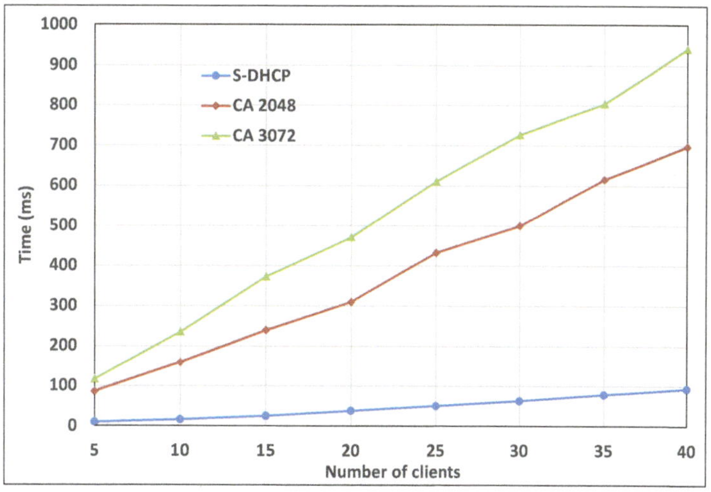

Figure 8. MAT versus *N* with in the case of using digital certificate and S-DHCP.

9. Conclusion

Dynamic host configuration protocol (DHCP) is one of the most used network protocols for host configuration that works in data link layer. DHCP is vulnerable to a number of attacks, such as the DHCP rouge server attack, DHCP starvation attack, and malicious DHCP client attack. This work introduces a new scheme called Secure DHCP (S-DHCP) to secure DHCP protocol. The proposed solution consists of two techniques. The first is the authentication and key management technique that is used for entities authentication and management of security key. It is based on using Diffie-Hellman key exchange algorithm supported by the difficulty of Elliptic Curve Discrete Logarithm Problem (ECDLP) and a strong cryptographic one-way hash function. The second technique is the message authentication technique, which uses the digital signature to authenticate the DHCP messages exchanged between the clients and server. Security analysis of the proposed scheme showed that it mitigates main security vulnerabilities that a DHCP server or client can encounter in wireless or wired LAN. In addition, performance analysis of the proposed scheme indicated that its performance is better compared to the digital certificate scheme introduced in literature.

References

[1] Hu, X.D., Gao, Z. and Li, W. (2009) Research on the Switched LAN Monitor Mechanism and its Implementation Method Based on ARP Spoofing. *International Conference on Management and Service Science*, Wuhan, 20-22 September 2009, 1-4. http://dx.doi.org/10.1109/icmss.2009.5304207

[2] Altunbasak, H.C. (2006) Layer 2 Security Inter-Layering in Networks. Georgia Institute of Technology, Atlanta.

[3] The TCP/IP Guide. DHCP Security Issues. http://www.tcpipguide.com/free/t_DHCPSecurityIssues.htm

[4] Droms, R. (1997) Dynamic Host Configuration Protocol. RFC 2131.

[5] Droms, R. and Lemon, T. (2002) The DHCP Handbook. 2nd Edition, Sams Publishing, Carmel, Indiana.

[6] Alexander, S. and Droms, R. (1997) DHCP Options and BOOTP Vendor Extensions. RFC 2132.

[7] Droms, R. and Arbaugh, W. (2001) Authentication for DHCP Messages. RFC 3118.

[8] Stevens, M.M.J. (2007) On Collisions for MD5. Master Thesis, Eindhoven University of Technology, Eindhoven.

[9] Lemon, T. and Cheshire, S. (2002) Encoding Long Options in the Dynamic Host Configuration Protocol (DHCPv4). RFC 3396.

[10] Xu, Y., Manning, S. and Wong, M. (2011) An Authentication Method Based on Certificate for DHCP. DHC Internet Draft.

[11] Glazer, G., Hussey, C. and Shea, R. (2003) Certificate-Based Authentication for DHCP.

[12] Duangphasuk, S., Kungpisdan, S. and Hankla, S. (2011) Design and Implementation of Improved Security Protocols for DHCP Using Digital Certificates. ICON, Singapore.

[13] De Graaf, K., Liddy, J., Raison, P., Scano, J.C. and Wadhwa, S. (2011) Dynamic Host Configuration Protocol (DHCP) Authentication Using Challenge Handshake Authentication Protocol (CHAP) Challenge. United States Patent Application 20110154440.

[14] Hornstein, K., Lemon, T., Adoba, B. and Trostle, J. (2001) DHCP Authentication via Kerberos V. IETF DHC Working Group.

[15] Ricciardi, F. (2007) Kerberos Protocol Tutorial. The National Institute of Nuclear Physics Computing and Network Services, LECCE, Italy.

[16] Ju, H. and Han, J.W. (2005) DHCP Message Authentication with an Effective Key Management. *Proceedings of the World Academy of Science, Engineering and Technology*, **8**.

[17] Komori, T. and Saito, T. (2002) The Secure DHCP System with User Authentication. *Proceedings of the 27th Annual IEEE Conference on Local Computer Networks*, Tampa, 6-8 November 2002. http://dx.doi.org/10.1109/lcn.2002.1181774

[18] Yang, C., Wang, R. and Liu, W. (2005) Secure Authentication Scheme for Session Initiation Protocol. *Computers & Security*, **24**, 381-386. http://dx.doi.org/10.1016/j.cose.2004.10.007

[19] Open SSL. Open SSL Project. http://www.openssl.org/

Permissions

All chapters in this book were first published in JCC, by Scientific Research Publishing; hereby published with permission under the Creative Commons Attribution License or equivalent. Every chapter published in this book has been scrutinized by our experts. Their significance has been extensively debated. The topics covered herein carry significant findings which will fuel the growth of the discipline. They may even be implemented as practical applications or may be referred to as a beginning point for another development.

The contributors of this book come from diverse backgrounds, making this book a truly international effort. This book will bring forth new frontiers with its revolutionizing research information and detailed analysis of the nascent developments around the world.

We would like to thank all the contributing authors for lending their expertise to make the book truly unique. They have played a crucial role in the development of this book. Without their invaluable contributions this book wouldn't have been possible. They have made vital efforts to compile up to date information on the varied aspects of this subject to make this book a valuable addition to the collection of many professionals and students.

This book was conceptualized with the vision of imparting up-to-date information and advanced data in this field. To ensure the same, a matchless editorial board was set up. Every individual on the board went through rigorous rounds of assessment to prove their worth. After which they invested a large part of their time researching and compiling the most relevant data for our readers.

The editorial board has been involved in producing this book since its inception. They have spent rigorous hours researching and exploring the diverse topics which have resulted in the successful publishing of this book. They have passed on their knowledge of decades through this book. To expedite this challenging task, the publisher supported the team at every step. A small team of assistant editors was also appointed to further simplify the editing procedure and attain best results for the readers.

Apart from the editorial board, the designing team has also invested a significant amount of their time in understanding the subject and creating the most relevant covers. They scrutinized every image to scout for the most suitable representation of the subject and create an appropriate cover for the book.

The publishing team has been an ardent support to the editorial, designing and production team. Their endless efforts to recruit the best for this project, has resulted in the accomplishment of this book. They are a veteran in the field of academics and their pool of knowledge is as vast as their experience in printing. Their expertise and guidance has proved useful at every step. Their uncompromising quality standards have made this book an exceptional effort. Their encouragement from time to time has been an inspiration for everyone.

The publisher and the editorial board hope that this book will prove to be a valuable piece of knowledge for researchers, students, practitioners and scholars across the globe.

List of Contributors

Avula Anitha
Department of Computer Science and Engineering, Keshav Memorial Institute of Technology, Hyderabad, India

Jungwan Hong, Seungju Baek, Heesoo Jung and Yongjin (James) Kwon
Department of Industrial Engineering, Ajou University, Suwon, South Korea

Suhwan Kim
1st Division, 3rd Department, Agency for Defense Development, Daejeon, South Korea

Liping Chen, Xiangyang Chen, Sile Wang, Wenzhu Yang and Sukui Lu
School of Computer Science and Technology, Hebei University, Baoding, China

Libo Cai, Yu Ma and Tangming Yuan
Department of Computer Science, University of York, Heslington, York, UK

Haifeng Wang
National Engineering Research Center of Rail Transportation Operation and Control Systems, Beijing Jiaotong University, Beijing, China

Tianhua Xu
Laboratory of Rail Traffic Control and Safety, Beijing Jiaotong University, Beijing, China

Minoru Sasaki, Muhammad Syaiful Amri Bin Suhaimi, Kojiro Matsushita and Satoshi Ito
Department of Mechanical Engineering, Gifu University, Gifu, Japan

Muhammad Ilhamdi Rusydi
Department of Electrical Engineering, Andalas University, Padang, Indonesia

Qinggong Ma and Bo Yang
Changzhou University Huaide College, Taizhou, China

Jungwan Hong and Yongjin (James) Kwon
Department of Industrial Engineering, Ajou University, Suwon, South Korea

Suhwan Kim
1st Division, 3rd Department, Agency of Defense Development, Daejeon, South Korea

Jia Shi and Kejian Yang
Department of Computer Science and Technology, Wuhan University of Technology, Wuhan, China

Yuqing Peng
Research Lab for Medical Imaging and Digital Surgery, Shenzhen Institutes of Advanced Technology, Chinese Academy of Sciences, Shenzhen, China
Shenzhen College of Advanced Technology, University of Chinese Academy of Sciences, Shenzhen, China

Xiaodong Zhang and Qingmao Hu
Research Lab for Medical Imaging and Digital Surgery, Shenzhen Institutes of Advanced Technology, Chinese Academy of Sciences, Shenzhen, China

Xiangyang Chen, Liping Chen, Sile Wang, Wenzhu Yang and Sukui Lu
School of Computer Science and Technology, Hebei University, Baoding, China

Qian Yang
School of Mathematical Science, University of Electronic Science & Technology of China, Chengdu, China

Hanbin Xi and Tiantian Chang
School of Science, Xi'an University of Posts and Telecommunications, Xi'an, China

Enyuan Jiang and Xuejun Wang
College of Communication Engineering, Jilin University, Changchun, China

Hua Li
Department of Information Engineering, Hangzhou Polytechnic College, Hangzhou, China

Mingzhu Wei, Ryad Chellali, Yang Yi, Ting Wang and Wen Qin
College of Electric Engineering and Control Science, Nanjing Tech University, Nanjing, China

Basheer Riskhan and Raza Muhammed
School of Computer Science, Huazhong University of Science and Technology, Wuhan, China

Shunrui Li, Jianjun Chen, Zuocheng Xing, Jinjin Shao and Xi Peng
Collage of Computer Science of National University of Defense Technology, Changsha, China

Aninda Majumder, Mohammad Raihan Ruhin, Tahsina Hashem and Md. Imdadul Islam
Department of Computer Science and Engineering, Jahangirnagar University, Dhaka, Bangladesh

Asif Shabbir and Fayyaz Khalid
Department of Computer Science, University of Gujrat, Gujrat, Pakistan

Syed Muqsit Shaheed
Department of Computer Science, University of Lahore, Lahore, Pakistan

Jalil Abbas and M. Zia-Ul-Haq
Department of Computer Science& IT, Government Collage University Faisalabad (Layyah Campus), Layyah, Pakistan

A. W. D. Udaya Shalika and Lasantha Seneviratne
Sri Lanka Institute of Information Technology, Malabe, Sri Lanka

Linlin Zhan
Heilongjiang Bayi Agricultural University Library, Heilongjiang Bayi Agricultural University, Daqing, China

Gamil Abdel Azim
College of Computers & Informatics Canal University, Ismailia, Egypt

Jingang Cao
Department of Computer, North China Electric Power University, Baoding, China

Hanaa Mohsin Ahmad and Bushra Jabbar Abdulkareem
Computer Science, University of Technology, Baghdad, Iraq

Linlin Zhang
Heilongjiang Bayi Agricultural University Library, Heilongjiang Bayi Agricultural University, Daqing, China

Abey Bruck and Tulu Tilahun
Department of Computer Science & IT, AMiT, Arba Minch University, Arba Minch, Ethiopia

Zvi Retchkiman Königsberg
Centro de Investigacion en Computacion, Instituto Politecnico Nacional, Mexico City, Mexico

Deshun Sun and Chongyu Wei
Qingdao University of Science & Technology, Qingdao, China

Janne Vatjus-Anttila, Mika Hoppari, Lance Fono, Kari Kolehmainen and Subayal Khan
Communication Systems, VTT Technical Research Centre of Finland, Oulu, Finland

Samia Allaoua Chelloug
Networks and Communication Systems Department, College of Computer and Information Sciences, Princess Nourah Bint Abdul Rahman University, Riyadh, KSA

Osama S. Younes
Faculty of Computers and Information, Technology Tabuk University, Tabuk, KSA
Faculty of Computers and Information, Menoufia University, Al Minufya, Egypt